T0262271

Biodiversity Enrichment: Socio-Economic Impacts and Genetics

Biodiversity Enrichment: Socio-Economic Impacts and Genetics

Edited by **Neil Griffin**

New York

Published by Callisto Reference,
106 Park Avenue, Suite 200,
New York, NY 10016, USA
www.callistoreference.com

Biodiversity Enrichment: Socio-Economic Impacts and Genetics
Edited by Neil Griffin

© 2015 Callisto Reference

International Standard Book Number: 978-1-63239-093-6 (Hardback)

Printed in the United States of America.

Contents

Preface

The book broadly covers agricultural aspects, socio-economic impacts and genetics. The complexity of the association has been approached from various angles, taking into account the interventions at various stages. A scientific approach has been adopted in this book which exhibits the inter-connectivity of all the three levels, emphasizing the need for conservation and protection of the systems if human existence is to continually benefit from it. This book is a valuable reference for researchers and students interested in biodiversity.

This book is the end result of constructive efforts and intensive research done by experts in this field. The aim of this book is to enlighten the readers with recent information in this area of research. The information provided in this profound book would serve as a valuable reference to students and researchers in this field.

At the end, I would like to thank all the authors for devoting their precious time and providing their valuable contribution to this book. I would also like to express my gratitude to my fellow colleagues who encouraged me throughout the process.

Editor

Agricultural Aspect

Amaranthaceae as a Bioindicator of Neotropical Savannah Diversity

Suzane M. Fank-de-Carvalho, Sônia N. Báo and Maria Salete Marchioretto

Additional information is available at the end of the chapter

1. Introduction

Brazil is the first in a ranking of 17 countries in megadiversity of plants, having 17,630 endemic species among a total of 31,162 Angiosperm species [1], distributed in five Biomes. One of them is the Cerrado, which is recognized as a World Priority Hotspot for Conservation because it has around 4,400 endemic plants – almost 50% of the total number of species – and consists largely of savannah, woodland/savannah and dry forest ecosystems [2,3]. It is estimated that Brazil has over 60,000 plant species and, due to the climate and other environmental conditions, some tropical representatives of families which also occur in the temperate zone are very different in appearance [4].

The Cerrado Biome is a tropical ecosystem that occupies about 2 million km² (from 3-24° Lat. S and from 41-43° Long. W), located mainly on the central Brazilian Plateau, which has a hot, semi-humid and markedly seasonal climate, varying from a dry winter season (from April to September) to a rainy summer (from October to March) [5-7]. The variety of landscape – from tall savannah woodland to low open grassland with no woody plants - supports the richest flora among the world's savannahs (more than 7,000 native species of vascular plants) and a high degree of endemism [6,8]. This Biome is the most extensive savannah region in South America (the Neotropical Savannah) and it includes a mosaic of vegetation types, varying from a closed canopy forest ("cerradão") to areas with few grasses and more scrub and trees ("cerrado *sensu stricto*"), grassland with scattered scrub and few trees ("campo sujo") and grassland with little scrub and no trees ("campo limpo") [3,9]. Among the grassland areas there are some flat areas with rocky soil, called "campos rochosos", which are considered Cerrado areas because of their flora, especially when located in Chapada Diamantina (Bahia State), a transition area between Cerrado and Caatinga Biomes.

Although the Cerrado is considered a Hotspot for the conservation of global biodiversity, with plant species completely adapted to survive adverse conditions of soil and climate, only 30% of this Neotropical Savannah biodiversity is reasonably well known [8,10]. Coutinho [11] believes that the frequent occurrence of fire is one of the most important factors to determine this Biome's vegetation, acting as a renewal element that selects structural and physiological characteristics. Nowadays it is believed that more than 40% of the original vegetation has already been converted into human-disturbed areas, due to the expansion of crops [12,13]. This process has accelerated the fragmentation of natural habitat, increasing the pressure on local biodiversity extinction and introducing exotic species, also amplifying soil erosion, water pollution and alterations in vegetation and hydrologic conditions [2,8,14].

The Amaranthaceae family is composed of 2,360 species and here will be listed those that occur in Brazil, emphasizing the Cerrado species and including information on endemism, endangerment and economic or potential use. We also provide a list of the most important bibliographical references for those who are interested in studying the species of this family. Some aspects of morphology, leaf anatomy and ultrastructure will be shown for six species found in the Neotropical Savannah core area (Chapada dos Veadeiros) and some of these aspects, as well as taxonomy and ecology, will be discussed in order to propose the use of this plant family as an indicator of the diversity in open areas of this Biome.

2. Methodology

2.1. List of the Brazilian Amaranthaceae species and those in RPPN Cara Preta

The Brazilian Amaranthaceae list (Table 1) was based on the research by the Brazilian taxonomists Marchioretto [15-21] and Siqueira [22-25] and on the most important taxonomic references to this Family both from the literature (Table 2) and Brazilian Herbaria (Table 3). All cited Herbaria are listed according to the Index Herbariorum [26,27].

The species to be detailed were collected in a Conservation Unit named Reserva Particular do Patrimônio Natural Cara Preta (RPPN Cara Preta), located in Alto Paraíso, Goiás State, Brazil. After obtaining authorization from the NGO Oca Brasil, random walks were done in order to locate, photograph and mark species with a Global Positioning System device, and to collect and make exsiccates for Herbaria deposits, from September 2006 until March 2009. Although some plant leaves were collected during the vegetative stage, these specimens were visited until flowering to identify them correctly. All exsiccates were deposited in Brazilian Herbaria as standard control material (prioritizing PACA, UnB and IBGE Herbaria) and these species are included in Table 1.

2.2. Leaf anatomy and ultrastructure

Completely expanded leaves, from 3rd to 5th node from the apex, of two to six specimens of each species were collected and sectioned. Part of the leaf medial region was fixed in ethanol, acetic acid and formaldehyde [28] for 24 hours and preserved in ethanol 70% until analysis to describe the anatomy and identify starch and crystal composition [28].

Some pieces of the leaf medial region were immediately submerged in a Karnovsky solution [29] of glutaraldehyde 2%, paraformaldehyde 2% and sucrose 3% in sodium cacodylate 0.05 M buffer for 12 to 24 hours and preserved in sodium cacodylate 0.05 M until processing for analysis under an electron microscope. For the latter analysis, these pieces were post-fixed in 2% osmium tetroxide and 1.6% potassium ferricyanide (1:1 v/v), followed by in-block staining with 0.5% uranyl acetate solution (overnight). These samples were then dehydrated in an acetone ascending series and slowly embedded in Spurr's epoxy resin. Semi-thin and ultra-thin sections were obtained in ultramicrotome with glass and diamond knives. Semi-thin sections were stained with toluidine blue and analysed under the optical Zeiss Axiophot, and ultra-thin sections were analysed under the transmission electron microscope TEM JEOL JEM 1011.

3. Results and discussion

The Amaranthaceae family is composed of 2,360 species and 146 of them are found in Brazil (Table 1). Ninety-eight species within the family are found in the Cerrado and 73 spp. are endemic to Brazil, of which 13 are endemic to the Cerrado Biome (Table 1). Twenty Amaranthaceae species are exclusive to the Cerrado (Table 1).

At least 22 Amaranthaceae species are referred to as being used in folk medicine (Table 1). In Brazil, only two of these species are already used as commercial drugs, as capsules containing their powdered roots, with studies to support their medicinal activity: *Hebanthe eriantha* (Poir.) Pedersen and *Pfaffia glomerata* (Spreng.) Pedersen (Table 1), both known as "Brazilian-ginseng". However, there is neither registered success in isolating or synthesizing their components nor any economic studies about the viability of this kind of pharmaceutical procedure.

Although the species *Gomphrena macrocephala* St.-Hil. is not cited as medicinal (Table 1), the fructan content in its roots has been determined [30] because this species was considered synonymous with *G. officinalis* Mart. [31]. Later, it was determined that *G. officinalis* was synonymous with *G. arborescens* L.f. and not with *G. macrocephala* [22]. Studying *G. arborescens*, fructan was also determined as the principal carbohydrate in its subterranean system [32]. This species is used in popular medicine to heal respiratory diseases (asthma and bronchitis), to reduce fever and as a tonic [33-35]. An *in vivo* study (in cats) with the use of fructans isolated from *Arctium lappa* L. (Asteraceae) reported a cough-suppressing activity [36], and the presence of fructan in *G. arborescens* roots can partially justify the use of this species as a medicinal plant.

Most members of Brazilian Amaranthaceae are only known by taxonomists and 42 species are in danger of extinction according to Brazilian regional lists; 14 of them are recognized as endangered by the Brazilian Ministry of the Environment (MMA – "Ministério do Meio Ambiente") (Table 1). Most of the endangered species are classified according to the IUCN Red List of vulnerability categories, some even with the same criteria, and there is a wide range of research still to be done.

Species	Bioma and level of endemism	Species Threat Level	Habit, popular name and species knowledge
Achyranthes aspera L.	Cerrado		Herb; plant used as indigenous medicine in Ethiopia with chemistry study [37]
Achyranthes indica (L.) Mill.	Cerrado		Herb
Alternanthera adscendens Suess.	Cerrado exclusive		Shrub
Alternanthera albida (Moq.) Griseb.			Subshrub; C4 photosynthesis physiology [38]
Alternanthera aquatica (D.Parodi) Chodat			Herb
Alternanthera bahiensis Pedersen	Cerrado, endemic to Brazil		Herb or subshrub
Alternanthera bettzichiana (Regel) G.Nicholson			Herb; popularly named "anador"; folk medicinal plant, used as analgesic and antipyretic [39]
Alternanthera brasiliana (L.) Kuntze	Cerrado, endemic to Brazil		Herb; called "perpétua-do-mato, periquito-gigante, penicilina" or Brazilian joyweed; folk medicinal plant, used as diuretic, digestive, depurative, bequic, astringent and antidiarrhoeal; ornamental plant; C3 photosynthesis structure [40-42]
Alternanthera decurrens J. C. Siqueira	Brazilian Cerrado endemic (Januária - MG)	CR [43]	Subshrub
Alternanthera dendrotricha C.C.Towns.	Cerrado, endemic to Brazil		Shrub
Alternanthera flavida Suess.			Subshrub
Alternanthera hirtula (Mart.) R.E.Fr.		EN [44]	Herb
Alternanthera januariensi J. C. Siqueira	Brazilian Cerrado endemic (Januária - MG)	CR [43]	Subshrub
Alternanthera kurtzii Schinz			Herb
Alternanthera littoralis P.Beauv.			Herb; called "periquito-da-praia" [45]
Alternanthera malmeana R.E.Fr.		EN [44]	Herb
Alternanthera markgrafii Suess.	Brazilian Cerrado endemic (Serra de Grão Mogol - MG)		Herb

Alternanthera martii (Moq.) R.E. Fries	Cerrado, endemic to Brazil		Subshrub
Alternanthera micrantha R.E.Fr.	Endemic to Brazil	VU [44]	Herb; called "periquito-da-serra" [45]
Alternanthera minutiflora Suess.	Endemic to Brazil		Herb
Alternanthera multicaulis Kuntze	Endemic to Brazil		Herb
Alternanthera paronychioides A.St.-Hil.	Cerrado	VU [44,46]	Herb; called "periquito-roseta, periquito"; C_3-C_4 intermediary photosynthesis structure; C_4 photosynthesis physiology; ornamental plant [38,41,42,45,47]
Alternanthera philoxeroides (Mart.) Griseb.	Cerrado		Herb; called "perna-de-saracura, carrapicho-de-brejo" and alligatorweed [45]
Alternanthera pilosa Moq.			Herb
Alternanthera praelonga A.St.-Hil.		CR [44]	Herb
Alternanthera puberula D.Dietr.	Cerrado exclusive		Herb
Alternanthera pulchella Kunth			Herb; C_4 photosynthesis physiology [38]
Alternanthera pungens Kunth	Cerrado		Herb; called "erva-de-pinto"; folk medicinal plant, used to treat syphilis and skin diseases; C_4 photosynthesis physiology [38,39]
Alternanthera ramosissima (Mart.) Chodat	Cerrado		Herb
Alternanthera regelii (Seub.) Schinz	Cerrado exclusive, endemic to Brazil		Herb
Alternanthera reineckii Briq.	Cerrado	VU [44]	Herb; called "periquito-de-reineck" [45]
Alternanthera rufa (Mart.) D.Dietr.	Cerrado, endemic to Brazil		Herb
Alternanthera sessilis (L.) R.Br.	Cerrado	LC [48]	Herb
Alternanthera tenella Colla	Cerrado	VU [44]	Herb; called "apaga-fogo, carrapichinho, corrente, folha-de-papagaio, periquito, periquito-figueira, perpétua-do-mato, sempre-viva"and joyweed; folk medicinal plant, used as diuretic; this species is naturally infected by a potyvirus; C_3-C_4 photosynthesis physiology and structure [38,39,40,45,47,49,50]

Alternanthera tetramera R.E.Fr.			Herb
Amaranthus blitum L.			Herb; called "caruru"; folk medicinal plant, used to fight anemia; C_4 photosynthesis physiology [38,39,51]
Amaranthus caudatus L.	Cerrado		Herb; called "rabo-de-gato, cauda-de-raposa, disciplina-de-freira, rabo-de-raposa"; folk medicinal plant, used to treat pulmonary diseases and as emoliente; C_4 photosynthesis physiology; ornamental plant [33,38,39,41]
Amaranthus cruentus L.	Cerrado		Herb; called "caruru-vermelho, veludo, bredo-de-jardim, crista-de-galo"; folk medicinal plant, used as emollient and laxative; C_4 photosynthesis physiology [33,38,39,51]
Amaranthus deflexus L.			Herb; called "caruru-rasteiro"; C_4 photosynthesis physiology [38,51]
Amaranthus hybridus L.	Cerrado		Herb; called "caruru, bredo" and smooth pigweed; C_4 photosynthesis physiology [38,51]
Amaranthus muricatus (Moq.) Hieron.			Herb;, C_4 photosynthesis physiology [38]
Amaranthus retroflexus L.	Cerrado		Herb; C_4 photosynthesis physiology [38]
Amaranthus rosengurtii Hunz.		EN [44]	Herb
Amaranthus spinosus L.	Cerrado		Herb; called "caruru-bravo, caruru-de-espinho, bredo-de-espinho, caruru-de-porco"; folk medicinal plant, used to combat eczema and as emollient, laxative and antiblenorragic; C_4 photosynthesis physiology [33,38,39]
Amaranthus viridis L.	Cerrado		Herb; called "caruru-bravo, caruru-verdadeiro, cururu, caruru-de-soldado, caruru-de-folha-miúda, amaranto-verde"; folk medicinal plant, used as emollient and diuretic desobstruente; C_4 photosynthesis physiology [33,38-40,51]
Blutaparon portulacoides (A.St.-Hil.) Mears	Cerrado	VU [44]	Herb; called "capotiraguá"; folk medicinal plant, used to combat leukorrhea; C_4 photosynthesis physiology [38,39]
Blutaparon vermiculare (L.) Mears	Cerrado		Herb; C_4 photosynthesis physiology [38]

Celosia argentea L.	Cerrado	Herb; called "celosia-branca, celósia-plumosa, crista-de-galo, crista-de-galo-plumosa, suspiro, veludo-branco"; folk medicinal plant, used to combat diarrhea and as anthelmintic and astringent; ornamental plant [39,41,52]
Celosia corymbifera Didr.	Endemic to Brazil	Subshrub
Celosia grandifolia Moq.	EN [44]	Herb, subshrub; called "bredo-do-mato" [45]
Chamissoa acuminata Mart.	VU [44]	Subshrub; called "mofungo-rabudo" [45]
Chamissoa altissima (Jacq.) Kunth	Cerrado	VU [44]	Subshrub; called "mofungo-gigante" [45]
Chenopodium album L.	Cerrado	Herb
Chenopodium ambrosioides L.	Cerrado	Herb; called "erva-de-santa-maria, erva-santa, quenopódio" [40]
Chenopodium murale L.	Cerrado	Herb
Cyathula achyranthoides (Kunth) Moq.		Herb
Cyathula prostrata Blume	Cerrado	Herb
Froelichia humboldtiana (Roem. & Schult.) Seub.	Cerrado	Herb; C_4 photosynthesis physiology [38]
Froelichia interrupta (L.) Moq.		Herb; C_4 photosynthesis physiology [38]
Froelichia procera (Seub.) Pedersen	Cerrado	Herb; called "ervaço"; C_4 photosynthesis physiology [38,41]
Froelichia sericea (Roem. & Schult.) Moq.		Herb
Froelichia tomentosa (Mart.) Moq.	Cerrado	Herb; C_4 photosynthesis physiology [38]
Froelichiella grisea R.E. Fries	Braziliann Cerrado endemic (Chapada dos Veadeiros - GO)	VU [43]	Herb; C_3 photosynthesis structure [42]
Gomphrena agrestis Mart.	Cerrado, endemic to Brazil	EN [46]	Herb
Gomphrena arborescens L.f.	Cerrado exclusive	Herb, subshrub; called "perpétua, perpétua-do-campo, perpétua-do-mato, paratudo-do-campo, paratudo-erva, raiz-do-padre"; folk medicinal plant, used as tonic, to reduce fever and against respiratory deseases; potential use as ornamentalplant; roots are fructan-rich; C_4 photosynthesis physiology/structure [32,35,38,40,42,49,53,54]

Gomphrena basilanata Suess.	Endemic to Brazil		Subshrub; C₄ photosynthesis physiology [38]
Gomphrena celosoides Mart.	Cerrado		Subshrub; C₄ photosynthesis physiology [38]
Gomphrena centrota E.Holzh.	Endemic to Brazil	VU [43]	Subshrub; C₄ photosynthesis physiology [38]
Gomphrena chrestoides C.C.Towns.	Brazilian Cerrado endemic (Chapada Diamantina - BA)	VU [43]	Subshrub
Gomphrena claussenii Moq.	Cerrado, endemic to Brazil		Subshrub
Gomphrena debilis Mart.	Endemic to Brazil		Subshrub; C₄ photosynthesis physiology [38]
Gomphrena demissa Mart.	Cerrado, endemic to Brazil		Subshrub; folk medicinal plant, used to combat the flu; C₄ photosynthesis physiology [38,49]
Gomphrena desertorum Mart.	Cerrado, endemic to Brazil		Subshrub; C₄ photosynthesis physiology [38]
Gomphrena duriuscula Moq.	Endemic to Brazil	EN [43]	Subshrub; C₄ photosynthesis physiology [38]
Gomphrena elegans Mart.	Cerrado, endemic to Brazil	VU [46]	Subshrub
Gomphrena gardnerii Moq.	Cerrado, endemic to Brazil		Subshrub; C₄ photosynthesis physiology [38]
Gomphrena globosa L.	Cerrado		Subshrub; called "gonfrena, perpétua, perpétua-roxa, sempre-viva, suspiro, suspiro-roxo"; folk medicinal plant used to fight respiratory diseases; C₄ photosynthesis physiology; ornamental plant [38,40,45,55,56]
Gomphrena graminea Moq.	Cerrado	VU [44]	Subshrub; called "perpétua-gramínea"; C₄ photosynthesis physiology [38,45]
Gomphrena hatschbachiana Pedersen	Cerrado, endemic to Brazil	VU [43]	Subshrub
Gomphrena hermogenesii J.C. Siqueira	Brazilian Cerrado endemic (Chapada dos Veadeiros - GO)		Subshrub; C₃ photosynthesis physiology; C₄ photosynthesis structure [38,42]
Gomphrena hillii Suess.	Brazilian Cerrado endemic (Paraíso do Norte - TO)		Subshrub; C₄ photosynthesis physiology [38]
Gomphrena incana Mart.	Cerrado exclusive, endemic to Brazil		Subshrub; C₄ photosynthesis physiology [38]
Gomphrena lanigera Pohl *ex* Moq.	Cerrado exclusive, endemic to Brazil		Subshrub; C₄ photosynthesis physiology and structure [38,42]
Gomphrena leucocephala Mart.	Endemic to Brazil		Subshrub; C₄ photosynthesis physiology [38]
Gomphrena macrocephala A.St.-Hil.	Cerrado exclusive, endemic to Brazil		Subshrub; roots are fructan-rich; C₄ photosynthesis physiology [30,38]

Gomphrena marginata Seub.	Brazilian Cerrado endemic (Diamantina - MG)		Subshrub
Gomphrena matogrossensis Suess.	Cerrado exclusive, endemic to Brazil		Subshrub
Gomphrena microcephala Moq.	endemic to Brazil		Subshrub
Gomphrena mollis Mart.	Cerrado, endemic to Brazil		Subshrub; called "erva-mole, erva-rosa"; folk medicinal plant, used as tonic and carminative [39]
Gomphrena moquini Seub.	Brazilian Cerrado endemic (Serra do Cipó - MG)		Subshrub
Gomphrena nigricans Mart.	Cerrado, endemic to Brazil	VU [43]	Subshrub
Gomphrena paranensis R.E.Fr.	Cerrado exclusive, endemic to Brazil		Subshrub, C_4 photosynthesis physiology [38]
Gomphrena perennis L.		VU [44]	Subshrub; called "perpétua-sempreviva"; C_4 photosynthesis physiology [38,45]
Gomphrena pohlii Moq.	Cerrado exclusive		Subshrub; called "infalível, paratudo, paratudinho, paratudo-amarelinho"; roots are used in folk medicine against respiratory deseases; C_4 photosynthesis physiology and structure [38,39,42,49]
Gomphrena prostrata Mart.	Cerrado, endemic to Brazil		Subshrub; C_4 photosynthesis physiology and structure [38,42]
Gomphrena pulchella Mart.		EN [44]	Subshrub; C_4 photosynthesis physiology [38]
Gomphrena pulvinata Suess.	Endemic to Brazil		Subshrub; C_4 photosynthesis physiology [38]
Gomphrena regeliana Seub.	Cerrado exclusive, endemic to Brazil		Subshrub; C_4 photosynthesis physiology [38]
Gomphrena riparia Pedersen	Endemic to Brazil	CR [43]	Subshrub; C_4 photosynthesis physiology [38]
Gomphrena rudis Moq.	Cerrado exclusive, endemic to Brazil		Subshrub; C_4 photosynthesis physiology [38]
Gomphrena rupestris Nees	Cerrado, endemic to Brazil		Subshrub
Gomphrena scandens (R.E.Fr.) J.C.Siqueira	endemic to Brazil	VU [43]	Subshrub
Gomphrena scapigera Mart.	Cerrado, endemic to Brazil		Subshrub; C_4 photosynthesis physiology [38]
Gomphrena schlechtendaliana Mart.		EN [44]	Subshrub; called "perpétua-schlechtendal"; C_4 photosynthesis physiology [38,45]

Gomphrena sellowiana Mart.	Endemic to Brazil	VU [44]	Subshrub
Gomphrena serturneroides Suess.	Endemic to Brazil		Subshrub; C₄ photosynthesis physiology [38]
Gomphrena vaga Mart.	Cerrado, endemic to Brazil	VU [44]	Subshrub; called "thoronoé"; folk medicinal plant, used as analgesic [57]
Gomphrena virgata Mart.	Cerrado, endemic to Brazil		Subshrub; called "cangussú-branco, vergateza"; folk medicinal plant, antiletargic; C₄ photosynthesis physiology and structure [33,38,42]
Hebanthe eriantha (Poir.) Pedersen	Cerrado	EN [44], VU [58]	Subshrub, shrub; called "corango-açu, ginseng-brasileiro, picão-de-tropeiro,solidonia, suma"; folk medicinal plant, used to combat colic and enteritis; most of its chemical constituents are known and roots of this plant are already used by pharmaceutical companies [40,59]
Hebanthe grandiflora (Hook.) Borsch & Pedersen	Cerrado		Bush scandentia
Hebanthe occidentallis (R.E.Fr.) Borsch & Pedersen	Cerrado		Subshrub scandentia
Hebanthe pulverulenta Mart.	Cerrado	VU [58]	Subshrub scandentia; called "corango-veludo" [45]
Hebanthe reticulata (Seub.) Borsch & Pedersen			Subshrub, shrub scandentia
Hebanthe spicata Mart.			Shrub erect or scadentia
Herbstia brasiliana (Moq.) Sohmer		EX [46]	Subshrub
Iresine diffusa Humb. & Bonpl. *ex* Willd.	Cerrado		Subshrub; called "bredinho-difuso" [45]
Iresine poeppigiana Klotzsch			Subshrub
Lecosia formicarum Pedersen	Endemic to Brazil		Subshrub
Lecosia oppositifolia Pedersen	Endemic to Brazil	CR [43]	Herb or subshrub
Pedersenia argentata (Mart.) Holub			Herb
Pfaffia acutifolia (Moq.) O.Stützer	Cerrado		Herb or subshrub
Pfaffia aphylla Suess.	Brazilian Cerrado endemic (Gouveia - MG)		Subshrub
Pfaffia argyrea Pedersen	Cerrado exclusive, endemic to Brazil	VU [43]	Herb or subshrub
Pfaffia cipoana Marchior. *et al.*	Brazilian Cerrado endemic (Itambé do Mato Dentro - MG)		Subshrub

Pfaffia denudata (Moq.) Kuntze	Cerrado exclusive, endemic to Brazil		Herb, subshrub, shrub
Pfaffia elata R.E.Fr.	Cerrado exclusive, endemic to Brazil		Subshrub
Pfaffia glabrata Mart.	Cerrado exclusive		Herb, subshrub; called "corango-sempreviva" [45]
Pfaffia glomerata (Spreng.) Pedersen	Cerrado	VU [44]	Herb, subshrub; called "anador, canela-velha, ginseng-brasileiro, finseng, páfia, paratudo, corango-sempreviva"; folk medicinal plant, most of its chemical constituents are known and roots of this plant are already used by pharmaceutical companies; butanolic extract showed antihyperglycemic potential in vivo; C_3 photosynthesis physiology and structure [38,40,42,45,60]
Pfaffia gnaphaloides (L.f.) Mart.	Cerrado	VU [44]	Herb, subshrub, called "corango-de-seda", C3 photosynthesis physiology and structure [38,42,45]
Pfaffia hirtula Mart.	Cerrado exclusive, endemic to Brazil		Herb, subshrub
Pfaffia jubata Mart.	Cerrado, endemic to Brazil		Herb, subshrub; called "marcela-branca, marcela-do-campo, marcela-do-cerrado"and kytertenim; roots are used in folk medicine against intestinal problems [39,49]
Pfaffia minarum Pedersen	Cerrado exclusive, endemic to Brazil	VU [43]	Subshrub
Pfaffia rupestris Marchior. *et al.*	Brazilian Cerrado endemic (Rio Pardo de Minas - MG)		Subshrub
Pfaffia sarcophylla Pedersen	Brazilian Cerrado endemic (Niquelândia - GO)		Subshrub; nickel hyperaccumulator, it is one of the first species to recolonize the ground with high concentrations of total Ni in the soil (>1%) [61]
Pfaffia sericantha (Mart.) Pedersen	Cerrado, endemic to Brazil		
Pfaffia siqueiriana Marchior. & Miotto	Cerrado, endemic to Brazil		Subshrub
Pfaffia townsendii Pedersen	Cerrado, endemic to Brazil	VU [43]	Subshrub; C3 photosynthesis physiology and structure [38,42]
Pfaffia tuberculosa Pedersen	Cerrado, endemic to Brazil		Herb, subshrub
Pfaffia tuberosa (Spreng.) Hicken	Cerrado		Herb, subshrub; called "corango-de-batata" [45]
Pfaffia velutina Mart.	Cerrado exclusive, endemic to Brazil		Subshrub

Pseudoplantago friesii Suess		PE [44]	Popular name is "caruru-açu" [45]
Quaternella confusa Pedersen	Cerrado exclusive, endemic to Brazil		Shrub
Quaternella ephedroides Pedersen	Cerrado, endemic to Brazil		Shrub
Quaternella glabratoides (Suess.) Pedersen	Endemic to Brazil	EN [44]	Subshrub; called "corangão" [45]
Xerosiphon angustiflorus (Mart.) Pedersen	Cerrado, endemic to Brazil		Subshrub
Xerosiphon aphyllus (Pohl ex Moq.) Pedersen	Cerrado, endemic to Brazil		Subshrub

Notes: The species threat level category is the same as that used in the IUCN Red List: CR (critically endangered), EN (endangered), EX (extinct), LC (Least Concern) and VU (vulnerable).

Table 1. Amaranthaceae species found in Brazil, identifying those endemics to Brazil and the ones found in the Neotropical Savannah (Cerrado), level of threat, habit, popular name (mostly in Portuguese) and some of the knowledge about the species.

3.1. Morphology, taxonomy challenge and species list of Brazilian Amaranthaceae

Taxonomy is the science that aims to identify and characterize species. It includes the study of the plant's behaviour in nature and is based on plant morphology. The use of other data, such as anatomy studies, genetic characters, ecology and geographic pattern, aims to include and define affinities and parental relations among plant groups. Only by knowing the species is it possible for Botany to contribute to other scientific areas, including to the conservation of species *in situ*, not only of plants but also of animals.

It is not easy to correctly identify Brazilian Amaranthaceae species. Different species can be very alike in habit and vegetative morphology. The correct identification depends almost exclusively on some flower details, whose small dimensions make it especially difficult to work in the field, demanding a highly specialized work, only partially carried out for this family (15-22).

Brazilian species of this family are predominantly herbs, shrubs or climbing plants. They can be annual or perennial, with erect, prostrate, decumbent or scandent stem. In species from the Neotropical Savannah or from rocky fields, the underground organ is thickened and composed of roots and a xylopodium – a portion of the subterranean system which is responsible for the re-sprouting after a fire or other environmental stress [62]. The leaf arrangement can be opposite, alternate or with a basal aggregation of leaves. Leaves are exstipulate, glabrous or pubescent, with entire lamina and margins. Inflorescences can be cymoses, in spikes, in heads, corymboses or paniculates, axillary or axial. Flowers are bisexual or monoecious and small. The perianth is undifferentiated, actinomorphic, with five distinct or partially connated sepals. Flowers are associated with dry and papery bracts. Fruits are dry, usually a single-seeded achene or capsules with few seeds [15-22]. A short list of the most important Brazilian Herbaria to visit in order to study Amaranthaceae taxonomy is presented on Table 2 and the literature used to identify the species of this family is presented in Table 3.

Index	Herbarium Name	Institution and municipality
ALCB	Herbário da Universidade Federal da Bahia	UFBA/Campus de Ondina, Salvador, Bahia, Brazil
BHCB	Herbário da Universidade Federal de Minas Gerais	UFMG, Belo Horizonte, Minas Gerais, Brazil
BOTU	Herbário da Universidade Estadual Paulista	UNESP, Botucatu, São Paulo, Brazil
CEN	Herbário da EMBRAPA Recursos Genéticos e Biotecnologia	EMBRAPA/CENARGEN, Brasília, Distrito Federal, Brazil
CEPEC	Herbário do Centro de Pesquisas do Cacau	CEPEC, Itabuna, Bahia, Brazil
CESJ	Herbário da Universidade Federal de Juiz de Fora	UFJF, Juiz de Fora, Minas Gerais, Brazil
CPAP	Herbário do Centro de Pesquisas Agropecuárias do Pantanal	CPAP, Corumbá, Mato Grosso do Sul, Brazil
ESA	Herbário da Universidade de São Paulo	ESALQ/USP, Piracicaba, São Paulo, Brazil
GUA	Herbário Alberto Castellanos	FEEMA/INEA, Rio de Janeiro, Rio de Janeiro, Brazil
HTO	Herbário da Universidade Federal do Tocantins	UFTO, Porto Nacional, Tocantins, Brazil
HUEFS	Herbário da Universidade Estadual de Feira de Santana	UFES, Feira de Santana, Bahia, Brazil
IAC	Herbário do Instituto Agronômico de Campinas	IAC, Campinas, São Paulo, Brazil
IBGE	Herbário da Reserva Ecológica do IBGE	IBGE/RECOR, Brasília, Distrito Federal, Brazil
JPB	Herbário da Universidade Federal da Paraíba	UFPB, Cidade Universitária, João Pessoa, Paraíba, Brazil
MBM	Herbário do Museu Botânico Municipal	Prefeitura Municipal/SMA, Curitiba, Paraná, Brazil
PACA	Herbarium Anchieta	Instituto Anchietano de Pesquisas/UNISINOS, São Leopoldo, Rio Grande do Sul, Brazil
RB	Herbário do Jardim Botânico do Rio de Janeiro	JBRJ, Rio de Janeiro, Rio de Janeiro, Brazil
SP	Herbário do Instituto de Botânica	Secretaria de Meio Ambiente, São Paulo, São Paulo, Brazil
SPF	Herbário da Universidade de São Paulo	USP, São Paulo, São Paulo, Brazil
UB	Herbário da Universidade de Brasília	UnB, Brasília, Distrito Federal, Brazil
UEC	Herbário da Universidade Estadual de Campinas	UNICAMP, Campinas, São Paulo, Brazil
UFG	Herbário da Universidade Federal de Goiás	UFG, Goiânia, Goiás, Brazil
VIC	Herbário da Universidade Federal de Viçosa	UFV, Viçosa, Minas Gerais, Brazil

Notes: Herbaria are cited according to the Index Herbariorum and all of them have good collections of Amaranthacae. The Institution/city where the Herbaria are located is also referred.

Table 2. List of the most important Herbaria references for researchers interested in studying the Brazilian Amaranthaceae

---

[15-21] Revisions of Brazilian *Froelichia, Froelichiella, Hebanthe* and *Pfaffia*; species list and phytogeography

[22-25] Revision of Brazilian *Gomphrena*; species list na phytogeography

[45] Amaranthaceae from Santa Catarina State, Brazil

[63] Restoring the *Hebanthe* genera

[64,65] Brazilian Amaranthaceae species and the Family in the World

[66] Revision of Amaranthaceae in the World

[67-71] Studies in South American Amaranthaceae

[72] Amaranthaceae in Flora Brasiliensis

[73] Studies of *Pfaffia* and *Alternanthera* genera

[74,75] Amaranthaceae in Central and South America

[51,52,56,76] Amaranthaceae from Rio Grande do Sul State, Brazil

---

Note: These references are ordereb by author and should be consulted in order to identify Brazilian species correctly.

Table 3. List of the most important bibliographical references for researchers interested in studying the Brazilian Amaranthaceae

The Reserva Particular do Patrimônio Natural (RPPN) Cara Preta, in Alto Paraíso, Goiás State, is a good representative of Neotropical Savannah vegetation, at about 1,500 meters of altitude and showing rocky slopes with Cerrado *sensu stricto* (Figure 1), grassland with scattered scrubs and few trees and grassland with few scrubby plants and no trees (Figure 2). The *Pfaffia* genus was restricted to a rocky slope and the other species were found in a level field of sandy soils, usually covered by Poaceae and Cyperaceae. It was very difficult to find all the species. It was only possible because of frequent visits to RPPN Cara Preta, using GPS to mark the local after finding any probable member of the family in order to be able to accompany them until the flowering stage. The area was monitored for one and a half year and only *Gomphrena hermogenesii* J.C. Siqueira and *Pfaffia townsendii* Pedersen (Figure 3) were localized, the first one always in vegetative stage. A key event to help finding all six species was a fire that burned out the vegetation in August of the year 2008: without the competition of the grasses, the Amaranthaceae species regrew and flowered rapidly, in order to spread their seeds before the grasses could fully recover (Figures 4-8).

Pfaffia townsendii is a shrub species with persistent aerial portions that flowers throughout the year (Figure 3). The herb *G. hermogenesii* is endemic to Chapada dos Veadeiros and also has permanent aerial portions (about 10-20 cm high), but it was commonly found in vegetative stage under the grass leaves; its flowering stage was stimulated by fire (Figure 4). *Froelichiella grisea* R.E.Fr. (Figure 5), *G. lanigera* Pohl. ex Moq. (Figure 6), *G. prostrata* Mart. (Figure 7) and *P. gnaphaloides* (L.f.) Mart. (Figure 8) species were recorded in the flowering stage at RPPN Cara Preta around 20 days after a fire that burned out all the vegetation in the area, which is evidence of the pirophytic behaviour of most Neotropical Savannah Amaranthaceae. Five of these species had never been recorded in this RPPN before and one of them was last recorded in 1966 (*F. grisea*), according to Herbaria data. Figures 1-8 are reproduced [77] with the authorization of the Biota Neotropica Editor, Dr. Carlos Joly.

Figure 1. (Figures 1-8) Photographs of the environment and of the studied species at Reserva Particular do Patrimônio Natural (RPPN) Cara Preta, Alto Paraíso, Goiás State, Brazil. **Fig. 1.** Rocky slope where were found the species *Pfaffia townsendii* Pedersen and *P. gnaphaloides* (L. f.) Mart. **Fig. 2.** Humid rocky grassland where were found the species *Froelichiella grisea* R.E.Fr., *Gomphrena hermogenesii* J.C. Siqueira, *G. lanigera* Pohl. ex Moq. and *G. prostrata* Mart. **Fig. 3.** *P. townsendii*. **Fig. 4.** *G. hermogenesii*. **Fig. 5.** *F. grisea*. **Fig. 6.** *G. lanigera*. **Fig. 7.** *G. prostrata*. **Fig. 8.** *P. gnaphaloides*.

Five species are herb to subshrub, and only *P. townsendii* is a shrub (Figure 3). Well-developed tuberous subterranean systems were found in *F. grisea* (Figure 5) and *G. hermogenesii*, while in *G. lanigera* and *P. gnaphaloides* the underground organ was less developed, also tuberous. *G. prostrata* and *P. townsendii* presented a well-developed and lignified underground organ. Leaves of *F. grisea* and *G. hermogenesii* are opposite and alternate in the other studied species. *F. grisea* and *G. lanigera* can present a basal aggregation of leaves. Leaves are always tomentose, with exception of the adaxial face of *F.*

grisea, which can be glabrous. Inflorescence is axial in all these species, spikes in *F. grisea* and *G. lanigera* and heads in the other studied species. Flowers are yellowish in *F. grisea, G. hermogenesii* and *G. lanigera,* with a tendency to turn red in the first and last one. In *G. prostrata, P. gnaphaloides* and *P.townsendii* flowers are white, turning beige in the last species. All the species have flowers associated with dry and papery bracts that persist alongside their dry fruits, usually a single-seeded achene, favouring anemocoric dispersion.

The fastest lifespan was observed in *G. lanigera,* which took around 20 days to regrowth and finish the flowering phase. In Figure 6, *G. lanigera* was about 20 days old and fruits were almost mature, indicating proximity to the seed dispersal phase. *Pfaffia townsendii* alone showed behaviour that was independent of fire, since even *G. hermogenesii* only flowered after being burned to the ground and regrowing from its xylopodium. The other four species were found after the occurrence of fire, all of them in the flowering stage.

In the Taxonomy and Morphology areas, studies of the genera Achyranthes, Alternanthera, Amaranthus, Blutaparon, Celosia, Chamissoa, Chenopodium, Cyathula, Iresine, Lecosia, Pedersenia, Pseudoplantago, Quaternella and Xerosiphon still need to be done, not only covering the revision of the Brazilian species, biogeography and morphological evolution, but also molecular biology to establish synonyms and to delimit variations among individuals of each species.

3.2. Leaf anatomy of Amaranthaceae species

Leaves of the six studied species have anatomical variation among the genera and are more similar between species of the same genus. Transverse sections show that *G. hermogenesii* (Figure 9), *G. lanigera* (Figure 10) and *G. prostrata* (Figure 11) have large nonglandular trichomes covering the single layered epidermis, dorsiventral mesophyll with upper palisade parenchyma and spongy parenchyma near the lower epidermis. All these three species are amphistomatic, and a complete well-developed parenchymatous sheath with thicker cell walls surrounds the vascular bundles (Kranz cells), in which starch accumulates. Calcium oxalate druses were found in the mesophyll. The leaf anatomy of the three *Gomphrena* spp. is compatible with the C4 photosynthesis pathway.

Pfaffia gnaphaloides (Figure 12) and *P. townsendii* (Figure 13) have more undulating surfaces and a thinner leaf blade in relation to the *Gomphrena* species. Trichomes are also more frequent and thinner and the mesophyll is dorsiventral. The parenchymatous sheath has thinner walls than the neighbouring cells in *Pfaffia* species. Both species had elevated stomata on the lower epidermis and only *P. gnaphaloides* had few stomata on the upper epidermis. Starch was distributed in all mesophyll cells and calcium oxalate druses were rare. The anatomy of *Pfaffia* spp. leaves is compatible with C3 photosynthesis metabolism.

Froelichiella grisea (Figure 14) has the only isobilateral mesophyll among the studied species, with palisade parenchyma near both upper and lower epidermis. Palisade cells are shorter near the lower epidermis. The parenchymatic vascular bundle is not conspicuous and organelles in these cells are positioned towards the outer cell walls, in the same way as they are found in the other mesophyll cells. Calcium oxalate druses were more common near the midrib, and the reaction to starch was similar to that of all the mesophyll cells. Its leaf anatomy is compatible with C3 photosynthesis metabolism. Figures 9-14 [77] were reproduced with the authorization of the Biota Neotropica Editor, Dr. Carlos Joly.

Gomphrena trichomes are similar to the ones described for *G. arborescens* [32,54,78]. Although it is expected that stomata are reduced on the upper surface of land plants, the Cerrado *Gomphrena* species *G. arborescens, G. pohlii* and *G. virgata* have a similar number of stomata on both surfaces [78], subjecting them to a greater water loss, which is compensated by the well-developed subterranean systems that guarantee water supply during the lifespan of their leaves. The size and number of stomata on both leaf surfaces of *G. hermogenesii, G. lanigera* and *G. prostrata* is still to be verified, but simple observation indicates that it should be similar to the phenomena observed in the first cited species, since they also have a relatively well-developed subterranean system.

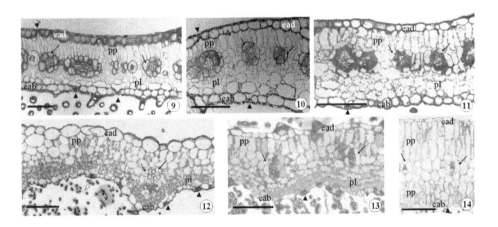

Figure 2. (Figures 9-14) Micrographies of the middle leaf transversal sections of the studied Amaranthaceae species. **Fig. 9.** *Gomphrena hermogenesii* - leaf blade thickness from medium to thick, dorsiventral mesophyll and complete parenchymatous bundle sheath, with thick cell walls and collateral vascular bundles. **Fig. 10.** *G. lanigera* - medium leaf blade, dorsiventral mesophyll and complete parenchymatous bundle sheath, with thick cell walls and collateral vascular bundles. **Fig. 11.** *G. prostrata* – thin to medium leaf blade, dorsiventral mesophyll and complete parenchymatous bundle sheath, with thick cell walls and collateral vascular bundles. **Fig. 12.** *Pfaffia gnaphaloides* – thin leaf blade, dorsiventral mesophyll, less defined parenchymatous bundle sheath and collateral vascular bundles. **Fig. 13.** *P. townsendii* – only hypostomatous leaf species, thin leaf blade, dorsiventral mesophyll, less defined parenchymatous bundle sheath and collateral vascular bundles. **Fig. 14.** *Froelichiella grisea* - thick leaf blade, isobilateral mesophyll with elongated palisade parenchyma under the adaxial epidermis, less defined bundle sheath and collateral vascular bundles. **Legend:** eab = abaxial epidermis; ead = adaxial epidermis; pl = spongy parenchyma; pp = palisade parenchyma; arrowhead = stoma; circle = druse; arrow = parenchymatous bundle sheath. Bar = 100 μm.

The leaf anatomy of the three *Gomphrena* spp. is similar to that of *G. arborescens* L.f. [54], *G. cespitosa*, *G. dispersa*, *G. nitida*, *G. sonorae* [79] and *G. conica*, *G. flaccida* [80] among others, most of them arranged in a *Gomphrena* atriplicoid-type of Kranz anatomy [81,82]. As expected, there was no significant variation in these species' leaf anatomy due to the life cycle stage, although older leaves collected during the vegetative stage have a thicker cuticle covering both epidermis surfaces, especially in *G. hermogenesii* species. The leaf anatomy observed in the two *Pfaffia* spp. is similar to that described in *P. jubata* [83], which also lacks the Kranz anatomy. There is no previous study about the anatomy of *F. grisea* leaves and its genus is monoespecif.

There are still a number of studies to be done in the field of anatomy and histology of Brazilian Amaranthaceae plants. Most of the medicinal species need to be analyzed and validated for their use as drugs, including anatomic description and an investigation of the secondary compounds of the used organs, by histology and by chromatography. Due to the difficulties in correctly identifying the species in the field, anatomical and morphological markers should be defined to guarantee these species' identity even during the vegetative stage. Besides that, anatomical studies can improve the taxonomy data and explain some morphological characters of this plant family, like the anatomical variations in the leaf that are connected to photosynthesis, or the secondary thickening and xylopodium development in underground organs, which is a character for the Cerrado species. The anatomy of few Brazilian Amaranthaceae species has been described, with the exception of some from the *Gomphrena* and *Pfaffia* genera [83-86].

3.3. Leaf ultrastructure of Amaranthaceae species

Leaves of the six studied species have less ultrastructural variation among the species of the same genus. *Froelichiella grisea* organelles are equally distributed among chlorenchyma tissues, usually near the cell walls. The chloroplasts of this species are always granal (Figure 15), even in the vascular cells, with large starch granules (usually one or two per organelle) in all tissues. Plastoglobuli are small and less numerous in mesophyll chloroplasts (Figure 15), but guard cell chloroplasts usually have just one large plastoglobulus and less conspicuous grana. Mitochondria and peroxisomes (Figure 15) were found in mesophyll and bundle sheath cells. Leaf ultrastructure is compatible with C_3 photosynthesis metabolism.

Mesophyll cell chloroplasts of *Gomphrena* species have conspicuous grana, rare starch granules and variable size of plastoglobuli: *G. hermogenesii* has larger ones in relation to *G. lanigera* and *G. prostrata*. Bundle sheath chloroplasts are completely devoid of grana or have few stacked thylakoids (Figure 16) in all studied *Gomphrena* species, but always have large starch granules and plastoglobuli. The larger the starch granules, the more deformed the chloroplasts' typical lens shape, as shown in *G. hermogenesii* (Figure 16). Mitochondria are usually numerous in bundle sheath cells and are always near chloroplasts, grouped next to the inner cell wall (towards the vascular bundle). Peroxisomes are rare, and a few were observed near chloroplasts in palisade and spongy parenchyma cells, but not in the bundle sheath cells. Phloem companion cells are mitochondria-rich in all *Gomphrena* species, as shown in *G. prostrata* (Figure 17). The presence of dimorphic chloroplasts, disposition of the organelles and the occurrence of Kranz syndrome seen in the leaf anatomy indicate that the C_4 photosynthesis pathway operates in the three studied *Gomphrena* spp.

Pfaffia species organelles are equally distributed among chlorenchyma tissues, usually near the cell walls. *Pfaffia* chloroplasts are granal even in the vascular cells, showing large starch granules and a similar size in all mesophyll cells, as can be observed in the palisade parenchyma of *P. townsendii* (Figure 18). Mitochondria and peroxisomes are common near chloroplasts (Figure 18). Phloem companion cells are mitochondria-rich and chloroplasts are smaller and granal, as in the other species of this study. Along with the aspects of *Pfaffia* anatomy described previously, their ultrastructure is compatible with the C_3 photosynthesis pathway.

Pfaffia gnaphaloides (Figure 19) and *G. hermogenesii* (Figure 20) leaves, collected during the flowering stage, were colonized by two distinct forms of microorganisms: (i) a smaller organism was found in the intercellular spaces (ics) of the spongy parenchyma (Figure 19); (ii) a larger and distinctly eukaryotic organism was found within distinct cells, with some morphological alterations suggesting an infectious process (Figures 19-20).

The external envelopae membranae system of the chloroplasts is disrupted in infected cells (Figure 20) and a size reduction was observed in the chloroplast plastoglobuli. All morphological characteristics observed in the intracellular microorganism suggest that it should be an obligate biotroph endophytic fungus belonging to the Ascomycete division (Figure 20). The invading fungus may be using the plastoglobuli lipids as its primary source of carbon and energy; the reduction of the plastoglobuli could also be due to its mobilization by the host plants in response to the stress caused by these biotic interactions. The complete identification of the fungus and its effect on the plants depends on its isolation from the environment/hosts and complementary studies.

The rare peroxisomes in *Gomphrena* spp. leaf cells and their presence among all chlorenchyma tissues of the *Pfaffia* spp. leaf cells is compatible with their possible photosynthesis metabolisms. Along with the presence of Kranz syndrome and dimorphic chloroplasts, the absence of peroxisome indicates that *Gomphrena* spp. perform photosynthesis via the C_4 pathway. In *Gomphrena* species, CO_2 concentration in the bundle sheath cells must be efficient, leading to a significant reduction in the oxygenase function of its RuBisCO enzyme. This leaves the species virtually free of the photorespiration process, aided by the large walls of the bundle sheath cells. Although a carbon isotope ratio study [38] indicates that *G. hermogenesii* is not a C_4 species, this species also has Kranz anatomy and ultrastructure compatible with C_4 metabolism, as do all the other studied *Gomphrena* spp. [42]. The distribution of its key photosynthetic enzymes will be carried out using immuno-cytochemistry, in our laboratory, in order to complete these data.

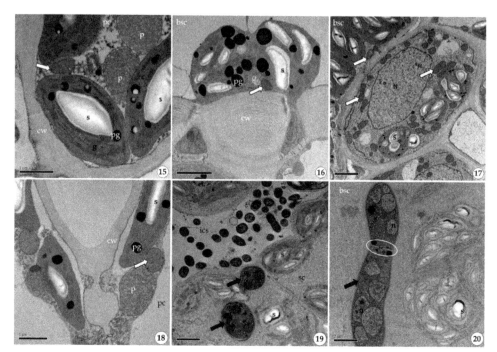

Figure 3. (Figures 15-20) Citological aspects of Amaranthaceae species as seen through a Transmission Electron Microscope. **Fig.15.** *Froelichiella grisea* palisade parenchyma cell. **Fig. 16.** *Gomphrena hermogenesii* bundle sheath cell. **Fig. 17.** *G. prostrata* phloem companion cell and bundle sheath cell on top. **Fig. 18.** *Pfaffia townsendii* palisade parenchyma cells.**Fig. 19.** *P. gnaphaloides* spongy parenchyma cells and invading microorganisms (black arrows). **Fig. 20.** *G. hermogenesii* bundle sheath cell and the invading Ascomycete fungus (black arrow) and the disrupted chloroplasts with smaller plastoglubuli. **Legend:** black arrow = invading organism; white arrow = mitochondria; ellipsis = septum with a simple pore; bsc = bundle sheath cell; cw = cell wall; ics = intercellular space; n = nucleus of the microorganism; N = nucleus of the plant species; p = peroxisome; pc = palisade parenchyma cell; pg = plastoglobulus in a chloroplast; s = starch granule in a chloroplast; sc = spongy parenchyma cell.

4. Conclusions and perspectives

This chapter presents data on Amaranthaceae species, with no pretension to explain the full potential of this plant family for scientific studies, but rather to provide a basic tool for those interested in amplifying studies on the species of this family. Based on our results, we are convinced of the importance of studying this family further, not only as a tool in the better preservation of endemic species, but also to explore its undoubted economic importance more fully. Basic research is still needed, with the aim of applying knowledge on these species to technological advances, especially in growing crops - since C_4 species have a faster metabolism and growing capacity, as observed in the species found in the RPPN Cara Preta - and to explore medicinal molecules of these plants. C_4 species are also important to balance CO_2 in the atmosphere because of their efficiency in the transformation of carbon into biomass; in Cerrado Amaranthaceae species, this storage is basically underground in their well-developed subterranean roots and xylopodium.

The number of medicinal plants among the Brazilian Amaranthaceae species may well be higher than already reported (Table 1), because Cerrado inhabitants are particularly interested in the highly developed subterranean systems of some medicinal species [34,49,58] which can be collected at any time of the year, even from species whose aerial portions are not persistent. Due to the morphological similarity among Amaranthaceae species in the Neotropical Savannah, their collectors can easily mistake one species for another during the vegetative stage, which confirms the need for further and more complete studies of the known medicinal and endangered species, at least.

Preparation of plant samples for transmission electron microscopy also proved to be useful in studying the morphology of fungi inside plant cells, as well as aspects of host-parasite interaction. This kind of study could be recommended for plants considered toxic to herbivores and to any medicinal plant consumed by humans, in order to give more information about the real source of poisoning or medicinal effect and for fine quality control. In both studied species (*G. hermogenesii* and *P. gnaphaloides*) the external macro aspects of the plants did not indicate the presence of the endophytic fungus.

RPPN Cara Preta is a small Private Conservation Unit (only 1.5% of the area of the Chapada dos Veadeiros National Park, a government-preserved area of 65,038 hectares). Both Conservation Units are separated only by a road, in Alto Paraíso municipality of Goiás State, Brazil. The latter site is registered by UNESCO as a natural protected Cerrado zone. RPPN Cara Preta has 245 species representing 47 family plants [75,86], which is 9.2% of the 2,661 plant species of Chapada dos Veadeiros [87], a good diversity of plants in relation to the occupied area. There are six Amaranthaceae species in the RPPN – 25% of the 27 species found in the National Park [87]. Considering that the RPPN Cara Preta Utilization Planning Report [86] indicated the presence of three endemic species, plus two Amaranthaceae species not reported initially [75], this Conservation Unit has 2% of endemic species – which is more than expected. According to [2,12], the Cerrado Biome is one of the priority hotspots for conservation because it has, among others, 4,400 endemic plants (1.5% of the Earth's 300,000 species). The Amaranthaceae family in RPPN Cara Preta can be considered a taxon indicator of the good diversity of the Neotropical Savannah. This taxon could be considered a plant diversity indicator in other works on flora in open areas of the Cerrado Biome. Because of the predominant habit (herbs and shrubs) and survival strategies, the presence of species from this family among the collected species clearly indicates a well performed collection effort.

There are a number of important factors indicating that this plant family deserves more studies for a greater understanding by researchers working in Brazil, and we recap them as follows: the Cerrado Biome holds 98 of the 146 Brazilian Amaranthaceae species (almost 70% of the total species) (Table 1); their pirophytic behavior and survival strategies (fast regrowth and seed dispersal before the complete recovery of grasses after fire) are coherent with the Biome's characteristics; their morphology shows exceptional adaptation to the seasonal climate and open areas (hairy aerial portions, partial or total loss of the aerial portions during the dry season, well-developed underground system with xylopodium, dry fruit dispersal by wind); their metabolism (evolution of C_4 and intermediary C_3-C_4 photosynthesis) may have importance for biomass conversion and CO_2 balance; and, finally, many of these plants are already used in medicines by Cerrado inhabitants and there may be much wider medicinal potential in other species of this family.

Author details

Suzane M. Fank-de-Carvalho*
National Council for Scientific and Technological Development - CNPq, Brasília, Distrito Federal, Brazil
Universidade de Brasília – UnB, Biological Institute, Electron Microscopy Laboratory, Brasília, Distrito Federal, Brazil

Sônia N. Báo
Universidade de Brasília – UnB, Biological Institute, Electron Microscopy Laboratory, Brasília, Distrito Federal, Brazil

Maria Salete Marchioretto
Instituto Anchietano de Pesquisas/UNISINOS, PACA Herbarium, São Leopoldo, Rio Grande do Sul, Brazil

Acknowledgement

We would like to thank CAPES, CNpq and FINEP for financial support; NGO Oca Brasil and Herbaria IBGE, UB and PACA for access authorization and research infrastructure; the aditional collectors for help in searching for and collecting the species at RPPN Cara Preta; and Susan Casement Moreira for the English review.

5. References

[1] Forzza, R.C.; Leitman, P.M.; Costa, A.F.; Carvalho Jr., A.A.: Peixoto, A.L.; Walter, B.M.T.; Bicudo, C.; Zappi, D.; Costa, D.P.; Lleras, E.; Martinelli, G.; Lima, H.C.; Prado, J.; Stehmann, J.R.; Baumgratz, J.F.A.; Pirani, J.R.; Sylvestre, L.; Maia, L.C., Lohmann, L.G.; Queiroz, L.P.; Silveira, M.; Coelho, M.N.; Mamede, M.C.; Bastos, M.N.C.; Morim, M.P.; Barbosa, M.R.; Menezes, M.; Hopkins, M.; Secco, R.; Cavalcanti, T.B. & Souza, V.C. (2010). Catálogo de Plantas e Fungos do Brasil – Vol. 1. Rio de Janeiro: Andrea Jakobsson Estúdio/Instituto de Pesquisa Jardim Botânico do Rio de Janeiro. 870 p.

[2] Myers, N.; Mittermeier, R.A.; Mittermeier, C.G.; Fonseca, G.A.B. & Kent, J. (2000). Biodiversity hotspots for conservation priorities. Nature, 403: 853-858.

---

* Corresponding Author

[3] Conservation International (2012). Cerrado. In: Biodiversity Hotspots. Available: http://www.biodiversityhotspots.org/xp/hotspots/cerrado/Pages/default.aspx   Acessed 2012 Jan 31.

[4] Harley, R.M. & Giulietti, A.N. (2004). Wild flowers of the Chapada Diamantina – illustrated botanical walks in the mountains of NE Brazil. São Carlos: RIMA Ed. 344 p.

[5] Dias, B.F.S. (coord.). (1992). Alternativas de desenvolvimento dos cerrados: manejo e conservação dos recursos naturais renováveis. Brasília: FUNATURA/IBAMA. 97 p.

[6] Simon, M.F. & Proença, C. (2000). Phytogeographic patterns of Mimosa (Mimosoideae, Leguminosae) in the Cerrado biome of Brazil: an indicator genus of high-altitude centers of endemism? Biological Conservation, 96/3: 279-296.

[7] Eiten, G. (2001). Vegetação natural do Distrito Federal. Brasília: SEBRAE/UNB. 162 p.

[8] Klink, C.A. & Machado, R.B. (2005). A conservação do Cerrado brasileiro. Megadiversidade, 1/1: 147-155.

[9] Marinho-Filho, J.; Machado, R.B. & Henriques, R.P.B.H. (2010). Evolução do conhecimento e da conservação do Cerrado brasileiro. In: Diniz, I.R.; Marinho-Filho, J; Machado, R.B. & Cavalcanti, R.B. Orgs. Cerrado: conhecimento científico quantitativo como subsídio para ações de conservação. Brasília: Thesaurus Editora. Pp. 13-31.

[10] Paiva, P.H.V. (2000). A Reserva da Biosfera do Cerrado: fase II. In: Cavalcanti, T. B.; Walter, B. M. T. editors. Tópicos atuais em botânica - palestras convidadas do 51° Congresso Nacional de Botânica. Brasília: Sociedade Brasileira de Botânica/Embrapa-Cenargen. Pp. 332-334.

[11] Coutinho, L.M. (2006). O conceito de bioma. Acta Botanica Brasilica, 20/1: 13-23.

[12] Mittermeier, R.A.; Myers, N.; Thomsen, J.B.; Fonseca, G.A.B. & Olivieri, S. (1998). Biodiversity Hotspots and Major Tropical Wilderness Areas: Approaches to Setting Conservation Priorities. Conservation Biology, 12/3: 516-520.

[13] Ratter, J.A.; Ribeiro, J.F. & Bridgewater, S. (2000). Woody flora distribution of the Cerrado Biome: phytogeography and conservation priorities. In: Cavalcanti, T. B.; Walter, B. M. T. editors. Tópicos atuais em botânica - palestras convidadas do 51° Congresso Nacional de Botânica. Brasília: Sociedade Brasileira de Botânica/Embrapa-Cenargen. Pp. 340-342.

[14] Miranda, H. & Miranda, A.C. (2000). Queimadas e estoques de carbono no Cerrado. In: Moreira, A.G. & Schwartzman, S. editors. As Mudanças Climáticas e os Ecossistemas Brasileiros. Brasília: Ed. Foco. Pp. 75-81.

[15] Marchioretto, M.S.; Windisch, P.G. & Siqueira, J.C. (2002). Os gêneros Froelichia Moench e Froelichiella R.E. Fries (Amaranthaceae) no Brasil. Pesquisas-Botânica, 52: 7-46.

[16] Marchioretto, M.S.; Windisch, P.G. & Siqueira, J.C. (2004). Padrões de distribuição geográfica das espécies de Froelichia Moench e Froelichiella R.E. Fries Amaranthaceae) no Brasil. Iheringia Série Botânica, 59/2: 149-159.

[17] Marchioretto, M.S. (2008a). Os gêneros Hebanthe Mart. e Pfaffia Mart. (Amaranthaceae) no Brasil. Instituto de Biociências, Doutorado em Botânica, Universidade Federal do Rio Grande do Sul, Porto Alegre, Brazil. (Thesis). 255 p.

[18] Marchioretto, M.S; Azevedo, F.; Josende, M.V.F & Schnorr, D.M. (2008b). Biogeografia da família Amaranthaceae no Rio Grande do Sul. Pesquisas-Botânica, 59/2: 171-190.

[19] Marchioretto, M.S.; Miotto, S.T.S. & Siqueira, J.C. (2009). O gênero Hebanthe (Amaranthaceae) no Brasil. Rodriguésia, 60/4: 783-798.

[20] Marchioretto, M.S.; Miotto, S.T.S. & Siqueira, J.C. (2010). O gênero Pfaffia Mart. (Amaranthaceae) no Brasil. Hoehnea, 37/3: 461-511.

[21] Marchioretto, M.S.; Senna, L. & Siqueira, J.C.de. (2012). Amaranthaceae. In: Lista de Espécies da Flora do Brasil. Available: http://floradobrasil.jbrj.gov.br/2010/FB000042 Accessed 2012 Jan 09.

[22] Siqueira, J.C. (1992). O gênero Gomphrena L. (Amaranthaceae) no Brasil. Pesquisas-Botânica, 43: 5-197.

[23] Siqueira, J.C. (1995). Fitogeografia das Amaranthaceae Brasileiras. Pesquisas-Botânica, 45: 5-21.

[24] Siqueira, J.C. (2002). Amaranthaceae. In: Wanderley, M.G.L.; Shepherd, G. & Giulietti, A.M. editors. Flora Fanerogâmica do Estado de São Paulo. São Paulo: Ed. FAPESP-HUCITEC. Pp. 11-30.

[25] Siqueira, J.C. (2004). Amaranthaceae: padrões de distribuição geográfica e aspectos comparativos dos gêneros Africanos e Sulamericanos. Pesquisas-Botânica, 55: 177-185.

[26] Holmgren, P. K.; Holmgren, N.H. & Barnett, L.C. (1990). Index Herbariorum - Part I: The Herbaria of the World. New York, IAPT. 693 p.

[27] Thiers, B.M. (2012). Index Herbariorum: A Global Directory of Public Herbaria and Associated Staff. In: NYBG Herbarium. Available: http://sciweb.nybg.org/science2/IndexHerbariorum.asp Accessed 2012 Feb 24.

[28] Kraus, J.E. & Arduin, M. (1997). Manual básico de métodos em morfologia vegetal. Rio de Janeiro: EDUR. 198 p.

[29] Souza, W.de. (Ed.) (2010). Técnicas básicas de microscopia eletrônica aplicadas às ciências biológicas, 3.ed. Rio de Janeiro: Sociedade Brasileira de Microscopia. 357 p.

[30] Shiomi, N.; Odera, S.; Vieira, C.C.J. & Figueiredo-Ribeiro, R.C.L. (1996). Structure of fructan polymers from tuberous roots of Gomphrena macrocephala (Amaranthaceae) from the cerrado. New Phytologist, 133/4: 643-650.

[31] Figueiredo-Ribeiro, R.C.L.; Dietrich, S.M.C.; Carvalho, M.A.M.; Vieira, C.C.J.; Isejima, E.M.; Dias-Tagliacozzo, G.M.; Tertuliano, M.F. (1982). As múltiplas utilidades dos frutanos – reserva de carboidratos em plantas nativas do cerrado. Ciência Hoje, 14/84: 16-18.

[32] Fank-de-Carvalho, S.M. (2004). Contribuição ao conhecimento botânico de Gomphrena arborescens L.f. (Amaranthaceae) - estudos anatômicos e bioquímicos. Instituto de Ciências Biológicas, Mestrado em Botânica, Universidade de Brasília, Brasília, DF, Brazil. (Dissertation). 139 p.

[33] Corrêa, M.P. (1931). Dicionário das plantas úteis do Brasil e das exóticas cultivadas, Vol. II. Rio de Janeiro: Ministério da Agricultura. 707 p.

[34] Barros, M.G.A.E. (1982). Plantas medicinais – usos e tradições em Brasília – DF. Oréades, 8/14-15: 140-149

[35] Almeida, S.P. de; Proença, C.E.B.; Sano, S.M. & Ribeiro, J.F. (1998). Cerrado - espécies vegetais úteis. Planaltina: Embrapa. 464 p.

[36] Kardošová, A.; Ebringerová, A; Alföldi, J.; Nosál'ová, G.; Fra˘nová, S. & Hˇr'ıbalová, V. (2003). A biologically active fructan from the roots of Arctium lappa L., var. Herkules. International Journal of Biological Macromolecules, 33: 135–140.

[37] Kunert, O.; Haslinger, E.; Schmid, M.G.; Reiner, J.; Bucar, F.; Mulatu, E.; Abebe, D. & Debella, A. (2000). Three Saponins, a Steroid, and a Flavanol lycoside from Achyrantes aspera. Monatshefte für Chemie, 131/2: 195-204.

[38] Sage, R.F.; Sage, T.L.; Pearcy, R.W.; Borsch, T. (2007). The taxonomic distribution of C4 photosynthesis in Amaranthaceae sensu stricto. American Journal of Botany, 94/12: 1992-2003.

[39] Siqueira, J.C. (1987). Importância alimentícia e Medicinal das Amaranthaceae do Brasil. Acta Biologica Leopoldense, 1: 99-110.

[40] Lorenzi, H. & Matos, F. J. A. (2008). Plantas medicinais no Brasil - nativas e exóticas, 2.ed. Nova Odessa: Editora Plantarum. 576 p.

[41] Lorenzi, H. & Souza, H. M. (2008). Plantas ornamentais no Brasil - arbustivas, herbáceas e trepadeiras, 4.ed.. Nova Odessa: Editora Plantarum. 1120 p.

[42] Fank-de-Carvalho, S.M. (2011). Contribuição ao conhecimento da anatomia, micromorfologia e ultraestrutura foliar de Amaranthaceae do Cerrado. Instituto de Biologia, Doutorado em Biologia Celular e Estrutural, Universidade Estadual de Campinas, Campinas, SP, Brazil. (Thesis). 107 p.

[43] MMA (2008). Lista Nacional das Espécies da Flora Brasileira Ameaçadas de Extinção. In: Instrução Normativa no. 6 do Ministério do Meio Ambiente, de 23 de setembro de 2008. Availale: http://www.mma.gov.br/estruturas/179/_arquivos/ 179_05122008033615.pdf Accessed 2012 Apr 10.

[44] RS (2003). Lista Final das Espécies da Flora Ameaçadas – RS. In: Decreto estadual n 42.099, Estado do Rio Grande do Sul, publicado em 01/jan/2003. Available: http://www.fzb.rs.gov.br/downloads/flora_ameacada.pdf Acessed 2012 Apr 10.

[45] Smith, L.B. & Downs, R.J. (1972). Amarantháceas. In: Reitz, R. editor. Flora Ilustrada Catarinense. Itajaí: Herbário Barbosa Rodrigues. Pp. 1-110.

[46] SP (2004). Lista oficial das espécies da flora do Estado de São Paulo ameaçadas de extinção. In: Resolução SMA 48, Diário Oficial do Estado de São Paulo - Meio Ambiente, 22/set/2004. Available: http://www.ibot.sp.gov.br/pesquisa_cientifica/ restauracao_ecologica/resolu%C3%A7%C3%A3o_%20sma48.pdf Accessed 2012 Apr 10.

[47] Rajendrudu, G.; Prasad, J.S.R. & Rama Das, V.S. (1986). C3-C4 species in Alternanthera (Amaranthaceae). Plant Physiology, 80: 409-414.

[48] IUCN (2011). The IUCN Red List of Threatened Species. In: International Union for Conservation of Nature and Natural Resources. Available: http://www.iucnredlist.org/ apps/redlist/search Accessed 2012 Jan 21.

[49] Siqueira, J.C. (1988). Planta Medicinais – identificação e uso das espécies dos cerrados. São Paulo: Edições Loyola. 40 p.

[50] Almeida, A.M.R.; Fukushigue, C.Y.; Sartori, F.; Binneck, E.; Marin, S.R.R.; Inoue-Nagata, A.K.; Chagas, C.M.; Souto, E.R. & Mituti, T. (2007). Natural infection of Alternanthera tenella (Amaranthaceae) by a new potyvirus. Archives of Virology, 152: 2095–2099.

[51] Vasconcellos, J.M. DE O. (1985a). Amaranthaceae do Rio Grande do Sul, Brasil-II. Roessléria, 7/2: 107-137.

[52] Vasconcellos, J.M. DE O. (1985b). Amaranthaceae do Rio Grande do Sul, Brasil-III - Gêneros Celosia L. e Chamissoa H.B.K. Roessléria, 7/3: 165-182.

[53] Botsaris, A.S. (2007). Plants used traditionally to treat malaria in Brazil: the archives of Flora Medicinal. Journal of Ethnobiology and Ethnomedicine, 3: 1-8. Available: http://www.ethnobiomed.com/content/3/1/18 Accessed 2012 Jan 9.

[54] Fank-de-Carvalho, S.M. & Graciano-Ribeiro, D. (2005). Arquitetura, anatomia e histoquímica das folhas de Gomphrena arborescens L. f. (Amaranthaceae). Acta Botanica Brasilica, 19/2: 377-390.

[55] Vasconcellos, J. M. DE O. (1986a). Amaranthaceae do Rio Grande do Sul, Brasil V. Gêneros Pfaffia Mart. e Gomphrena Mart. Roessléria, 8/2: 75-127.

[56] Balbach, A. (1957). As plantas curam, 5.ed. São Paulo: Ed. Missionário. Pp. 299-300.

[57] Silva, I.M. (1983). A flora na vida do índio Karajá. Atas da Sociedade Botânica-secção RJ, 2/3: 21-32.

[58] ES (2005). Lista de Espécies Ameaçadas de Extinção no Espírito Santo. In: Decreto Nº 1.499-R, Diário Oficial do Estado do Espírito Santo, 14/jun/2005. Available: http://www.meioambiente.es.gov.br/web/flora.htm Accessed 2012 Apr 10.

[59] Siqueira, J.C. (1981). Utilização popular das plantas do cerrado. São Paulo: Editora Loyola. 60 p.

[60] Sanches, N.R.; Galletto, R.; Oliveira, C.E.; Bazotte, R.B. and Cortez, D.A.G. (2001). Avaliação do potencial anti-hiperglicemiante da Pfaffia glomerata (Spreng.) Pedersen (Amaranthaceae). Acta Scientiarum - Biological Sciences, 23/2: 613-617

[61] Reeves, R.D.; Baker, A.J.M.; Becquer, T.; Echevarria, G. & Miranda, Z.J.G. (2007). The flora and biogeochemistry of the ultramafic soils of Goiás state, Brazil. Plant and Soil, 293: 107–119.

[62] Appezzato-da-Gloria, B. (2003). Morfologia de sistemas subterrâneo - histórico e evolução do conhecimento no Brasil. Ribeirão Preto: A.S. Pinto Editor. 80 p.

[63] Borsch, T. & Pedersen, T.M. (1997). Restoring the Generic Rank of Hebanthe Martius (Amaranthaceae). Sendtnera, 4: 13-31.

[64] Martius, C.F.P. v. (1825). Beitrag Zur Kenntnis der natürlichen familie der Amaranthaceen. Bonn: Eduard Weber's Buchhandlung. 321 p.

[65] Martius, C.F.P.v. (1826). Amaranthaceae. In: Martius, C.F.P.v. editor. Nova genera et species plantarum – in itinere per Brasiliam, Vol.II. Munich, Typis C. Wolf. Pp. 1-64.

[66] Moquin-Tandon, A. (1849). Amaranthaceae. In: Candolle A. De editor. Prodromus Systematis Naturalis Regni Vegetabilis, Vol.13, part.2. Paris: Victoris Masson Ed. Pp. 419-423.

[67] Pedersen, T.M. (1967). Studies in South American Amaranthaceae. Darwiniana, 14/2-3: 430-462.

[68] Pedersen, T.M. (1976). Estudios sobre Amarantáceas sudamericanas – II. Darwiniana, 20/1-2: 269-303.

[69] Pedersen, T.M. (1990). Studies in South American Amaranthaceae - III (incluinding one amphi-Atlantic species) - Bull. Mus. Natl. Hist. Nat., B. Adansonia, Sér.4, 12/1: 69-97.

[70] Pedersen, T. M. (1997). Studies in South American Amaranthaceae – IV - Bull. Mus. Natl. Hist. Nat., B. Adansonia, Sér.3, 19/2: 217-251.

[71] Pedersen, T.M. (2000). Studies in South American Amaranthaceae – V. Bonplandia, 10/1-4: 83-112.

[72] Seubert, M. (1875). Amaranthaceae. In: Martius, C.F.P. Von, Endlicher & Urban editors.Flora Brasiliensis, Vol.V/part. 1. Munich, Typografia Regia. Pp.188-202.

[73] Stützer, O. (1935). Die Gattung Pfaffia mit einem Anhag neur Arten von Alternanthera. Feddes Repertorium Specierum Novarum Regni Vegetabilis, 88: 1-49.

[74] Suessenguth, K. (1934). Neue und kritische Amarathaceen aus Süd und Mittelamerika. Feddes Repertorium Specierum Novarum Regni Vegetabilis, 35: 298-337.

[75] Suessenguth, K. (1938). Amaranthaceen-Studien – I - Amaranthaceae aus Amerika, Asien, Australien. Feddes Repertorium Specierum Novarum Regni Vegetabilis, 44: 36-48.

[76] Vasconcellos, J. M. DE O. (1986b). Amaranthaceae do Rio Grande do Sul, Brasil – IV - Gêneros Pseudoplantago Suess., Iresine BR. e Blutaparon Rafin. Roessléria, 8/1: 03-15.

[77] Fank-de-Carvalho, S. M.; Marchioretto, M.S. & Báo, S.N. (2010a). Anatomia foliar, morfologia e aspectos ecológicos das espécies da família Amaranthaceae da Reserva Particular do Patrimônio Natural Cara Preta, em Alto Paraíso, Goiás, Brasil. Biota Neotropica, 10: 77-86. Available: http://www.biotaneotropica.org.br/v10n4/en/download?article+bn01310042010+abstract Accessed 2012 Apr 10.

[78] Fank-de-Carvalho, S.M.; Gomes, M.R.A.; Silva, P.I.T. & Báo, S.N. (2010b). Leaf surfaces of Gomphrena spp. (Amaranthaceae) from Cerrado biome. Biocell (Mendoza), 34/1: 23-35.

[79] Welkie, G.W. & Caldwell, M. (1970). Leaf anatomy of species in some dicotyledon families as related to the C_3 and C_4 pathways of carbon fixation. Canadian Journal of Botany, 48/12: 2135-2146.

[80] Carolin, R.C.; Jacobs, S.W.L. & Vesk, M. (1978). Kranz cells and mesophyll in the Chenopodiales. Austalian Journal of Botany, 26/5: 683-698.

[81] Kadereit, G.; Borsh, T.; Weising, K. & Freitag, H. (2003). Phylogeny of Amaranthaceae and Chenopodiaceae and the evolution of C_4 photosynthesis. International Journal of Plant Sciences, 164/6: 959-986.

[82] Sage, R.F. (2004). The Evolution of C_4 photosynthesis. New Phytologist, 161/2: 341-370.

[83] Estelita-Teixeira, M.E. & Handro, W. (1984). Leaf ultrastructure in species of Gomphrena and Pfaffia (Amaranthaceae). Canadian Journal of Botany, 62/4: 812-817.

[84] Gavilanes, M.L. (1999). Estudo anatômico do eixo vegetativo de plantas daninhas que ocorrem em Minas Gerais. 1. Anatomia foliar de Gomphrena celosioides Mart. (Amaranthaceae). Ciência e Agrotecnologia, 23/4: 881-898.

[85] Handro, W. (1964). Contribuição ao estudo da venação e anatomia foliar das amarantáceas dos cerrados. Anais da Academia Brasileira de Ciências, 36/4: 479-499.

[86] Handro, W. (1967). Contribuição ao estudo da venação e anatomia foliar das amarantáceas dos cerrados. II – Gênero Pfaffia. Anais da Academia Brasileira de Ciências, 39/3-4: 495-506.

[87] De Souza, A. (2004). Plano de Utilização RPPN Cara Preta. Nativa Proteção Ambiental, Alto Paraíso, Brazil. (Report). 327 p.

[88] Felfili, J.M., Rezend, A,V. & Silva-Júnior, M.D. (orgs). (2007). Biogeografia do Bioma Cerrado – vegetação e solos da Chapada dos Veadeiros. Brasília: Editora Universidade de Brasília/FINATEC. 256 p.

Chapter 2

Impact of Land-Use and Climate on Biodiversity in an Agricultural Landscape

Andrzej Kędziora, Krzysztof Kujawa, Hanna Gołdyn, Jerzy Karg,
Zdzisław Bernacki, Anna Kujawa, Stanisław Bałazy, Maria Oleszczuk,
Mariusz Rybacki, Ewa Arczyńska-Chudy, Cezary Tkaczuk,
Rafał Łęcki, Maria Szyszkiewicz-Golis, Piotr Pińskwar,
Dariusz Sobczyk and Joanna Andrusiak

Additional information is available at the end of the chapter

1. Introduction

The term "biodiversity" was used for the first time by wildlife scientist and conservationist [1] in a lay book advocating nature conservation. The term was not adopted by more then decade. In 1980 use of the term by Thomas Lovejoy in the Foreword to the book "Conservation Biology" [2] credited with launching the field of conservation biology introduced the term to the scientific community. There are many definitions of biodiversity. One of them, formulated in Millennium Ecosystem Assessment [3] reads: " Biodiversity is the variability among living organisms from all sources, including terrestrial, marine, and other aquatic ecosystems and the ecological complexes of which they are part; this includes diversity within species, between species, and of ecosystems". **Biodiversity forms the foundation of the vast array of ecosystem services that critically contribute to human well-being.** Biodiversity is important in human-managed as well as natural ecosystems. Decisions humans make that influence biodiversity affect the well-being of themselves and others [3].

Biodiversity is the one of basic driving forces which influence and determine most of ecosystem services [3]. Interdisciplinary and long-term (50 years) researches carried out by Institute for Agricultural and Forest Environment were focused on recognition of factors, that brought about impoverishment of biodiversity in agricultural landscape and finding the ways to counteract these negative changes. During a few last centuries the rapid decrease of biodiversity has been observed all over the world [4]. Human activity brought to worsening the habitat ability to ensure the conditions for living rich plant, animal and fungi population. Conversion of more stable ecosystems like forests, meadows, and wetlands into less stable ones like arable land causes increase of threats to such fundamental processes like energy flow and cycling of matter in the environment [5]. Such factors like climate changes, simplification of landscape structure and changes in farming practices have resulted in worsening habitat conditions (water conditions, environment pollution, soil degradation etc.) leading in consequence to impoverishment of biodiversity. Agriculture is commonly considered to be one of the main threats to the biological diversity. Farmers try to eliminate all organisms which could diminish crops to increase agricultural production. In order to channel the solar energy and nutrients into products useful for man, farmers simplify plant cover structure both within cultivated fields (selection of genetically uniform and productive cultivars, elimination of weeds), and within agricultural landscape (eradication of hedges, patches of trees, mid-field wetlands and ponds, riparian vegetation strips). Increasing usage of fertilizers and pesticides caused environmental pollution that threats all organisms. Apart from this, the unfavorable climate changes were observed during the last century. Increase of air temperature and wind speed without clear increases of precipitation brought about worsening water conditions.

The agricultural landscape located in Wielkopolska (western Poland) around the Turew village (currently protected by law as landscape park) is a distinct example of such management which allows keeping intensive crop production without causing serious degradation of habitat and impoverishment of biodiversity. A network of diverse linear habitats (shelterbelts, hedgerows, roadside verges, tree alleys etc.) and small wood patches established 200 years ago, stretches of meadows, small mid-field water reservoirs, or wetlands provide good refuge for many organisms and unique facility for studying the effects of farming intensification at stable spatial arrangement of non-farmed habitats. In the Turew agricultural landscape near 850 species of vascular plants, 2600 insect species, 120 species of breeding birds, and about 700 species of macrofungi were detected, including many woodland species, as well as a variety of invertebrates including rare, threatened, protected by law and umbrella species [6]. As many, as 60-100% of animal taxa showed by regional lists of taxa live in this agricultural landscape.

The crucial factor for maintenance numerous and favorable habitats for various groups of animals is well developed structure of landscape. Guidelines for shaping of landscape structure towards ensuring maintenance end enhancement biodiversity as a crucial factor for sustainable development of agricultural landscape. For instance, such elements of landscape like shelterbelts, hedges, and strips of meadows, allows to survive animals (like bees), which population is reduced in the result of intensive use of pesticides. The old trees with a lot of hollows, existing within shelterbelts, are necessary for keeping rich population of many species of birds. In the chapter we summarized our knowledge on quantitative relationships between landscape features and diversity (species richness and abundance) of taxa mentioned above and to present the spatio-temporal pattern of species distribution and their abundance.

It is possible to reconcile very high level of intensive agriculture with protection and even enhancement of biodiversity. Creating a very mosaic landscape (biotic and abiotic component of landscape) is a toolkit for solving this problem. The integrity of biological and physical or chemical processes is a basic foundation of modern ecosystem or landscape ecological approaches. Recognition of this functional relationships leads to the conclusion that biodiversity cannot be successfully protected only by isolation from hostile surrounding, but its conservancy should rely on the active management of the landscape structures in a direction of their diversification. The very important guidelines for enhancement of sustainable system resistance and resilience to threats are diversification of its structure Generally speaking, the results of our study are in line with very recent findings, that due to the great diversity of climatic and physiographic conditions as well as customary regulatory and management of the landscape across Europe, makes it very much needed regional approach to biodiversity protection and management. For example, application of agri-environmental schemes, with the rules elaborated mostly in Western Europe (where agricultural landscape is in general extremely simplified, and where climate is mild, e.g. winters are warmer) to CEE countries seems to be inaccurate [7, 8].

2. Description of Study areas (Turew mosaic landscape and Kościan Plain)

The "Turew Mosaic Landscape" in this chapter is the part of the larger study area called "Kościan Plain" is located about 40 km south from Poznań – the capital of Wielkopolska region and its geographical coordinates are 16°45′ to 17°05′ E and 51°55′ to 52°05′ N (Figure 1). The Field Station of Institute is situated in the middle of the landscape near a small village called Turew. Therefore, the name Turew is used to identify the landscape. Wielkopolska region is known as the "bread basket". Agriculture is the dominant activity of the region.

Figure 1. Location of Turew Landscape

The study area "Kościan Plain" (area about 200 km²) is located within radius of 5 – 10 km around Turew, in the area of the West Polish Lowland, which is a ground moraine created during the Baltic glaciation, that terminated about 10 000 year ago. Although the differences in altitude are small (from 75 m to 90 m a.s.l.) and the area consists of a rolling plain made up of slightly undulating ground moraine there are many drainage valleys. In general, light textured soils (Hapludalfs, Glossudalfs and less frequently met Udipsamments) with favorable water infiltration conditions are found in uplands. Deeper strata are poorly permeable and percolating water seeps to valleys and ditches or main drainage canal. In depressions Endoaquolls, poorly drained, collect water runoff and discharge water to surface drainage system [9].

The climate of the region is shaped by the conflicting air masses from the Atlantic, Eastern Europe and Asia (arctic 6%, polar martitime 59%, polar continental 28%, tropical 7%), which are modified by strong Arctic and Mediterranean influences. It results in a great changeability of weather conditions and the predominance of western winds brings strong oceanic influence that manifest in milder winters and cooler summers in comparison to the centre and east of Poland. Within Poland, this area is one of the warmest, with an annual mean temperature above 8°C (range from 6.9 to 8.5°C). Mean annual global radiation amounts 3700 MJ/m² and, mean annual net radiation equals to 1315 MJ/m².

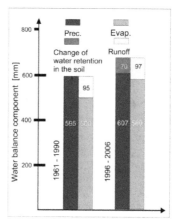

Figure 2. Water balance in Wielkopolska in 1961 – 1990 reference period and In 1996-2006 warm and dry period

Thermal conditions existing in the Turew landscape are favorable for cultivated plant growth. The mean plant growth season with temperatures above 5°C lasts from 21 March till 30 October. In reference 1961 – 1990 period (Figure 2) the mean annual precipitation is equal to 595 mm, of which 365 mm falls between April-September, and 230 mm in the period of October-March. Although the amount of precipitation in the spring-summer period is higher than in autumn and winter, a shortage of water occurs frequently in the plant growth season. This situation is aggravated by the dominance of light soils with poor water storing capacities. Average annual evapotranspiration amounts to 500 mm (485 mm in whole country) and water runoff is equal to 95 mm (212 mm in whole country). But in warm and dry period of 1996 – 2006 evapotranspiration was as high as 580 mm, which resulted in decreasing water retained in soil by 70 mm. It caused worsening of water conditions for plants.

In the land-use structure of whole catchment arable land makes 62.2%, forests and shelterbelts (mid-field rows of trees) cover 17.9%, meadows and pastures 12.5 %, water bodies 3.5 %, and villages, roads the rest of the area. There are no industrial facilities. The mean density of inhabitants equals to 55 individuals per 1 km².

- The natural forest complexes were replaced by woods planted by man or converted into cultivated fields that constitute up 70% of the total area, shelterbelts and small forests 16 % and grasslands 9 %.

- The structure of crops at the beginning of the 21 century was as follows: cereals (mainly, wheat and *Triticale*) including maize made 76.7%, vegetables 16%, potato, seed-rape, and sugar beets 6%.

The specific landscape of Turew neighborhood located in Wielkopolska was shaped during the twentieths of XIX century by general D. Chłapowski who farmed on 10 thousands hectares. He introduced essential changes in farming system as well as in the field today called landscape engineering. Conversion of the open and uniform agricultural landscape in mosaic which is rich in stable element like shelterbelts small mid-field water reservoirs was the results of his activity. Many wooded patches, shelterbelts, tree lines, clumps of trees were planted in the landscape. They were designed as a shelter for domestic animals and the measures against wind erosion. Since 1950s last century the investigation of agricultural landscape functioning has been carried out within that areas. About of one hundred km of linear and 10 hectares of new shelterbelts and woody patches were planted during the two last decades.

The "Turew Mosaic Landscape" (Figure 3A) is still abundant in various wooded patches located in upland parts of the landscape or along banks of the drainage water system as well as along other non-crop habitats, such as small water reservoirs, marshy habitats, bounds and so on. All together there are more than 800 shelterbelts forming the network in an area of 17200 ha. Cultivated fields make 70 % of the total area, forests and shelterbelts -16 % and grasslands - 9 %. The majority of farms are small. Only 27% are larger than 10 ha.

Comparative studies have been carried out in uniform agricultural landscape (Figure 3B) composed of large fields located about 10 km apart characterized by the same climatic conditions and similar soil types. This uniform landscape is almost entirely devoid of shelterbelts and the drainage system is mostly operating as an underground system. The cultivated fields are bigger than those in the mosaic landscape, and are ranging from 15 to 150 ha. A similar crop structure is appears in both mosaic and uniform landscape. The structure of the crops at the end of 1990s was as follows: cereals (68.1%), maize (9.6%), vegetables (16%), potatoes, seed-rape and sugar beets (6%). The substantial increase of cereal cultivations were observed over the period of 1984-2004. In 1985 cereals covered 48.1% of total arable land, in 1997 their contribution increased to 63.5% and in 2002 cereals were cultivated on 73.9% of arable land. Among the cereals wheat and triticale cultivations dominated.

(a)

(b)

Figure 3. A. Mosaic landscape. Phot. K. Kujawa; B. Uniform landscape. Phot. K. Kujawa

3. Methods

Due to a great variety of biological and ecological features of studied taxa, their distribution, abundance and diversity patterns were studied with the use of many methods differing strongly between the taxa. Diverse goals of former research projects carried out in the area described above involved various scales (approximately from hectares to tens of hundred square kilometers) and diverse habitats. Generally speaking, we refer our results mostly to three different objects:

a. landscape in broad sense (ca. 200 km², called in the chapter "Kościan Plain"), i.e. comprising all found habitat types, such as crop fields, meadows, semi-natural grasslands, woodlands, small wooded clumps (tree or shrub lines, belts or clumps), wetlands, roadside verges, water bodies and courses,

b. agricultural land in narrow sense (ca. 20 km², called "Turew Mosaic Landscape"), i.e. consisted of cultivated fields, meadows and grasslands, field margins (with all kinds of plant cover, roadside verges etc.), drainage ditches, small wooded patches, small water bodies and courses,

c. habitats (with area of tens hectares), called just wheat field, arable land, small wooded patches, drainage ditches, etc.

In other cases, described object (piece of landscape other than "Kościan Plain" and "Turew Mosaic Landscape") is described separately to distinguish it from the above listed. During the data analyzing, various statistical tests and procedures were applied, which are cited in appropriate place with commonly used names (t-test, Mann-Whitney test, Wilcoxon test, Analysis of Variance - ANOVA, Principal Component Analysis - PCA, General Linear Model - GLM, etc.). The significance level used for all tests was 0.05. For estimation total bird species richness, the Jack-Knife 2 estimator was used [10]. Arythmetic mean is given with standard deviation (SD).

3.1. Plants

Studies on spontaneous flora of the "Kościan Plain" have been carried out since 1975. First they were collected mainly in aquatic and marsh ecosystems and only occasionally apart from them. From 2000 they included the entire landscape and were taken with the method of mapping of all the species in the network of equal basic fields. Lists of the plant species of every investigated plot were analyzed. Material for the study on relationships between the floristic diversity of agricultural landscape and the diversity of its structural elements was collected on 37 square plots (1 km² each) [11]. Landscape diversity was assessed by number of elements of landscape (i.e. spatial units that differ in land use).

The most important measures used in floristic research include: species richness, floristic distinction (i.e. mean coefficient of species rarity), number of environmentally valuable species (recessive i.e. declining species, species included in red lists for Poland or Wielkopolska, protected species, and those rare regionally and locally), sociological groups, and geographic-historical groups. The status of species in the flora follows reports on Polish and regional floras. To diagnose the sociology of the taxa [12] elaboration was used. Names of alien geographic–historical groups follow [13,14]. Those are: archaeophytes (naturalized alien species introduced before ca. 1,500), naturalized neophytes (naturalized alien species introduced after ca. 1,500) and casual neophytes (casual alien species introduced after ca. 1,500 occurring sporadically or for a short time within the studied area). Among native species nonsynanthropic spontaneophytes (species which do not show permanent trends in occupying transformed anthropogenic habitats) and apophytes (indigenous species, permanently occupying strongly transformed anthropogenic habitats) are distinguished according to [15]. Species names follow [16].

Field inventory work on dendroflora was carried out during vegetative periods from 2001 to 2003. Prior to field work there were preparatory work on maps and aerial photos conducted. The inventory and evaluation encompassed all wooden plants assembles of open landscape, the area of which was smaller than 5 ha: shelterbelts, woodlots, avenues, tree lines and hedgerows. Desk work, which followed the field survey, was aimed at evaluation of wooden flora located in different landscape context among other by number of species and share of native and alien species [17].

The study of the flora and plant associations in the lake Zbęchy, peat bogs and Wyskoć Canal were carried first in 1976-1979. They were repeated in 2006-2007 [18 - 21]. Between these periods occasional studies were still performed. The commonly used Braun-Blanquet method was applied. The associations were distinguished and the taxonomy of syntaxa was adopted after [22]. The data for both investigated seasons were compared and valorized regarding to they threat and frequency of occurence according to [22] elaboration. The same method was used in the research on the flora of ponds that started in 1985 [23].

3.2. Insects

The basic apparatus (called "biocenometer" – Figure 4) used in the study of above-ground entomofauna in vegetation season is composed of cap in the form of a truncated pyramid with trapping surface 50x50 cm and height 70 cm. It is randomly barraged on the study area and trapped invertebrates are taken by suction apparatus, powered by a generator [24]. In all years (from 1970s till now) and habitats (crop fields, meadows, small wooded areas) the samples were collected with a frequency no less than three samples during one season (usually monthly, from May to October), in the series including 10 samples along a transect located in a given sample plot (habitat studied). Samples were separated by hand and the gained material was stored dried and identified to families (all insects) or genus and species (some groups of insects)[25].

Figure 4. Biocenometer. Phot. J. Karg.

The samples of insects wintering in litter and soil were taken from 1990s till now in shelterbelts and on adjacent crop fields. As many as 10 soil monoliths (without litter) with dimensions 10x10x10 cm were taken in each series three times per winter. Soil was manually separated. Obtained material was preserved in 75% alcohol and identified to the level of family or order (some larvae and pupae). Simultaneously, on the same dates and on the same places, litter samples from the shelterbelts were collected. Frames with dimensions 25x25 cm were used, and 20 samples were taken in each series. Further procedures were the same as in the case of soil samples.

The butterflies were studied mostly with the use of a transect method [26]. Eleven transects (with total length of 10.25 km) were located along the roads adjacent to various habitats (wooded areas, crop fields and meadows, young shelterbelts etc.) and the butterflies were usually counted and identified to species level in 5 m wide belt. The gathered data were used for the estimation of total or species abundance index estimation.

Since beginning of 1960s the observation of nocturnal butterflies were carried out using light trap method.

3.3. Spiders

Spider communities in winter cereals, sugar beet and alfalfa crops were studied in two types of agricultural landscape within the "Kościan Plain": mosaic one with the net of shelterbelts and the uniform one with lack of shelterbelts, where the majority of the terrain is covered by large sized crop fields. The results obtained from three fields in each landscape within three years were compared. To recognize an effect of wooded habitats on spider community occurring in crop fields, the spiders were collected in the distances of 10 m, 50 m and 100 m from shelterbelts (in case of mosaic landscape) or from a road (in case of uniform landscape). To assess the spider density, the samples were taken with biocenometer twice a year (in May and July). Foliage-dwelling spiders were taken by sweet-netting (50 sweeps were a sample), 2-3 times per season (from April to September).

3.4. Fishes

The results have been obtained by using an electrofishing method with IUP-12 device [27], used for catching fish in Wyskoć Canal. The most recent data were gathered in two 100 m long sections of this water course in 2007 and earlier – in 1997-1999 in 1000 m long section. The information on Zbęchy Lake was obtained from the user (Polish Angling Association, the Fishing Company in Osieczna).

3.5. Amphibians

Investigations of amphibians were carried out in the "Kościan Plain" mostly in the years 1995-2000, since March to October. Investigation covered mainly reproducing amphibian populations as well as water bodies in the spring because amphibians aggregate there for reproduction. The first step was cartographical analysis of the study area on the basis of topographic maps (1:10000) and aerial photographs. Then, location of ponds and all wet areas was checked in the field. As many as 150 water reservoirs were found. They were divided into 4 types, depending on impact of man activity: village ponds (located up to 100 m from buildings), reservoirs located on the meadows, in the fields and in the forests. In each water body the number of species and its abundance were studied. The presence of species was verified on the basis of direct observations of adults, juveniles or tadpoles. In some cases voices of males or unique characters of egg deposits were taken into account. Abundance of populations was estimated on the basis of direct observations of adults, counting of male voices during their high vocal activity (at night only) and numbers of egg clumps (for the brown frogs *Rana temporaria* and *Rana arvalis* only). The threats to amphibians (pollution, presence of fishes, ducks) were identified for each water body. Each reservoir was classified with respect to its usefulness to amphibian breeding (type of water plants, grade of shading, depth etc.). Moreover, in 32 reservoirs some chemical features of water (pH, ammonium, nitrates, and phosphates) were analyzed [28, 29].

3.6. Birds

At first, it is worth to mentioning that during the study species diversity in heterogenic habitat or landscape, serious methodological problem is defining the "representative" both study area and sampling effort. In case of birds, the representative agricultural landscape was defined as a mosaic of cultivated fields, meadows, pastures and small wooded area with the area not exceeding 4 ha and it was located in the "Turew Mosaic Landscape". To control biases related to sampling effort and to avoid formulation of biased conclusions, we applied here triple approach: a) description of species richness, b) comparisons to other comparable data in Poland, c) analysis of species richness indicators.

The distribution and population density of birds was studied with the aid of a variety of techniques, depending on the goal for a given research project.

a. Most data have been gathered with the use of a combined version of cartographic methods [30] , which was used in 1964-66, 1984, and since 1988 till now (with some time gaps). In all cases (plots and years) at least 6 counts/per plot were done (in most years – 9 counts) and number of breeding pairs was established on the basis of "paper territories" (clumps of at least 3 records of territorial males) or other observations of breeding behavior. The method was applied in all kinds of small wooded areas – clumps of trees, tree alleys and tree belts (in total of ca. 100 plots) as well as in few fragments of open landscape.

b. Transect method [31] was used for recognition of bird community patterns in an open landscape, among others for assessment of red fox impact on distribution and abundance of birds. The birds were censused on 200-300 m wide transects (several tens km), in the morning, twice a breeding season.

c. Point count methods [31] was applied when relationships between landscape structure and birds abundance and species richness had been studied in scale of tens of square kilometers (hundred points). Birds were counted twice a breeding season in the morning, up to 100 or 150 m from place of standing.

In next step, for understanding the effect of environmental factors on the bird communities in space and time, habitat structure was quantified with the aid of several variables in three scales: within the plot (tree belt, tree clump etc.), in close neighborhood (adjacent area) and in landscape around a plot (i.e. within a radius of 1500 m from given wooded patch).

a. Within the plot: tree, shrub, herb percentage cover, tree stem density, tree and shrub species diversity etc. [32, 33].
b. Area adjacent to studied plots : crop diversity, crop patch density etc. [33].
c. Landscape: woodiness index, percentage cover of crop fields, density of ecotones etc. [33].

In the next step the number of independent variable was reduced with the aid of PCA if needed, and the relationships between habitat or landscape structure were verified with the use of a various regression models.

3.7. Mammals

Small mammals were studied in 1999-2001 in two shelterbelts of different age located in crop fields in the "Turew Mosaic Landscape". The young one (16 m wide) was planted in autumn 1993 year and consisted of oaks, pines, birches, poplars, elms, maples, beeches and some other species in addition. The old one (36 m wide) was created with the shelterbelts' network in XIX century. The tree stand consisted of false acacia (with admixture of several oaks). Catch-Mark-Release (CMR) method was used. Traps were located in consistent lay out in both studied shelterbelts. 16 transects each 20 meters were set, 4 points including 2 traps were located in each transect. The traps were inspected every morning for 10 days within each season series (spring, summer, autumn).

In former years also other groups of mammal were studied. Carnivores were studied with trapping (small mustelids), recording of burrows and snow tracking. Night counting (with searchlight attached to a car) was also used. Belt assessment method with a line of beaters moving through selected area was used for estimation of hares. Roe deer and wild boar were counted by direct observation in late winter and early spring seasons [34].

3.8. Macrofungi

The study on macrofungi species diversity in the "Kościan Plain" was carried out basically with the use of three methods :

1. Route methods, enabling a basic (rough) recognition of species diversity in a large area.
2. Permanent plots, allowing for detailed description of fungal communities in various habitat types as well as for making comparison between the communities.
3. Transect method, which enables gathering the information from relatively large areas, comparing between them and detecting changes in species richness and diversity in a various ecosystems.

Additionally, in 2000-2003 study was carried out in 50 permanent plots (area of 400m²) located in managed forests, village parks, road tree alleys, tree belts and clumps of trees.

The route method has been used systematically since 1997 in various habitats. The rate of colonization of windbreak introduced in arable field is studied from 1998 [35], and preliminary study on the mycobiota of ephemeral habitats characteristic for a Polish lowland agricultural landscape (manure and straw piles) are realized from 2009. In result, in 1997-2010 numerous data on species diversity of a fragment of agricultural landscape have been collected.

3.9. Microfungi

Due to specificity of biological features of fungal pathogens, i.e. demanding special research techniques, the study was designed and carried out in other way than previously described taxa. The study of the fungal pathogens (infecting insects (Insecta), true spiders (Araneae) and mites (Acari)) was aimed at (1) recognition of their real resources and diversity in particular habitats, (2) – estimation of their effects on noxious and beneficial arthropod populations. They were conducted in German, France, and Poland during 10 vegetation seasons (1995-2004) in agroecosystems and neighboring forests. In agroecosystems three groups of elements were studied: annual crops (cereals and row plants), perennial crops (including meadows and pastures) and non-farmed clumps or strips of wild vegetation, including rushes and arborescent plants. This allowed for synthesis of the results for simplified (uniform) and diversified (mosaic) landscape types. The cases of epizootic appearance of particular pathogens' species were treated with special attention independently on the noxiousness or usefulness of their host species [36]. Before undertaking the extended studies in three countries, the methods were elaborated and tested in the "Kościan Plain" and the Wielkopolski National Park – both in the central Wielkopolska region (Poland). Both these areas have been included into the referred researches.

The sample plots were located along transects running across a chosen landscape fragments differing in plant cover, habitat structure or/and human impact. Material from 3-5 randomly scattered sample plots (2x2 m), where plant cover up to the height of 2.5 m and the litter including superficial layer of soil was carefully searched, served as basic samples [37]. The invertebrates were sampled two times during every growing season – in the turns of spring/summer and summer/autumn. Additional checking searches were made in other dates, but only in Poland. The applied methods give the best results in the recognition, sampling, and species isolation of these pathogens group, often allowing to discover spatial distribution of epizootic appearance and new species for science.

Apart from the study on pattern of occurrence, for some habitats and countries the investigations on the pathogens of the subcortical and wood-boring insects were carried out. Mortality causes of these insects – together with associated with them invertebrates in their feeding sites - have been investigated by the immediate stereo-microscopic searches of samples from invaded trees (logs and branches) and laboratory rearing of them, for periodical isolation of appearing pathogens [38, 39].

Occurrence and diversity of entomopathogenic fungi in the soil from different habitats of agricultural landscape, including the "Turew Mosaic Landscape" and closely situated Farm Karolew applying no-tillage system have been investigated in th period 1996 – 2006. In total

296 soil samples from 74 locations in Poland were collected in the years 1996-2006 mainly from agrocenoses (arable fields, meadows, pastures), semi-natural biotopes (shelterbelts, mid-field afforestations, balks) and forest biotopes. Fungi were isolated from soil samples by means of the "Galleria bait method" [40] and selective agar medium [41]. The infective potential of particular soil samples has been related to the mean colony forming units (CFU) (colony forming units) per 1 g of checked soil. Standard methods for χ^2 tests were used for comparison of their quantitative relation in soils samples from different habitats.

4. Characteristic of biodiversity

The methods used and goals of investigations conducted in the "Kościan Plain" differed strongly between the studied taxa or groups. That is, why some data included in this chapter should be regarded rather as the approximate information about "species pool" of organisms occurring there and as a general picture of wildlife taxonomical diversity and richness, but not in the term of detailed, complete datasets on species richness, concise across the taxa in scale of the landscape. Presumably, only the data on vascular plants, butterflies, amphibians and birds represent the groups for which we can evaluate or estimate the number of species for farmland (arable fields, meadows, field margins, small water bodies, drainage ditches) as well as for the whole landscape in broad sense. Other groups can be used for evaluation of biodiversity mostly in respect to farmland ("Turew Mosaic Landscape") or to selected habitats occurring in farmland (for example the fishes – for evaluation of small water bodies and water courses importance).

4.1. Plants

The total number of the spontaneously occurring vascular plant species, which have been noted until now within the "Kościan Plain" amounts to 848. It comprises just 54 % of the number of species in the Środkowa Wielkopolska Region. There are 773 naturalized species and 75 casual neophytes. Natives are 72 % and all alien species 28% of the total flora. Among the alien species the most important group are archaeophytes that present 39% of the total alien species number, whereas naturalized neophytes are 30% and casual neophytes 31% [11]. Among the naturalized neophytes there are 4 invasive species (*Heracleum sosnowskyi, Echinocystis lobata, Impatiens glandulifera, Reynourtia japonica*) from the ministerial list of alien species that especially threat the biodiversity in Poland. As many as 6 of the neophytes that are noted in the Landscape Park (*Echinocystis lobata, Elodea canadensis, Impatiens glandulifera, Prunus serotina, Reynourtia japonica, Robinia pseudoacacia*) are included in the list of 100 WORST in Europe [42]. The species typical for meadows show the highest share (22%) among the sociological groups of plants. The smallest is the group of thicket plants (5%). The share of other groups (aquatic, forest, xerothermic grassland, ruderal, segetal) is 11-14%.

Currently, the total landscape flora consists of 828 species due to the extinction of 20 species in the last decades. Nowadays 85 plant species of special care (rare, threatened, declining, protected) occur spontaneously in the "Kościan Plain". Among them, there is one species

(*Ostericum* palustre) protected in UE that is in the II Annex to the Habitats Directive, and two species (*Botrychium matricariifolium, Ostericum palustre*) listed in the "Polish Red Data Book of Plants" [43]. As many as 44 species from this group of special care plants are protected in Poland. Despite their exposure to extinction the other species from this group are not protected. Apart from them 12 valuable vascular plant species had been noted in the past, but they disappeared in the last decades.

The flora of fields that dominate in the landscape, consists of 224 plant species. It is 74% of the segetal flora of the Wielkopolska Region. Native species are 56% and alien species 44% of the total field flora. Archaeophytes that presents 65% of alien species are the biggest group among them. 21% are naturalized neophytes and 14% casual neophytes. In the group of naturalized neophytes there are often noted invasive species as *Amaranthus retroflexus, Conyza canadensis, Galinsoga parviflora* and *Veronica persica* and less frequently *Anthoxanthum aristatum, Bromus carinatus, Bidens frondosa* and *Galinsoga ciliata*. Among the sociological groups of plants the highest share show segetal species (34%) and from other sociological groups the biggest are meadow (20%) and ruderal species groups (19%). The share of remaining groups (aquatic, forest, xerothermic grassland, thicket) is 3-8%. 23 plant species of special care were noted in the fields. Among them there were species threatened in Poland and in the Wielkopolska Region (e.g. *Myosurus minimus, Valerianella locusta, Valerianella rimosa, Conium maculatum*), vulnerable archaeophytes (e.g. *Agrostemma githago, Anthriscus caucalis, Melandrium noctiflorum, Valerianella rimosa*) and only one species protected in Poland (*Ornithogalum umbellatum*) among them [11].

A studies conducted in the mid-1970s showed 131 plant communities in the "Kościan Plain", but actually 122 plant communities were confirmed. Number of aquatic associasions was reduced by 9 only in last four decades.

Forest communities represent less than 15% of the land cover and most of them are of anthropogenic origin. Pine monoculture stands are the most common forest communities in the area. Natural forest are represented by riparian (elm-oak) and alder stands – which covers narrow areas along water courses. Elm-ash riparian forests are noticed in the vicinity of the alder stands – characterized by lower level of groundwater.

As a result of thousands years of agricultural use of this area, dominating Middle European lowland oak-hornbeam forest habitats (*Galio-Carpinetum*) have been substituted with arable fields, whereas lowland ash-elm floodplain forest habitats (*Ficario-Ulmetum*) mostly with intensively cultivated meadows. Predominant deciduous forest habitats have been almost destroyed, and only small forest patches like manor park in Turew remained. As early as in the XIX century this area was among the most intensively used in this part of Europe. Arable fields occupied 65.2%, meadows – 13.5% , pastures – 3.4%, forests – 13.2%. In order to prevent the soil erosion, in 1820s the then holder D. Chłapowski introduced into fields a network of windbreaking shelterbelts which consisted mainly of *Robinia pseudoacacia* with some addition of *Quercus robur* and with hedgerows that consisted of *Crataegus monogyna*. Consequently, although most of them have been destroyed, the studied area is particularly rich in those species and shelterbelts and woodlots with *Robinia pseudoacacia* are distinctive features of the Park's landscape, distinguishing it from other parts of Poland. This species is not only predominating species or even the only woody species in many shelterbelts, but it was also recognized as the species which was most often found in all woody plant assembles surrounded by arable fields in the investigated area (found in 37% of all the studied tree lines, belts and clumps growing among crop fields – former Oak-Hornbeam habitat). The other frequently found species were: *Quercus robur*, *Pyrus pyraster* and *Acer platanoides*. In contrast, on meadows the most abundant species were *Salix alba* and *Alnus glutinosa* [17, 44].

Characteristic feature of "Turew Agricultural Area" is a network of tree or shrub lines or belts, which develop mostly along roads or ditches, and much less frequently constitute border line between adjoining crop fields. Tree and shrub lines constituted 80% of all woody vegetation clumps in this area [17]. However in most field margins, especially in the area consisted of very small (several hectares) pieces of arable land, there are no perennial (including woody) plant communities. Also thicket communities are very rare in the area. Usually, the anthropogenic habitats gives conditions for grassland communities which enriches biodiversity by creating favorable conditions for many rare plant species in the landscape.

Research of 50 small water reservoirs located in the Kościan Plain showed presence of 40 plant communities [23]. Those communities were often represented by pleustonic plants of Lemnetea class. Most of studied reservoirs with submerged vegetation formed hornwort phytocenoses - *Ceratophylletum demersi* while very rarely were observed stonewort communities (Characeae). Among the rush vegetation the *Phragmitetum australis* and *Typhetum latifoliae* were dominating (poor floristically). Sedge rushes - *Magnocaricion* were better formed.

In the investigated ecosystems There were noted 180 species of vascular plants, 3 of stoneworts and 1 moss were noted. The analysis of identified plant communities showed that 38 of them are of native origin. Communities of alien species that occurred in small mid-field ponds are formed by Canadian waterweed (*Elodea canadensis*) and sweet flag (*Acorus calamus*). The majority of plant communities in the mid-field ponds (29) were distinguished as native communities, which increase their areas as effect of the anthropogenic influence. Phytocoenoses in this category usually inhabit very fertile waters as mid-field ponds. Low diversity of plant species within phytocoenoses show the adverse changes in ecosystems as a result of human activities.

Mid-field ponds are marginal habitats in the agricultural landscape. Regardless of their small area they create conditions in which natural vegetation finds refuge, especially where the intensification of agricultural activities reduces ecological quality of landscape. Therefore, it is useful to enrich the landscape through creation of new and renovation of neglected water ponds.

One of the objects of investigations in the Turew landscape was the pond (1800 square meters) which was dug in the summer 1995 in natural, periodically flooded area. In this pond aquatic and marsh plant succession was observed in 1995-2005. During the first and second vegetation seasons the stonewort meadows covered 91% of bottoms area. In 1998 the stonewort domination finished. In that period there were not observed any phytoplankton blooms. After the third year of the pond existence the plants with floating leaves covered large area. The shade the bottom area was the reason of charophytes disappearance. In 2000 the communities of emerged plants dominated in the pond. Since 2003 the patches of hornwort are growing among the other submerged plants. It means the end of the early development stage of the pond. In the pond species rare in Poland and Wielkopolska Region occurred, for example: *Chara fragilis, Chara vulgaris, Ceratophyllum submersum* and *Teucrium scordium*.

The most valuable natural plant communities in the "Kościan Plain" are represented by aquatic (peat pits, mid-field ponds, lake and water courses) and meadow phytocenoses. Among the meadow communities the most important for biodiversity are marshy and swamp meadows endangered by drying and intensification of agricultural activities. The sedge (*Carex*), purple moorgrass (*Molinia caerulea*) and thistle (*Cirsium*) meadows were recognized as the most vulnerable to mentioned threats.

It is worth emphasizing that among 122 plant communities noticed in the whole "Kościan Plain", 54 are endangered in Wielkopolska. The most valuable communities are the 23 associations recognized in "V" category - at risk of extinction (rare or very rare communities or communities consisting of vulnerable species; those are also communities with simplified and poor species composition and with decreasing area of occurrence. Existence of these associations can be prolonged only by preserving actual conditions and reduction of anthropogenic pressure. In addition, 31 associations have been reported in high risk of withdraw - in the "I" category (communities of indeterminate threats because their distribution, dynamic tendency and systematics are weakly recognized). In the investigated area there have been identified 39 that are recognized as indicators of habitat listed in II Annex of the Habitats Directive.

4.2. Invertebrates

4.2.1. General characteristics of invertebrate diversity

„Turew Mosaic Landscape" favors to preserve high biodiversity level. According to data gathered by numerous authors, there were recorded ca. 3500 species of invertebrates, of that 2600 insects species since 1970s (Table 1), which represent a majority of Central European insect families. The most abundant are the species strictly related to agrocenoses (agrophags) and their natural enemies (predatory and parasitoid species). Some of the species are rare in Poland, many of them (carabid beatles, bumble bees, some butterflies) are protected by law or are listed in Red Lists [45]. Unfortunately, extremely strong differences in sampling efforts between the studied groups and qualitative character of much data do not enable to approximate total species richness.

Taxon/group	No of. species	Source		Taxon/group	No of. species	Source
Nematoda	40	[46]		Microlepidoptera	~150	Karg (unpubl.)
Enchytraeidae	16	[47]		Coleoptera	~500	Karg (unpubl.)
				Hetero- and		
Lumbricidae	7	[48]		Homoptera	~200	Karg (unpubl.)
Acarina	146	[49]		Hymenoptera	~600	[52]
Araneae	224	[50]		- Apoidea	260	
				Thysanoptera	40	[53, 54]
Water insects	~190	[51]		Orthoptera	~30	Karg (unpubl.)
- Odonata	36			Dermaptera	>5	Karg (unpubl.)
- Heteroptera	41			Blattoptera	>2	Karg (unpubl.)
- Ephemeroptera	10			Diptera	~300	Karg (unpubl.)
- Coleoptera	>90			Others	~50	Karg (unpubl.)
Macrolepidoptera	~500	Karg (unpubl.)				
- butterflies	51	Sobczyk (unpubl.)				

Table 1. Number of species in invertebrate taxa.

4.2.2. Insects

The studies that have been carried out for above 40 years on above-ground insects of agrocenoses and small wooded patches, tree belts and lines, enable evaluation of their status and changes they are undergoing. The results of these investigations indisputably show the crucial role of semi-natural habitats in preserving insect diversity in an agricultural landscape. Biomass, abundance and diversity of above-ground insect communities in small wooded patches and in ecotone zone are on average significantly bigger than in open arable land. What is more, the abundance of predatory and parasitoid species, which play important biocenotic function, is in small wooded patches markedly higher, too.

Small wooded patches constitute refuge area as many species of insects can find there a favorable condition for wintering. The differences in density and biomass of insects between wooded area and crop fields are significant. In the first habitat and in ecotone zone (0.5 m wide) the values were several tens higher than in adjacent crop fields. More detailed information on distribution and abundance pattern of insects in relation to habitat and landscape structure are included in Chapter 5.2.

As many as 51 butterfly species were recorded in the "Kościan Plain". Most of them were found in two ecotones - between wooded area and crop fields (27 species) and between 'wooded area and meadows (41 species), where the butterfly abundance was 4-5-fold higher than in the previous ecotone. There were observed a few species listed in Red List: *Lycaena dispar*, *Colias myrmidone* and two other species worth to conserve regionally: *Heteropterus morpheus* and *Polyommatus amandus*. *Lycaena dispar*, hydrofilous species, inhabited meadows with drainage ditches, where larvae of the species could easily find their food - *Rumex hydrolapatum* Huds. Additionally, some observations of second generation of the species evidence for a favorable conditions, which characterize some habitats of the studied landscape. The population density of the species in wooded patches surrounded by crop fields amounted to 0.1 ind./km and in ecotones at wooded areas and meadows – 0.8 ind./km. The presence of dense network of shelterbelts, tree alleys, etc., favors relatively easy dispersion of the rare species from the breeding habitat across the landscape.

4.2.3. Spiders

In the "Kościan Plain" as many as 224 spider species (27% of Polish spider fauna) were observed and 17 species are listed in "The Red list of threatened and endangered animals" – most species with VU (vulnerable) status, one with EN (endangered) status. Of that, 204 species were found in the "Turew Mosaic Landscape" and 19 other species – exclusively in forest island constituted by a village park in Turew were noted [55,56].

In total, 21 spider families were noted, what constitutes 56 % of all spiders families of Poland. The first three spider families, which are most numerous in species, were in the same sequence as in the whole Polish spider fauna. Linyphiidae was the most species-rich spider family and it was represented by 63 species (32 % of the pool of species occurring in crop fields and shelterbelts), and the next ones were: Theridiidae – 19 species (30 %) and Lycosidae - 18 species (28%), respectively. The other frequently represented families were: Gnaphosidae -17 species, Salticidae -14, Araneidae -13, Thomisidae and Philodromidae – 9, Tetragnathidae – 7, Clubionidae and Dictynidae – 6 species. The rest of families was represented by 1-4 spider species.

4.3. Fish

Zbęchy Lake, the biggest basin in the "Kościan Plain" is inhabited by 11 species (bream *Abramis brama*, roach *Rutilus rutilus*, silvery crucian carp *Carassius gibelio*, common crucian carp *Carassius carassius*, tench *Tinca tinca*, carp *Cyprinus carpio*, ide *Leuciscus idus*, pike *Esox lucius*, perch *Perca fluviatilis*, pikeperch *Sander lucioperca*, eel *Anguilla anguilla*) from four families: *Cyprinidae, Anguillidae, Percidae* and *Esocidae* (by Polish Angling Association, the Fishing Company in Osieczna). Fish in the „Wyskoć Canal" are represented by six families (*Cyprinidae, Anguillidae, Percidae, Esocidae, Gasterosteidae, Gadidae* and *Cobitidae*) and 14 species (roach, rudd *Scardinus erythrophtalmus*, bream, gudgeon *Gobio gobio*, common crucian carp, silvery crucian carp, tench, three-spined stickleback *Gasterosteus aculeatus*, bleak *Alburnus alburnus*, perch, pike, eel, burbot *Lota lota*. Particularly valuable species, listed in Annex II of the Habitats Directive (Council Directive 92/43/EEC), is mud loach (*Misgurnus fossilis*).

4.4. Amphibians

The "Kościan Plain" is inhabited by 12 amphibian species. It is 67 % of all amphibian species living in Poland and 86 % of lowland amphibian species. The two species not recorded here, are rare (natterjack *Bufo calamita*) or very rare (agile frog *Rana dalmatina*) in Poland. The most common amphibians in the agricultural landscape belong also to the most common in Poland: water frogs (edible frog *Pelophylax esculentus* and pool frog *P. lessonae*) – inhabited

94% of all water bodies, brown frogs (common frog *Rana temporaria* and moor frog *R. arvalis*) - found in 89% of reservoirs. To the rarest species, inhabited less than 10 % reservoirs, belong marsh frog *Pelophylax ridibundus* and the European tree frog *Hyla arborea* [57, 28, 29]. One species (crested newt *Triturus cristatus*) is included in the Polish Red Book of Vertebrates (with the statuts NT - near threatened) and two other ones (crested newt and fire-bellied toad *Bombina bombina*) in the Polish Red List, with the status NT and DD (data deficient), respectively. The same two species are included in Appendix II of EU Habitat Directive. Crested newt and fire-bellied toad are not common in the study area – they were found in 20-30% of all investigated water reservoirs.

All types of water bodies and landscapes within the "Kościan Plain" were inhabited by similar number of species (11-12), but they differed in frequency and abundance of species. The highest number of species was noted in water bodies located in forests (mean 6.6 per reservoir) and in ponds surrounded by cultivated fields (6.2) and the lowest one – in reservoirs in meadow complexes (4.8). The species that occurred mostly in forest are: pool frog (70 % sites), moor frog (65%) and crested newt (35%). In the ponds in cultivated fields bred mainly (60-70% of sites) common spadefoot (*Pelobates fuscus*) and common frog. One species – green toad (*Bufo viridis*) – was found mostly (40%) in village ponds. The most abundant amphibian populations (> 500 individuals) were observed in forest (17% of all sites) and in meadow (8%) reservoirs.

4.5. Birds

4.5.1. Regional species pool

Total number of breeding species recorded in the "Kościan Plain" (in all habitats) amounted to *ca.* 120 [58]. Complete study on breeding bird community in the "Turew Mosaic Landscape" was carried out in 1991-1994 near Turew on the area of 1380ha (crop fields - 80 %, grasslands - 12 %, small wooded patches and lines - 5 %, roads - 2 %, water bodies and courses - 1 %) [59]. As many as 76 species were found to nest (with the population density of 140 pairs/100 ha) in that mosaic landscape. Small wooded patches were the most important as they covered only 5% of studied area, while constituted breeding habitat for as much as 88% of species (N=67) species and 54% of all breeding pairs. Although the studied community consisted mostly of species known as common and abundant in Poland, population density of several species which are considered to be threatened and/or rare in Europe was relatively high. It concerns e.g. ortolan bunting (*Emberiza hortulana*), corn bunting (*Emberiza calandra*) and red-backed shrike (*Lanius collurio*). The community included also some species from Polish Red List (DD – quail *Coturnix coturnix* and European turtle dove *Streptopelia turtur*) and from Appendix 1 of EU Bird Directive (white stork *Ciconia ciconia*, marsh harrier *Circus aeruginosus*, black woodpecker *Dryocopus martius*, barred warbler *Sylvia nisoria*, red-backed shrike, ortolan bunting, tawny pipit *Anthus campestris*).

4.5.2. Variability in landscape scale

Species number in the "Turew Mosaic Landscape" differed strongly between its fragments (with the area of 35-55ha). The minimal number of species (4) was observed in a plot without wooded areas, with distinctive predominance of crop fields (98.8%), the maximal one (33) was recorded in the mosaic-like area, rich in a variety of tree lines and patches. The population density was much less variable and ranged from 13 to 33 pairs/10 ha [59].

4.5.3 Variability of bird communities in small wooded patches and lines.

The analyses presented here were done for the data on 74 various tree lines (length < 3 km) and patches (area < 10 ha) studied in 1991-1994 [32], and for 68 patches (<4 ha) studied in 2005-2007 [60]. In 53 tree lines as many as 54 breeding species were observed. However, the use of species richness estimators for that dataset suggests that total number of species could be even higher – ca. 60 (ca. ¼ of breeding avifauna in Poland). The species number per single

plot (i.e. tree line) ranged between 1 and 28, with arithmetic mean amounting to 12±6.3 (SD). Total density amounted 1-134 pairs/km, on average 26±22 pairs/km. In small wooded patches as many as 56 species (3-23 per patch, on average 12.1±5.4) were observed. According the Jack-Knife2 species richness estimator one may expect more than 80 species (Figure 5) in all types of wooded patches, tree lines etc. Among dominating species (>5% of community) were finch (*Fringilla coelebs*), blackcap (*Sylvia atricapilla*), yellowhammer (*Emberiza citrinella*), great tit (*Parus major*), nightingale (*Luscinia megarhynchos*), blackbird (*Turdus merula*) and icterine warbler (*Hippolais icterina*). Total population of breeding birds amounted to 125 pairs/10ha.

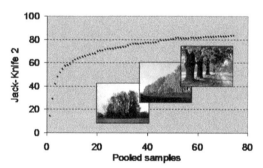

Figure 5. Species richness in wooded patches. Photo K. Kujawa.

4.6. Mammals

A list of big-sized mammal species in the "Kościan Plain" is based on field observations in the years 1998-2012 and references. Rodents were diagnosed on the basis of research carried in recent years by the CMR method, while the presence of other species was based on direct observation. Most of the recorded species are commonly encountered in this area commonly. There are no species for which the "Kościan Plain" would provide a key role in preserving their populations. However, the presence of protected species indicates the attractiveness of the area for those groups of animals. As many as 32 species (with the exception of bats), representing 14 families, were recorded in the area: mole *Talpa europaea* (partially protected), common shrew *Sorex araneus* (strictly protected), Eurasian water shrew *Neomys fodiens* (strictly protected), brown hare *Lepus capensis*, red squirrel *Sciurus vulgaris* (strictly protected), beaver *Castor fiber* (partially protected), water vole *Arvicola terrestris* (partially protected), red vole *Myodes glareolus*, common vole *Microtus arvalis*, field vole *Microtus agrestis*, house mouse *Mus musculus*, field mouse *Apodemus agrarius*, yellow-necked mouse *Apodemus flavicollis*, wood mouse *Apodemus sylvaticus* (partially protected), harvest mouse *Micromys minutus*, brown rat *Rattus norvegicus*, red fox *Vulpes vulpes*, raccoon dog *Nyctereutes procyonoides*, Eurasian lynx *Lynx lynx* (seen once), badger *Meles meles*, European otter *Lutra Lutra* (partially protected), pine marten *Martes martes*, beech marten *Martes foina*, least weasel *Mustela nivalis*, American mink *Neovison vison*, common raccoon *Procyon lotor*, wild boar *Sus strofa*, roe deer *Capreolus capreolus*, red deer *Cervus elaphus*, fallow deer *Dama dama*, Eurasian elk *Alces alces* (seen once), mountain sheep *Ovis ammon* [61, 34]. Two species – beaver and European otter – are listed in the annexes of the Habitats Directive. Traces of beaver were found mainly along of watercourses, e.g. Wyskoć Canal, but in melioration ditches, too. Fauna of bats requires additional research, but it is estimated that there are approximately 12 species of bats.

Mid-field shelterbelts are an important element of agricultural landscape for small mammals. Shelterbelts are mid-field refuges, food source and ecological corridors. They also join scattered elements of environment in mosaic landscape. Small mammals in the shelterbelts were studied e.g. by [62, 63]. Rodents can migrate for the longest distances along shelterbelts. In the shelterbelts, many species find the conditions for the reproduction and rearing, the overwintering or survival in the case of unfavorable weather conditions. Also wooded areas can be a food source for many species of mammals, often in the winter decisive for survival, or a supplemental food source. Seven species of small rodents (47% of Polish fauna) were recorded in the studied shelterbelts. In the old shelterbelt, yellow-necked mouse *Apodemus flavicollis* dominated markedly (55% of captured animals). Also numerous individuals of field mouse *Apodemus agrarius* (18%) and common vole *Microtus arvalis* (19%) were found. Additionally, individuals of wood mouse *Apodemus sylvaticus*, house mouse *Mus musculus*, harvest mouse *Micromys minutus* and bank vole *Myodes glareolus* were captured. In the young shelterbelts, small mammals were much more abundant but species structure was different than in the old ones. Field mouse dominated here (45%). Common vole was also numerous (32%). Other species (yellow-necked mouse, wood mouse, house mouse, bank vole) were much less frequent. According to classification by [64, 65], in which small mammals can be divided into "forest" species (yellow-necked mouse, bank vole), "field" species (house mouse, harvest mouse, common vole) and "intermediate" species (wood mouse, field mouse), our study indicates, that the old shelterbelt is an environment similar to the forest. In the young one, domination of "field" and "intermediate" species was observed. Small mammals were also studied in manorial park in Turew. Species typical for forest environment strongly dominated there (bank vole 54%, yellow-necked mouse 15%). Field mouse was also observed here (30%). Common vole was captured here only occasionally.

4.7 Fungi

4.7.1. Macrofungi

At first, it is worth to note that an agricultural landscape, till now, has been only rarely considered as the area, on which preservation of national species pool strongly depends. Only few studies on species diversity of Macromycetes (species, which can be observed with unarmed eye) have been carried out in agroecosystems and exceptionally in scale of landscape. The results of a study conducted since 1997 in the "Kościan Plain" indicate that mosaic-like agricultural landscape, rich in non-farmed habitats such as village parks, small wooded patches, menaged tree stands, is inhabited by a variety of fungal species, including rare and protected ones. In 1997-2011 as many as 687 species were found (Kujawa A., unpubl.), in that 99 from Ascomycota and 588 from Basidiomycota. According to substrate on which fungi grow, most frequent (55% of all species) were terricolous species (Figure 6A). Apart main type of substrates (soil, wood and litter), the fungi were found also on dung, dead fruitbodies, wood charcoal) and some others were parasites of herbs and insects. With respect to trophy, most species belongs to saprotrohic group (Figure 6B).

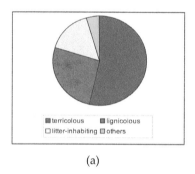

terricolous   lignicolous
litter-inhabiting   others

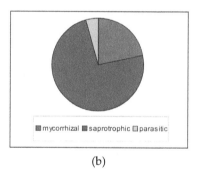

mycorrhizal   saprotrophic   parasitic

(a)   (b)

Figure 6. Share of species according to: a) substrate inhabited; b) their trophy.

Most of the studied elements of the „Kościan Plain" play a role of substitute habitats for some rare species and species protected by law. It is worth underlining, that 26 % of species observed in the "Kościan Plain" study area belong to the group of special concern species, i.e. these species which follow at least one of the below listed criteria:

a. Species protected by law [66] – 17 species, e.g. *Sarcocypha austriaca, Grifola frondosa, Sparassis crispa, Verpa conica, Fistulina hepatica, Geastrum striatum, G. berkeleyi, G. coronatum.*

b. Threatened species, listed in "Red list of the macrofungi in Poland" [67] – 98 species, e.g. *Coprinus bisporus, Crepidotus luteolus, Entoloma rhodocylix, Hygrocybe insipida, Inocybe calospora, Lepiota brunneoincarnata.*

c. Threatened species, listed in "European red list of the macrofungi" (Ing 1993) – 18 species, e.g. *Cordyceps capitata, Entoloma excentricum, Lepiota fuscovinacea, Mycenastrum corium, Perenniporia fraxinea*

d. Rare species, not protected by law and included in red lists, but observed in Poland only in 1-2 sites, in that the species found exclusively in the "Kościan Plain", according to checklist of Polish larger Basidiomycetes [68], checklist of Ascomycetes in Poland [69, 70] - 79 species, e.g. *Conocybe fuscimarginata, Entolona araneosum f. fulvostrigosum, E. cephalotrichum, E. incarnatofuscescens, Marasmius anomalus, Peziza ampliata,Xylaria oxyacanthae* .

e. Species found first time in Poland (in that, recorded exclusively in the "Kościan Plain") – 25 species, e.g. Desmazierella *Desmazierella piceicola, Entoloma parasiticum, Gammundia striatula,Hohenbuehelia cyphelliformis, Melanomphalia nigrescens, Pustularia patavina.*

4.7.2. Microfungi

The total number of the species of entomopathogenic fungi collected in the "Kościan Plain" amounted to 88. As many as 60 species were found in the "Turew Mosaic Landscape". The poorest resources of that group of organisms were regularly noted in both kinds of annual crops (cereal and row crops), where only 20-25 species occurred and the majority of them was connected with the roadside, balk or ruderal vegetation. However, even in such inhospitable conditions occur not rare epizootics in aphid colonies or adult anthomyid flies – mostly of the genus *Hylemya* - caused by *Zoophthora, Pandora* and *Entomophthora* species on big areas of cereals and rape seeds [71].

The main taxonomical units grouping the majority of obtained arthropod pathogenic species of fungi were the orders of Entomophthorales and Hypocreales. The greatest significance of the first order for the restriction of agrophagous and forest pest arthropods results from their spontaneous dispersal in dense host populations, ending often by almost total their mortality. Only few of representatives of the second order are able to form the perfect (ascomycetous) fructification form, but they produce as a rule abundant and strongly differentiated vegetative (conidial) sporulation forms (called anamorphs), allowing their overall dispersion and permanent restriction of arthropods. Their most common genera, as *Beauveria, Isaria, Lecanicillium, Metarhizium, Simplicillium* are polyphagous, non-selective with respect to a host species, permanently sustaining in soil conditions [72] and in "aeroplankton". Moreover, many of their strains have been applied as "active agents" in commercial biopesticides.

As many as eight entomopothogenic fungal species were found in soil of agrocenoses and semi-natural habitats [72]. Generally, three species of fungi: *B. bassiana, M. anisopliae* and *I. fumosorosea*, dominate in Polish soils. The dominance of particular species depended on habitat. In the soils from arable fields *M. anisopliae* dominated, with *I. fumosorosea* and *B. bassiana* as subdominants. *M. anisopliae, B. bassiana* and *I. fumosorosea* predominated in the soils from meadows and pastures, whereas *B. bassiana* in samples from forest soil and litter. It was found that infective potential and density of CFU of entomopathogenic fungi were significantly higher in soils from old or medium age shelterbelts than in adjacent large-area arable fields. Moreover, the soils from shelterbelts were characterized by more abundant fungal species composition (6 vs. 3 species respectively), being specific refuges for resources, diversity and persistence of entomopathogenic fungi in the agricultural landscape. Particularly valuable in this respect seem to be scrubby shelterbelts of rich species composition.

5. Impact of landscape structure and land-use

5.1. Plants

Floristic diversity of the whole agricultural landscape depends on landscape complexity. Results of the study on relationships between the floristic diversity of farmland and the diversity of its structural elements have shown a strong relationship between them [11]. The flora value increases with the number of spatial elements in the landscape (Figure 7).

Results of the study have proved that in a homogeneous landscape, composed of only arable fields and linear mid-field elements, the species number is half as high as in a landscape with numerous habitat islands. In fields, 224 plant species were noted. It was only 27% of the total landscape flora. The number of weed species in each field depended on method of cultivation and type and size of crop. The greatest species richness is characterized by small fields of farmers applying the extensive methods of cultivation.

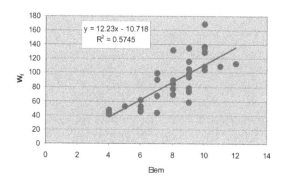

Figure 7. Relationship between the floristic value (*Wf) and the number of landscape elements (Elem) * Wf – the floristic value is the sum of the rarity coefficients of the naturalized plant species that are noted on the plots (1 km² each)

The studied landscape consists of various spatial elements with various land use. Evaluation of these elements allowed to point out those with the highest conservation value, which increase floristic diversity in agricultural landscape most significantly. Among them there are: water reservoirs and ditches, meadows, forests and manor parks. Nearly all the elements of the agricultural landscape, both patch-like and linear, are refuges for threatened and other environmentally valuable species.

The research on the flora carried on in the previous century was summarized by [6]. The study indicated that water reservoirs and meadows located in agricultural landscape are the habitats that provide refuge site for the greatest number of threatened species.

Lake differs from other landscape elements in plant cover, which is the most natural, and in numerous occurrences of species that are especially vulnerable to human impact. Rare species for the flora of Poland and Wielkopolska Region and taxa included in red lists of plants are noted more often in the water column and in the belt of rush of the lake than in other water reservoirs (e. g. *Hippuris vulgaris, Lathyrus palustris, Calamagrostis stricta, Carex disticha, Cladium mariscus, Dactylorhiza majalis, Juncus ranarius, Lotus tenuis, Nuphar lutea, Schoenoplectus tabernaemontani, Tetragonolobus maritimus, Teucrium scordium, Valeriana dioica*). They represent 32% of the total flora of the lake. The share of the native species group is very high (nowadays it is 95%) [11].

Species that are protected by law are the most numerous in peat pits. The flora of those biotopes is most similar to the flora of lakes in respect to natural value. Among threatened species, the following are worth mentioning are: *Achillea ptarmica, Calamagrostis stricta, Carex disticha, Cladium mariscus, Dactylorhiza majalis, D. incarnata, D. maculata, Hydrocotyle vulgaris, Lathyrus palustris, Menyanthes trifoliata, Nuphar lutea, Nymphaea alba, Pedicularis palustris, Schoenoplectus tabernaemontani, Teucrium scordium, Utricularia vulgaris, Viola palustris*. About 27% of peat pits total flora are valuable species.

High naturalness has distinguished also the flora of linear elements of landscape that are included in meliorative systems. Those are canals and ditches lying among meadows. Native species are 97% of this flora. Species typical for aquatic ecosystems and meadows have dominated in the sociological plant groups. The only records of *Batrachium fluitans, Sagittaria sagittifolia* and *Sparganium emersum* were found in canals and the only locality of *Alisma lanceolatum* in a ditch.

Ponds and midfield ditches are often the only aquatic ecosystems within rural areas. Thus the flora of them is especially valuable for biodiversity despite of their anthropogenic origin. This concerns particularly areas with big fields and intensive management. Those biotopes enrich poor agricultural landscape with a group of native species, also of the ones which are typical for natural ecosystems. Those are most of all aquatic, forest and ticket taxa. The flora of midfield ditches consists mainly of common species, but sometimes species from a vulnerable group can be observed. Those are e. g. protected by law *Batrachium trichophyllum, Centaurium pulchellum, Hedera helix* or sufficiently rare in Wielkopolska Region *Carex cuprina* and *Sagina nodosa*. Among all tree and shrub lines in the studied area, the dendroflora of the ones that were situated along ditches was the most natural and the share of native species was the highest [17].

Ponds are refuges for many aquatic and marsh species in agricultural landscape and therefore they are very important in maintenance of biodiversity. The only records of such species as *Potamogeton gramineus* and *Potamogeton trichoides* were observed in ponds. The flora of ponds is characterized by a higher share of alien species than the other aquatic biotopes. It is highest in case of naturalized neophytes [11].

The intensity of agriculture of the studied area rose during the last 30 years what resulted in the increase of the significance of agricultural nonpoint pollution leading to the eutrophication of ecosystems.

Thus changes in the flora of aquatic and marsh ecosystems (lake, peat pits, Wyskoć Canal) under increasing human impact were analyzed [20]. This was achieved by comparing the present flora of these habitats with their flora reported in the late 1970s. Vascular plant species richness during the last 30 years increased by 40, as 15 species disappeared and 55 new appeared. However, 6 moss and 4 stonewort species disappeared and only 3 new moss species were found (Table 2). The Wilcoxon test, taking into account changes in the dynamics of individual species, indicated statistical significant differences between the studied periods (Z = 2.531, P = 0.011). In the case of apophytes they were significant both in their number in both periods (sign test Z = 6.019, P<0.001) and their dynamics (Wilcoxon test Z = 5.453, P<0.001). Also the redundancy analysis (RDA) revealed significant differences in flora between the two study periods as the variables 'Present' and 'Past' were highly significant (F = 7.7; P<0.0001). In the case of apophytes they were significant both in their number in both periods (sign test Z = 6.019, P<0.001) and their dynamics (Wilcoxon test Z = 5.453, P<0.001).

Item	The number of plant species	
	1976-1980	2006-2007
Total flora	152	185
Vascular plants	129	169
Stoneworts	9	5
Mosses	14	11
Endangered from the red lists	25	18
New species of vascular plants	-	55
Disappeared vascular plants	-	15
Disappeared stoneworts	-	4
Disappeared mosses	-	6
New mosses	-	3
Nonsynanthropic native species	87	87
Apophytes (synanthropic native species)	40	73
Archaeophytes	0	2
Naturalized neophytes	2	6
Casual neophytes	0	1
Share of valuable species in the total flora (%)	51	37

Table 2. Transformations of the aquatic and marsh flora

Nearly all undetected species were rare in the Wielkopolska region. They are e.g. Batrachium trichophyllum, Carex diandra, Carex rostrata, Gentiana pneumonanthe, Hippuris vulgaris, Pedicularis palustris, Potamogeton friesii. However, complete disappearance of the water soldier (Stratiotes aloides) which was a very common species in peat pits in 1976-1980 was very spectacular. It is also interesting that hornwort (Ceratophyllum submersum) which was very rare in the Wielkopolska Region between 1976 and 1980 and only one record of it was known in the Kościan Plain (in a mid-field pond), widespread in the last 30 years. At present, patches of this species are often found in various types of water body.

Among the new species, the most common were apophytes (i.e. the synanthropic native species capable of occupying the habitats transformed by human activity). Among the species that disappeared from the flora, 84 % are nonsynanthropic spontaneophytes (i.e. the nonsynanthropic native species). An increase in the percentage share of apophytes (from 31% up to 43.2%) and of alien species (from 1.6% up to 5.3%) in the flora caused a decrease in the naturalness index (the percentage share of nonsynanthropic spontaneophytes in the flora) of aquatic and marsh flora from 67.4% to 51.5%.

In the group of new species there are 2 species of invasive neophytes. They are: *Bidens frondosa* and *Echinocystis lobata*. *Bidens frondosa* widespread already in the Kościan Plain landscape and outcompeted the native *Bidens* species, which completely disappeared from the flora of studied ecosystems. *Echinocystis lobata* has appeared in the natural vegetation patches in the last decade and its invasion in the next years is very probable.

In the group of species that disappeared there are only those characteristic for aquatic and meadow communities. New components of the flora are species from various sociological groups. Segetal and ruderal species, which mostly decrease the naturalness index, are 14% of the new species group.

14 moss species were found in aquatic and marsh ecosystems in 1976 - 1980. Their number decreased to 11 (Table 1). Two species protected by law disappeared (*Drepanocladus sendtneri* and *Leptodictyum humile*) and a new one appeared (*Amblystegium radicale*). Three calciphilous species vanished from the moss flora (*Drepanocladus sendtneri, Campylium polygamum, Campylium stellatum*). *Fontinalis antipyretica,* which in 1976-1980 was a very common component of the aquatic flora in the lake, currently belongs to the group of potentially endangered species. The frequency of occurrence of this species decreased by over 90%.

The flora of stoneworts (Characeae) decreased in number by 4 species (*Chara aculeolata, C. contraria, C. polyacantha, C. vulgaris*). All of them were included in the red list of stoneworts in Poland. *Chara polyacantha* was an especially valuable species because it is very rare in Polish flora and is protected by law.

A very significant decrease in the frequency of *Chara fragilis* and *Nitellopsis obtusa* (by 75% and 65%, respectively) was observed in the time interval of the investigations. These species, especially *Nitellopsis obtusa*, belonged to the most important components of submerged plant associations in the lake in 1976-1980.

Phytosociological studies on the vegetation of aquatic ecosystems were also carried out (Tab. 3). The total number of plant communities that were distinguished was 77. Half of them are mentioned in the list of threatened plant communities in Wielkopolska Region. The most valuable is the vegetation of peat pits – the highest number of species threatened, rare and vulnerable to human impact communities occurs in those biotopes. The highest phytocoenotical diversity characterized ponds and ditches lying among meadows. The smallest is the diversity of plant communities in midfield ditches.

Plant communities of the lake, peat pits and Wyskoć Canal were studied from the half of 1970, so it was possible to follow the transformation of them during the thirty-year period

	Plant communities		
Kind of ecosystem	Number	R	V
Lake	22	4	12
peat pits	36	8	23
Canals	25	2	8
ditches among meadows	44	7	18
Ponds	48	7	19
ditches among fields	20	2	4
all ecosystems	77	15	38

Table 3. Phytocoenotic diversity of aquatic ecosystems (Explanations: R - rare, V – vulnerable)

(1976-2006). Nine communities of aquatic plants and one of bulrush community have perished in all the studied ecosystems. Only one of them is not endangered in the Wielkopolska Region. Almost all are very rarely met in the Kościan Plain landscape. During the thirty years the number of plant associations (mainly components of the rush belt), as well as the number of the species that constituted them, simultaneously increased.

In Lake Zbęchy 5 associations of submerged plants disappeared [18].The maximal depth of plant occurrence decreased from 3.6 m in 1976 to 2 m at present. This caused a reduction of total phytolittoral area by 50 percent. The area overgrown by submerged macrophytes decreased from 13 to 2 ha. Valuable plant associations like *Nitellopsidetum obtusae*, *Myriophylletum spicati*, *Najadetum marinae*, *Potametum lucentis*, *Cladietum marisci*, and *Scirpetum maritimi*, considered as endangered in the Wielkopolska Region, disappeared in the lake, or the area of their occurrence decreased.

At the same time cyanobacteria blooms which were not observed in the 1970s appeared. The appearance of blooms at present in the lake can be linked to the disappearance of submerged macrophytes, especially the extinction of *Nitellopsidetum obtusae* that dominated this area thirty years ago.

The most important change in Wyskoć Canal was the complete disappearance of patches of *Sagittario-Sparganietum emersi* association, which dominated in 1970s [19]. Three aquatic plants associations (*Charetum aculeolatae*, *Myriophylletum verticillati*, *Stratiotetum aloidis*), which nowadays are included in the list of endangered communities, disappeared from the peat pits [21].

The study on the flora of meadows has shown that they are very valuable for biodiversity of the whole agricultural landscape. They are distinguished by a high total number of species that is 45% of the total landscape flora, the highest native species diversity and the highest number of vulnerable species [6, 11]. However, during the last 30 years serious changes of the meadow flora were observed, particularly regarding marsh meadows, which are habitat of many endangered and protected plants. As a result of the changes, vegetation of meadows becomes homogeneous and the same everywhere. Rare plant species disappear and common grasses grow in their place. *Alchemilla monticola*, *Crepis praemorsa*, *Eleocharis quinqueflora*, *Eriophorum latifolium*, *Euphrasia rostkoviana*, *Pulicaria vulgaris*, *Viola stagnina* were among plant species that vanished during the last decades. Sites of many valuable species such as *Carex davalliana, Dactylorhiza incarnata, D. majalis, D. maculata, Dianthus superbus, Gentiana pneumonanthe, Parnassia palustris, Pedicularis palustris, Polygala amarella, Tetragonolobus maritimus* and *Triglochin maritimum* are threatened and they are constantly decreasing. As yet, *Ostericum palustre* occurs numerously and in many places, but it is also endangered because of a decrease of wet meadows area and intensification of meadow management.

Forests and manor parks are very important for biodiversity of agricultural landscape despite occupation of small surface. Their species richness is almost the same as in meadows (species noted in all the kinds of forests were 40% of the whole landscape flora). The number of vulnerable species is bigger than in forests only in meadows. The most interesting of those species is daisy leaf grape fern (*Botrychium matricariifolium*). It was found in habitats under big human pressure – economically used monocultures of oak and ash, where some typical practices linked with cultivation of trees were regularly carried out [73]. The species is very rare in flora of Poland and strictly protected. Apart from this species, in the group of vulnerable plants of forests and parks also such species as *Calamagrostis stricta, Campanula latifolia, Cucubalis baccifer, Gagea arvensis, G. minima, Leucoium vernum, Listera ovata, Lycopodium annotinum, Ophioglossum vulgatum, Platanthera bifolia, Teucrium scordium* etc. were noted.

Majority of forests on the explored area are of anthropogenic origin. Separate studies, carried out on the flora of 15 planted shelterbelts lying among fields, showed that 100 species of vascular plants occurred there – they were 30 tree and shrub and 70 herb species. 83% of all the species were native and 17% were alien species. It was more than in natural alder forest, where the share of alien species was only of 5%, but equal or less than in all the forests in the landscape, where the mean share of this species group was of 18% [11]. In the group of alien species there were archaeophytes (8% of all species noted in shelterbelts) and naturalized neophytes (9%). The share of alien species was bigger in the tree and shrub layer (27%) than in the herb layer (13%). The group of alien species in the herb layer in majority constituted from archaeophytes (10% of all species), while among tree and shrub alien species naturalized neophytes dominated (24%). Three of the naturalized alien species – common robinia (*Robinia pseudoaccacia*), black cherry (*Padus serotina*) and box elder (*Acer negundo*) – are mentioned as invasive species that endanger the biodiversity in Poland [42]. Common robinia and black cherry are included in the list of 100 worst alien species in Europe (DAISIE – 100 of the worst http://www.europe-aliens.org/ speciesTheWorst.do). The most recently found expansive and potentially dangerous alien woody taxon was *Amelanchier* sp. [17].

Out of 86 segetal weed species which occurred in the flora of all the landscape 76 species were observed on arable fields and only 9 in shelterbelts. In addition, only 5 of those noted in shelterbelts were met in the flora of fields. The other 4 species were very rare in the flora of the Turew landscape e.g. sharp-leaved fluellen (*Kickxia elatine*) which is a vulnerable archaeophyte in Poland [74] and is also published on the red list of plants of the Wielkopolska Region [75]. Shelterbelts were the only localities of this and several other species in the studied area. Therefore these anthropogenic biotopes may also constitute refuges of rare plant species. According to [76] segetal weeds have appeared in masses only in young shelterbelts. The vegetation of older and stabilized shelterbelts is peculiar and they are not a place from which weeds widespread on the fields.

As a result of investigation (2001-2003) of woody species in tree lines, avenues, shelterbelts, hedgerows and woodlands surrounded by arable fields or intensively used meadows in the central part of the Kościan Plain in 467 mature objects 86 species have been found [17]. The most often noted species was *Robinia pseudoacacia*. The other considerably often noticed species were: *Quercus robur*, *Pyrus pyraster* and *Acer platanoides*. In contrast, on meadows *Salix alba* and *Alnus glutinosa* most often occurred. The number of wooded species in each object varied from 1 to 22.

Generally, tree lines among arable fields were richer than the ones on meadows. On the other hand, the alien species were more common there and had higher share. Total number of alien and cultivated woody species in tree and shrub lines surrounded by arable fields was similar to the number of native species, whereas in those on meadows number of native species was higher. As a matter of fact, around arable fields there was no wooden vegetation patch (wood) which tree layer consisted of native species only. On the contrary, on meadows there were 29% such small woods.

5.1.1. Factors determining plant diversity

The key factor determining richness of plant species and diversity is the landscape structure. In the agricultural landscape all the elements that are different from arable fields, both patch-like and linear, are refuges for environmentally valuable species. Thus, the most serious threat to biodiversity is the loss of appropriate habitats, what results from increase of intensity of agriculture. The intensity of agriculture particularly increased in the 1970s and 1980s. The most serious threat was the increase in the nitrogen content in surface and groundwater. It resulted from increasing non-point pollutions of agriculture landscape and intensive nutrient leaching from dried up peatbogs. Both aquatic and marsh ecosystems are very important refuges for plant species and their communities which are the most vulnerable to human impact.

In recent years, probably as the result of temperature increase and eutrophication, increase of frequency of cyanobacterial toxin-producing blooms have been noticed in various water bodies [77]. Cyanobacteria are very expansive and their massive appearance, regardless if they produce harmful compounds or not, it is an unfavorable phenomenon. First of all, it is a sign of impaired balance of the ecosystem [78]. The effects of the cyanobacterial blooms are a large concentration of biomass, loss of biodiversity, reduction in biocenotic stability system, the presence of large numbers of heterotrophic bacteria (especially in the phase of cyanobacteria decay), inhibition of photosynthesis of planktonic algae which are associated with cyanobacteria as well as profound oxygen deficits in the lower layer of water column [79]. The most sensitive to cyanobacterial toxins are warm-blooded Vertebraters.

Another threat for aquatic ecosystems biodiversity is an incorrect fishery management. For example, large stock of grass carp (*Ctenopharyngodon idella*) results in adverse changes in the littoral of aquatic ecosystems. This species eats vegetation, leading to the elimination of spawning areas of fish and fry regrowth of the native valuable species. Small bodies of water rich in grass carp may be completely devoid of vegetation.

Common carp (*Cyprinus carpio*) at high densities can significantly influence on environmental conditions, accelerating the rate of eutrophication of natural waters and the same have a negative impact on native fish species. The foraging behavior of carp can also negatively affect the structure and dynamics of benthic macroinvertebrates community, as well as the physical condition of the substratum [80]. These disturbances may also intensify the algal blooms due to the re-suspension of sediments [81] and release of nutrients [82, 83]. Serious problem for biodiversity of aquatic environments and marshes results from overdrying of peatlands and changes in their management.

The transformations of vegetation in the Wyskoć Canal reflect in the eutrophication of waters and changes in water flow rate. Water management measures undertaken in the 1980s caused a lowering of both the river bed and water table of the ditch. As a consequence, there was observed drainage from surrounding meadows, what caused an accelerated peat mineralization and intensive leaching of nutrients. The observed transformations of reed and sedge communities in the Kościan Plain have partially natural and partially anthropogenic character, because the natural plant succession also accelerated due to the human pressure as well as due to climate changes, especially low precipitation. The increase in the number of identified plant associations is a result of human impact. However this process is jointed with withdrawal of plant species distinctive for waters with low nutrient concentrations and by expansion of plant associations that are indicators of high nutrient levels in waters.

The natural values of wet and marsh meadows that belong to the richest elements of the studied agricultural landscape are threatened not only by drainage but also by changes in the usage patterns. The extinction of these habitats results from the intensification of management (fertilization and sowing of common grass species) but on the other hand from the abandonment of meadows cultivation, which leads to a complete overgrowth of them by willow shrubs.

5.2. Insects (above-ground insects and butterflies)

As the importance of landscape (and habitat) for insect diversity was studied in the "Kościan Plain" in many various research projects, here we focus on a comparison between cultivated fields and small wooded patches (wide shelterbelts in that case). Such approach allows underlining the role of non-farmed habitats presence as a key factor for preserving high biodiversity level of an agricultural landscape. The results obtained during the last three years (2009–2011) show that the density, biomass, and diversity (expressed as number of above-ground insects families) in the shelterbelts are significantly bigger than in arable fields. Also in narrow ecotone (0–0.5 m from the shelterbelt) higher values of those parameters were noted. Thus, the results showing differences in density, biomass and diversity of entomofauna, depending on shelterbelt's age, which were reported earlier [84, 85] have been confirmed. The highest values of those parameters were observed in young (several years old) shelterbelts, in early stages of ecological succession. The lowest diversity (47–61 families) was observed in the several years old shelterbelt and the highest diversity (63–74 families) in a few years old one. In the oldest (over 100 years old) shelterbelt moderate values were noted (58–69 families). Similar pattern was found for density and biomass of insects. Also in the ecotone zone these differences were similar, but they completely disappeared in the arable areas distant 100 m from the shelterbelt, values of studied parameters are much lower than in shelterbelts and ecotones (Figure 8). That pattern have been constant for many years (several decades), while the proportions were changing and during recent years the mean individual weight and biomass of insects in the open fields (away from shelterbelts) increased.

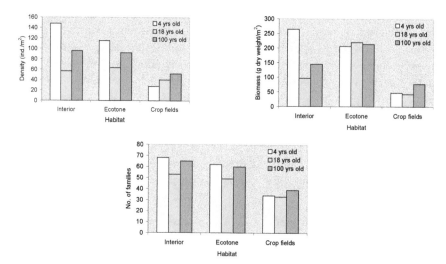

Figure 8. Relationships between insect communities (density, biomass, number of families) in vegetation season and age of shelterbelt with respect to three habitats: shelterbelt interior, ecotones between shelterbelt and cultivated fields and open fields.

The importance of non-farmed habitat for insect diversity was also studied in winter. As early as in fifth or sixth winter after introduction of the shelterbelts, they became an important place of wintering for numerous insect species. The diversity measured by the number of families reaches values (over 30 families) stable during next years of shelterbelt's growth (Figure 9). In the case of insect density, after rapid and strong increase of its value in 6th – 7th year of shelterbelt's growth (up to over 1300 ind./m²), some decline and stabilization at the level about 200-1300 ind.m² was observed. Similar pattern of changes was noted for biomass. All studied parameters (density, biomass, number of families) reached the highest value in the oldest shelterbelt and the lowest – in the youngest one (Figure 10).

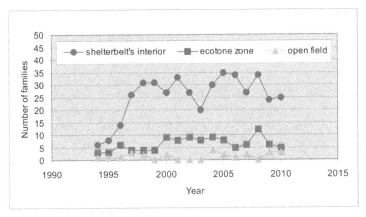

Figure 9. Number of insect families wintering in young shelterbelts and in adjacent crop fields.

Ongoing process of increasing share of cereals in crop structure creates favorable conditions for existence of species primarily living in grasslands, mainly steppes.

5.3. Spiders

Effect of landscape structure on spider assemblages is till now unclear and understanding it needs further investigations. However, gathered data also suggest positive effect of mosaic-like structure of landscape. Mean spider population density in cereal and sugar beet crops was similar across the landscape. In cereals it amounted to 7.5 ind./m² in mosaic landscape ("Turew Mosaic Landscape") and 7.3 ind./m² in homogeneous landscape (in the "Kościan Plain"). In

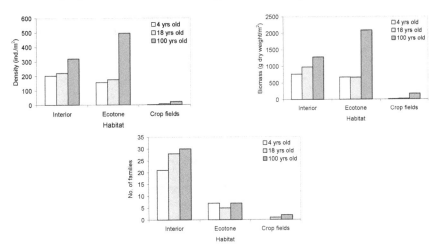

Figure 10. Relationships between insect communities (density, biomass, number of families) in winter and age of shelterbelt with respect to three habitats: shelterbelt interior, ecotone between shelterbelt and cultivated fields and open fields.

sugar beet crop it was equal to 3.1 and 2.9 ind./m², respectively. However, greater differences in alfalfa crop were noted. In homogeneous landscape spider density amounted 4.1 ind./m² and in the mosaic one - 10.8 ind. /m².

Spider species diversity in cereal and alfalfa crops was near twice higher in the mosaic landscape than in the homogeneous one. On the other hand, in sugar beet crop the same numbers of spider species were stated in both landscapes. In all studied crops in homogeneous landscape aeronautic spider species (small-in-size Linyphiidae) were more abundant than in the mosaic landscape. Foliage-dwelling spider assemblages structure in the studied crops were more diversified taxonomically in the mosaic landscape – higher numbers of spider families (cereals) or species (sugar beet and alfalfa) were noted. Moreover, the differences in share of particular spider families were stated. Non-web spiders were more abundant in these assemblages: ambush (families Thomisidae and Philodromidae) as well as actively hunting ones: Pisauridae, Salticidae, Mimetidae and Lycosidae.

In the cereal crops situated in homogeneous landscape the vast majority (75%) of the assemblage was composed of Araneidae – spiders hunting with orb webs, whereas in the mosaic landscape they constituted slightly more than 50% of the whole assemblage. In the last mentioned landscape type higher abundance of theridiid spiders was noted, which built three dimensional tangle webs. In the sugar beet crops the most abundant spider families were Theridiidae and Araneidae. The first one composed of over 40 % of assemblage in the homogeneous landscape, and share of Araneidae was similar (45 %) in the mosaic landscape. Among foliage-dwelling spiders in alfalfa crops linyphiid spiders (in majority Erigoninae) composed of 60% of assemblage in the homogeneous landscape and 25 % in the mosaic landscape. In the last one the next 25 % composed of Araneidae. The share of families Theridiidae and Tetragnathidae was similar in both landscapes.

The crop fields adjacency to the shelterbelts resulted in increasing of spider species diversity in above-ground layer of all studied crops. In the distance of 10 m from shelterbelts the highest numbers of spider species were stated and with the increasing distance the species numbers decreased until near twice lower in 100 m from shelterbelts.

5.4. Birds

The "Turew Mosaic Landscape" was in 1991-1994 inhabited by 76 breeding species [59] as mentioned in the chapter 4.5.1. The number may be evaluated as relatively high in Poland when agricultural areas are considered and the study area size is taken into account (Figure 11A). Moreover, regression analysis (GLZ) showed, that number of years had no significant effect on total number of species in this set of data. Thus, the Figure 11A evidences for high bird species richness in the "Turew Mosaic Landscape", presumably as a pure (not methodologically-biased) effect of habitat quality.

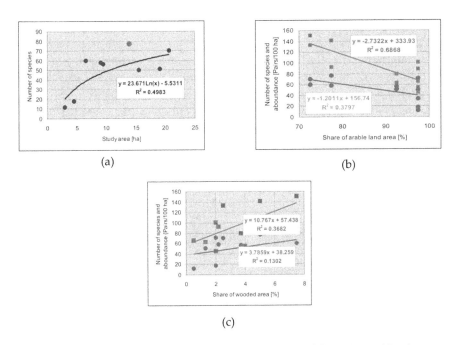

(a)

(b)

(c)

Figure 11. a) Species - area relationship for Polish farmlands [86]. Red dot – "Turew Mosaic Landscape"; b) Effect of cultivated fields share on breeding avifauna [86]. Red - species number, blue - density; c) Effect of wooded patches share on breeding avifauna [86]. Red - species number, blue - density

The data gathered and presented by [86] confirm clear relationships between landscape structure and bird richness and abundance. The values of both variables were negatively correlated with the share of arable land (Figure 11B) and positively correlated with the percentage of land covered by small wooded patches, lines etc. (Figure 11C). These two variables were weakly correlated (r=−0.43, p=0.19), so presumably both influenced the avifauna independently. It is worth to underline that species number of bird community in the "Turew Mosaic Landscape" is as high as observed in Podlasie, in eastern part of Poland [87], which is characterized by much less intensive farming practices than in Wielkopolska region and one could expect higher bird species richness just in E Poland. What is more, the species composition similarity between the "Turew Mosaic Landscape" and Podlasie was high (incidence-based Sörensen index = 82%, after [86], what indicates high efficiency of landscape structure (mostly wooded patches, lines etc. in that case) as a tool for mitigation of the impact of farming intensification on biota. Strong relationships between species richness and landscape structure is reflected also by the analyses performed for the data gathered in 25-55 ha subplots within the "Turew Mosaic Landscape" and in other places in Wielkopolska, which confirmed crucial importance of the saturation of the landscape with the wood patches and woodland-cropland ecotones [88].

The relationships between landscape structure and species richness (i.e. Jack-Knife 2) were studied with the use of the data from point count census presented by [89]. The analysis indicates similar overall species richness (65-80 species) in various landscape pieces with no respect to landscape structure (Figure 12), but the shape of estimated curves shows significant differences in species richness spatial distribution between the studied plots. In more simplified plots (arable fields – 95%) it is necessary to visit >25 points to observe 60 species, while in most diverse (arable fields – 52%) – only 15 points is needed to record 60 species It is also worth to underline that the landscape consisted of tree line, belts and patches (arable fields – 75%) had similar overall species richness to the area (arable fields – 52%) with relatively big forest complexes (Incidence- and abundance-based Sörensen index amounted to 76±2% and 94±3% S.E, respectively).

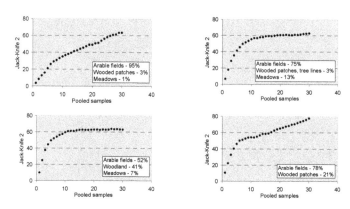

Figure 12. Bird species richness in various landscapes estimated with the use of Jack-knife 2.

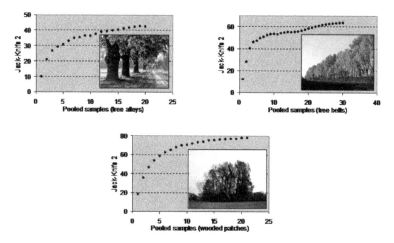

Figure 13. Bird species richness in various wooded habitats estimated with the use of Jack-knife 2. Photo K. Kujawa.

Due to a study focused on birds in small wooded patches, it could be possible to study the relationships between the bird communities and the variability in habitat structure (i.e. in scale of several hectares). The first issue was studied in small wooded patches, tree lines etc. Using the data presented by [32], the highest potential overall species richness (ca. 80 species) was estimated for wooded patches (Figure 13) and the value is close to the total species richness estimated for all kinds of patches (clumps, belts and lines) built of woody species.

Observed differences may be surely related to the differences in habitat structure between the three wooded habitat types as the analyses presented by [32] showed crucial importance of such habitat features as tree stand age, percentage cover of tree and shrub layer and the area (or length in case of line habitat), which all influenced species number positively. Indeed, the differences in bird species richness were in line with the differences in habitat structure between three habitat types [32].

However, the knowledge on relationships between individual habitat structure usually allows for explaining only a part of bird community variability in wooded areas. The analyses performed for the data on 66 islands located in the "Turew Mosaic Landscape" showed that taking into account landscape structure around studied habitats allows for understanding avifauna variability much better, especially for species strictly related to woodland, for which positive role for small wood island colonization played the presence of other wooded patches and tree lines in close adjacency [33]. Seemingly, these habitats significantly reduce the open landscape "resistance" for migrating woodland species and, as a result they enhance probability of colonization of given wood island by the birds.

5.4.1. Effect of land-use changes on bird communities

The quantitative investigations on birds in various mid-field wooded patches were conducted for the first time in 1964-1966 and were repeated in some plots in the years 1984, 1991-1994 [32], 1999-2002 [33] and 2005-2006 [89]. During the last 50 years, farming practices have been much intensified thus gathered data on birds enabled for testifying the effect of farming intensification on birds. Uniqueness of that research relies on dissecting the effect of landscape structure simplification and the effect of farming intensification. Commonly in Europe, landscape homogenization follows the intensification of agricultural techniques, but not in the "Turew Mosaic Landscape", where the spatial arrangement of non-farmed habitats is stable (even enriched with new tree lines and belts recently) due to establishing the Dezydery Chłapowski's Landscape Park, which successfully protects historically formed land mosaics (see "Study area").

First analysis of long-term changes in the breeding avifauna in the "Turew Mosaic Landscape" was done by [90]. The number of breeding species has increased from 44 in the 1960s to 51 in the 1990s (which can be at least partially explained by somewhat bigger sampling effort in the later period).

Total density of breeding birds remained unchanged in wooded patches (23–24 pairs/ha) with obvious exception of few areas where tree stand was cut. However, there was noticeable decrease in population density of some 'farmland specialists' like corn bunting and ortolan bunting. The analyses were then concluded that although the structure of bird community has changed only slightly in the study area, the population trend analysis suggests that agriculture intensification affects the avifauna [90].

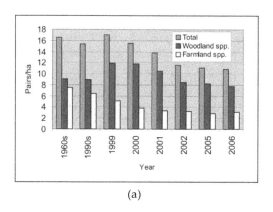

(a)

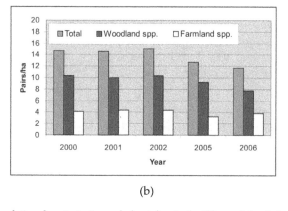

(b)

Figure 14. a) Bird population density in 6 wooded patches in the "Turew Mosaic Landscape" between 1964 and 2006 [90]; b) Bird population density in 55 wooded patches in the "Turew Mosaic Landscape" between 2000 and 2006 [90].

Recently, the changes that avifauna of small wooded patches is undergoing have been proved again [91]. The results of comparisons between the study periods (1960s, 1990s, 1999-2002 and 2005-2006) show large and significant different pattern of the changes for two main guilds of birds: woodland species (i.e. birds living in breeding season only in wooded areas, e.g. woodpeckers, robin *Erithacus rubecula*, nightingale, and spotted flycatchers *Muscicapa striata*) and farmland species (i.e. birds, which occur in wooded areas but using adjacent cultivated fields as feeding area or as place for building of nest, e.g. ortolan bunting, red-backed shrike, and goldfinch *Carduelis carduelis*). In the last 50 years the woodland species abundance and species number has been changing irregularly (increase in 1990s, some decline in XXI century), while the farmland species guild tends to decline permanently (Figure 14A). In consequence, the structure of the community has changed markedly. In 1960s the share of farmland species amounted to 45%, and it decreased in 2005-2006 to 27%.

Recently some decline in species richness has been observed too. Mean number of species per wooded patch decreased from 8.6±4.3 (SD) to 7.5±3.9 (SD) and bird abundance declined, too (Figure 14B). The differences between 2000-2002 and 2005-2006 were statistically significant for both guilds of species [91]. The observed pattern of changes shows that the intensification of agriculture is the main factor responsible for some bird impoverishment in small wooded patches in the "Turew Mosaic Landscape". Thus, it may be concluded, that mosaic of diverse habitat is necessary but not sufficient condition for successful protection of bird diversity in the agricultural landscape. To be as sufficient as possible, mosaics of diverse habitat has to accompany less intensive and more diversified (in terms of number of crops) farming.

5.4.2. Factors influencing avian diversity

The analyses and findings presented above show, that the bird communities in an agricultural landscape are influenced by a complex set of factors, which act in both habitat scale and landscape scale. The mechanisms involved are summarized below.

5.4.3. Habitat scale

- **Patch size**. The number of breeding species increases with the length and width of tree alleys and tree or hedge belts [32] and with the patch area [32, 33]. Birds are highly plastic species. Many species is able to breed in very small-sized patches. As many as 45 species were found to nest in wooded patches smaller than 1 ha [33].
- **Patch shape**. More complicated shape affects positively total species number as favors species preferring ecotonal and woodland edge species [33].
- **Patch vegetation structure**. More species breed in wooded patches with deciduous species, with old, diverse tree stands, with higher percentage cover of shrub layer, which means that more rich in species are wooded patches similar in structure to natural forest typical (but rarely occurring) for lowland regions in moderate climate [32, 33].
- **Diversity of crop fields**. This factor affected not only the number of species breeding in open area (more species in more fragmented and more diversified cultivated area) but also the birds occurring is small wooded patches (bird abundance and species number higher when the patch surrounded by diverse, fragmented crop fields) [33].

5.4.4. Landscape scale:

- **Landscape context (for wooded patches)**. Although isolation of studied was on average small (distance from woodland amounted to maximum few kilometers, presence of dense network of tree line and belts) some bird species, especially those typical to woodland) were more abundant in wooded patches, which were located in a landscape rich in other wooded patches [33]. It means, that the knowledge on relationships between birds and inside habitat structure is not enough to understand the distribution of birds. It is necessary to take into account the landscape structure as well.
- **Landscape heterogeneity**. The analysis of species richness indicates that avian diversity in a mosaic-like agricultural landscape may be kept at same level as in a landscape with high

woodland dominance (in case of considering managed forest). If we take into account, that avifauna in our study included also rare and protected species (with abundant populations of e.g. Ortolan Bunting, Corn Bunting), the conclusion clearly arise that key way for preserving high bird species richness is making landscape as heterogenic as possible. Even though non-farmed elements cover only 5-7%, it is enough to preserve almost all species from the woodland generalists and woodland edge species. The observation can me generalized with respect to the issue of fragmentation. Although in ecology the habitat fragmentation is considered as negative phenomenon, in agricultural landscape that process has rather positive role for species diversity, when we discuss about small (<10 ha) forest patches. Concluding with the use of simple, model example, it is more profitable to keep e.g. five forest islands with the area 3 ha each, connected to each other by several kilometers of tree belts, than one "large" forest island with the area of 20 ha.

The studied area, especially the "Turew Mosaic Landscape", is characterized by most of features listed above. There are numerous, diverse non-farmed habitat patches bodies and wetlands. Moreover, most of wooded patches have old tree stand, they have diverse vegetation structure. Due to these features and mosaic-like structure, the "Turew Mosaic Landscape" constitutes a distinct example of an agricultural landscape, in which relatively rich avifauna occurs despite of high agricultural pressure [86]. Such agricultural landscape is favorable for at least 100 species, i.e. ca 40% of breeding species observed in Poland. It is striking number and we claim that the "Turew Mosaic Landscape" may play a role of a reference point for the studies on diversity of birds in lowland agricultural landscape in Central Europe.

However, recent publications on long-term changes in avifauna showed that even strongly stable amount and arrangement of non-farmed, semi-natural habitats is not enough to effectively preserve species richness, when farming practices become more and more intensive. Long-term decline in ecotonal species populations [89, 90] clearly show, that stable amount and structure of wooded patches, tree lines and tree belts does not mitigate the effects of farming intensification for all bird species guilds. Farming extensification seems to necessary to preserve total bird diversity.

5.5. Fungi

In chapter 4.7.1 it was mentioned that 26 % of species observed in the "Kościan Plain" study area belong to the group of special concern species (Species protected by law, threatened species listed in "Red list of the macrofungi in Poland" , threatened species listed in "European red list of the macrofungi", rare species).

Occurrence of these species in the study area provides great facilities in protection of fungal diversity in Poland (including agricultural area), due to proper landscape management and planning. Moreover , it may be useful for evaluation of response of given species to habitat alteration as well as for elaboration on nationwide strategy for protection of rare and threatened (according to IUCN's criteria) species of fungi. Proper landscape structure has significant influence on species diversity of fungi all over Poland. It allows for surviving of many woodland species under strong, unfavorable human pressure. Additionally, some rare

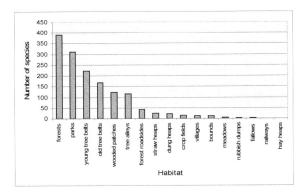

Figure 15. Number of fungi species in habitats located in the "Kościan Plain".

species dependent on traditional agriculture (growing in pastureland or organically fertilized fields) find appropriate niches in strongly diversified, mosaic agricultural landscape.

A key component for incorporating natural enemies, including fungi, in pest management in sustainable farming is the application of appropriate agricultural practices that improve conditions for these organisms in the agroecosystems and surrounding habitats. Agroecosystems suffer from mechanical disturbances due to necessary tillage regimes of various kinds, and these practices may negatively affect occurrence of entomopathogenic fungi in the soil [92, 93]. This was confirmed in [72] studies conducted in the years 2002-2004 in the area of Borek Wielkopolski directly adjacent to the "Kościan Plain".

6. Impact of climate change

Impact of climate change on biodiversity can be both direct and indirect. The direct influence lies in the fact that climate change will lead to changes in habitat conditions (mainly temperature and moisture) that can be favorable for some species and unfavorable for others. This influence is revealed by disappearing of existing species and by appearing of alien species, among which some can be invasive. The indirect impact is the fact that changes in habitat conditions are forcing farmers to change cropping patterns. For example, growth of thermophilic plants that require less rainfall (increase of cereals at the expense of root crop cultivation). This in turn influences the change of plant community and animals living in the agricultural landscape.

There is no doubts that climate change is not new in geological history of Earth, and organisms have adapted to most serious changes over evolutionary timescales. But the key question today is how will organisms respond to the current apparently rapid rate of anthropogenic climate change? [94].

Kędziora defined the process of climate changes that occurred in Poland as mediterraeanization. It's expressed by gradual increase of average air temperature with simultaneous absence of annual precipitation increase and move rainfalls to cold months. Similar trend for climate changes was observed in the neighboring state of Brandenburg. Such course of changes leads to deterioration of higrothermic conditions, especially during vegetative season.

It is obvious that impact of climate and land use changes is working synergistically. Therefore, the analysis of changes of land-use in the landscape during long period has been done in order to to distinguish the impact of climate from the impact of land use,. The study of a landscape structure which was conducted in the area of 16 km² in the vicinity of Turew, where the total area of meadows, pastures, and woody vegetation patches (the most important ecological elements in the agricultural landscape) was assessed at four time horizons: 1890, 1940, 1989, and 1996. Share of meadows, pastures and woody vegetation in total area of the landscape showed significant increase in the period 1890 – 1940 and remained slightly changed in the period 1940 - 1996 (Figure 15A). Although only mild increase of tree patches (from 64.5 to 69.5%) and decline of tree lines (from 35.2 to 30.5) and small decline of pastures was observed during the period 1940-1996 (Figure 15B), disadvantageous trend of changing their distribution occurred: disappearance of single trees and small strips of meadows distributed between fields [44].

Simplification of agricultural landscape may not only occur together with climate change, but it can also be a result of those changes. Deterioration of hydrothermal condition may cause disappearance of moist habitats, like swamps, marshes, and bogs. The presence of ecologically differentiated water reservoirs in landscape is also the key factor for maintenance of amphibian populations. Protection of such habitats in the agricultural landscape is especially difficult because this area is usually very (sometimes extremely) small and particularly vulnerable to influences. But the most important change, threatening biodiversity, is observed in the structure of the crop. Area of cereal crops increases rapidly at the expense of perennial and root crops. Until 1990s, [95] reported following structure: cereals 50%, row crops (including rapeseed) 20%, perennial fodder crops 10% and others 20%, in first decade of 21st century [96] noticed: cereals (including maize) 78%, legumes 16%, potato, rapeseed and sugar beets 6%.

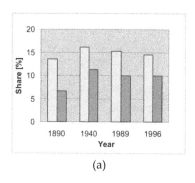
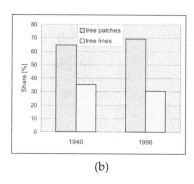

(a)   (b)

Figure 16. a) Changes of share of meadows and pastures (light green) and woody vegetation (dark green) in total area of investigated landscape [44]; b) Changes of share of tree patches (light green) and tree lines (yellow) in total area of woody vegetation in investigate area [44].

In global scale temperature increase may cause serious changes in species distribution which will expand their ranges towards higher latitudes. Such processes were just observed in the case of fungi and lichens in UK, Netherland and Baltic See region. Herk [97] found, basing on a long-term monitoring, that many tropical and subtropical species of epiphytic and terrestrial lichen species are invading the Netherlands while 50% of the arctic-alpine and boreomontane species already shows a decline. In Baltic See basin, 39 distinguished species of Macromycetes extends range to the northern latitudes, for example from Central Europe to Scandinavia [98, 99]. Expected dramatic decline of forests and tundra biome towards desert and grasslands. This prediction was supported by [100]. Some results in literature, especially these made on climatic transects [101] showed that in many regions of the world we should expect increase of α diversity (species or highest taxonomic units number) as climate change will progress [102]. On the other hand due to dramatic speed of change we will lose mainly vulnerable species with narrow ecological niches [103, 104]. Such species are rare and characteristic for unique regions or habitats [105]. But then again common species having wide amplitude of distribution will move towards highest latitudes [106]. Increasing ranges of alien species might mean also the invasion of pests, both animals as fungi. Widely known is the case of Harlequin frog in mountain areas of tropical Central America (Monteverde (Costa Rica)). [107] showed that ongoing global warming process changes microclimate condition of Monteverde (Costa Rica) towards optimum for chytrids (warmer nights and increased daytime cloud) (*Batrachochytrium dendrobatidis*). This fungi infects amphibians causing Chytridiomycosis - an infectious disease of amphibians responsible for extinctions of amphibian species in western North America, Central America, South America, eastern Australia, and Dominica and Montserrat in the Caribbean. Classic example is also horse chestnut leafminer (*Cameraria ohridella* - invasive species of uncertain origin, first observed in Macedonia in 1984 [108]. Pest invaded central and western Europe at an approximate rate of 60 km year. In Poland the species was discovered in 1998 for the first time in Lower Silesia [109]. The march of this insect, from south west to north east (according to the line of spring progress in Poland) suggests that its expansion is connected with climate changes. As far as plants are concerned changes in species distribution may lead to disturbances of balance between plants with different photosynthetic pathways. The system of interrelated factors influencing photosynthesis (CO_2, temperature, water availability) is very complicated and it is difficult to predict the direction of changes. For example, the increase of CO_2 concentration in the atmosphere will stimulate C3 plants [110-113] while in the case of C4 plants, biomass growth and competitive potency will be stimulated by temperature increase. Apart from that, C4 plants have better efficiency of water use (WUE) [114]. Therefore deterioration of humidity conditions caused by increase of temperature will cause better conditions for C4 plants than for C3 plants. Last observations suggest that despite of CO_2 elevation we may expect increasing share of C4 plants, in Central Europe. For example, between 372 species of exotic plants in Poland until 7.5% (28) consist C4 plants, while in native flora of Poland only 5 species represent C4 photosynthetic pathway. Some of them are aggressive invasive plants, (genus: *Setaria, Digitaria, Echinochloa, Eragrostis, Amaranthus,*). Many of them are common weeds, brought to the area before industrial era, with the crop plants, originated from the Middle East, but presently occurring of these species are not restricted to crop fields, some of them (*Echinochloa crus galli, Amarantus retroflexus*, different species from *Eragrostis* genus are aggressive invasive plants.

Changes in entomofauna of agricultural landscape, observed in last decades, are probably an effect of global climatic changes (since eighties). Increase of temperature without increase of precipitation, especially in summer is favorable for termophilous species. There are herbivorous species among them, so they are potential or real pests, mainly some of the beetles, bugs and butterflies. Several years ago, chemical protection from genus *Eulema*, strongly exceeding so-called threshold of harmfulness, started to be applied. Near this threshold there are tortoise bugs (*Eurygaster*), cutworms (*Agrotis*) and cereal ground beetle (*Zabrus tenebrioides*). The last species is an imposing beetle (family: Carabidae), feeding (larvae and adult forms) on cereals. High density of beetles may cause considerable loss, when they climb to ears and eat grain. In the last decade, number of Thysanoptera has also instantly grown. In 2002, in Wielkopolska region, they exceeded the harm level (mainly on oat). They were not chemically exterminated, and in last decades they were considered as potential pests Significant increase of Simulidae has also been observed. In the 70's and 80's years, only single individuals were found in large samples, containing several thousands of other Diptera, taken by motor net [25]. Increasing number of south-european (Mediterranean) species, especially moths (Sphingidae, Noctuidae) and Cicadidae, e.g. *Cicada orni* has been observed. For several years, the invasion of Asian ladybugs (*Harmonia axyridis*), is also observed.

The number of alien insect species will probably grow up. Some of them, with high invasive potential, are already present here, e.g. agrofagical western corn rootworm (*Diabrotica virgifera*), recorded in Poland since 2005. It is hoped that those species can not reproduce on a mass scale, because of natural, regulating biocenotic processes which efficiently act in an agricultural landscape. Such effect concerning horse chestnut leafminer (*Cameraria ohridella*) has been observed since the begin of the present century. Number of natural predators of this species grows from year to year and the presence of this pest decreases, what can be easily observed. On the contrary, 20-years study on insect fauna in Turew landscape showed graduating changes in its structure. During this period mean individual body mass of insects increased significantly (Figure 17A) as well as total biomass of insect community (Figure 17B). However biodiversity (expressed by number of families) significantly decreased much quickly in uniform landscape than in mosaic (Figure 18). This is probably an effect of the increase of the number of xerophillic and thermophillic insects on arable fields, including invasive large agrophagus, not connected with seminatural parts of agricultural landscape (*Zabrus, Agriotes, Eulema, and Eurygaster*) and predators (*Nabis*).

This phenomenon is a result of synergetic impact of increasing of cereal area and climate changes which leads to invasion of alien species having high body mass. Investigation showed that after 1990s a new species of butterflies have been observed. These species originated mainly from south and east Europe where climate is more dry and warm. It were thermofilic and xerophilic. For instance: *Ephesia fulminea, Syntomis phegea, Catephia alchymista, Arcthia villica* and more frequency inflying Mediterranean species (*Macroglossum stellatarum, Agrius convolvuli, and Phlogophora meticulosa*). Similar changes have been observed in other groups of insects (coleoptera, homoptera, hymenoptera, and heteroptera. Increase number of these species is especially evident in agrocenosis in uniform landscape, so diversity of insects living in cereals decreases slowly than in the mosaic landscape (Figure 18). Increasing share of maize in crop structure, which is the effect of climatic changes, brought about occurrence many efemeral habitats, like silage heap, which causes that some forest species migrates into agricultural landscape. For example: european rhinoceros beetle (*Oryctes nasicornis*)

Another change in insect community structure observed during 20 years was gradual vanishing of differences between communities in two type of landscape. In the eighties, mosaic landscape was characterized by higher biomass, higher individual body mass of insects and higher number of families. In first decade of 21st century those differences vanished and number of families in uniform landscape was even higher then in mosaic landscape. These data suggest declining role of landscape structure together with progress of climate change and simplification of crop structure. Long term study of bird assembles in Turew area confirms the results concerning insects. Since1964 up to 2006, in birds assembles in afforestations in gen. D. Chłapowski Landscape Park number of species declined of 20-25%, and this trend begun in the end of 20th century. Furthermore, permanent decrease of density of birds which use of agricultural fields during breeding period and typical forest species in 21st century (preceded by increase at the end of 20th century) has been observed. Comparison of the above results concerning birds rarely occurring in shelterbelts and hedges, typical for agricultural landscape with study results carried out 10 - 40 years ago leads to conclusion that farmland avifauna decreased approximately by 30 to 40%.

Data collected at the beginning of this century, together with those which were collected earlier indicate that the population density and species richness in the group of species living in afforestations, but using agricultural fields are decreasing, and this decrease albeit has been slowly is continued for many years. Moreover, recent data (from the years 2000-2006) indicate that the forest species begin to acquire negative changes. This is a new phenomenon, because so far the forest birds were at least stable, both in density as species richness, and in some periods showed even upward trends. It is hard to explain, but it should be emphasized that the phenomenon of declining of some species typical to forest, even common, like poor tit, robin and creepers, was noted also in the country scale [115].

In present climate change conditions amphibians are especially vulnerable animal group. The simplest measure of the scale decreasing amphibian population is change in number of water reservoirs as amphibian breeding sites through years. Probably the first time in Europe this method was used by Danish, which made the picture of changes in amphibian populations in years 1940-86 [116, 117]. In the Wielkopolska region in the end of 19th century there were 11 068 small water bodies (area less than 1 ha), but in 1960 only 2490 remained – decrease by 77% [118]. This process is ongoing. In the "Turew Landscape" between year 1960 and 1995 disappeared 55 out of 287 (19%) small reservoirs. The main causes of this process were direct impacts of man activity, such as drainage and filling 36 artificial reservoirs, but disappearance of remaining 19 reservoirs (35%) was caused by climatic changes – low precipitation and lowering of groundwater level [29].

Disappearance of these reservoirs together with biological degradation of many other caused decreasing of abundance of many amphibian populations (even up to 55-95 %), total disappearance of some other ones and drop of their diversity. The examples of such „climatic disasters" in Turew landscape are three large ponds situated in forest, far away from agricultural drainage system. All of them belonged in mid 1990s to the 10 most important amphibian breeding sites [28]. Pond 1 (Rabinek forest – 0,4 ha in 1988) – after lowering of water level number of water frogs droped from 1330 (in 1988) to 320 (in 1997)

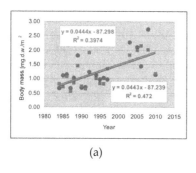
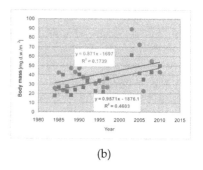

(a)   (b)

Figure 17. a) Individual body mass of epigeic insects on grain crops in two types of landscape; b) Total biomass of epigeic insects on grain crops in two types of landscape.

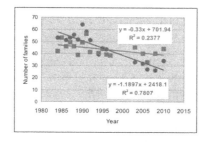

Figure 18. Diversity of epigeic insects (number of families) on grain crops in two types landscape.

individuals [51] at present there are ca. 50 frogs. Pond 2 and 3 (Błociszewo forest – 4 and 1,3 ha in 1995) – two biggest forest ponds in the Turew landscape inhabited by the largest populations (more than 100 individuals) of two amphibian species from Appendix II of the EU Habitat Directive: crested newt (*Triturus cristatus*), fire-bellied toad (*Bombina bombina*) and European tree frog (*Hyla arborea*). In the beginning of the 21st century these ponds were completely dried up and all amphibians disappeared.

In conclusion: long term observations evidenced that changes lead to uniformization of flora, fauna (and probably funga) of agricultural landscape. We cannot precisely distinguish changes caused by landscape (and crop) structure simplification and those caused by global change but there is no doubt that agricultural landscape diversity deteriorates. In recent 20 years it became obvious that we cannot stop decline of biodiversity only by use of protection of landscape heterogeneity, because the scale of changes in crop structure and that which are not directly connected with landscape, e.g. climate change or in the case of birds occurring in wintering areas is too high.

7. Conclusions

It seems obvious that human activity transformed nature in pursuit of safe food supplies, assertion of more comfortable housing conditions, exploitation resources and making transportation of people and goods easier and so on. The main change is transformation of stable ecosystems like forests, meadows, wetlands into unstable, mainly in farmlands. In order to obtain high yields, farmers must eliminate weeds, control of herbivores and pathogens, ensure that nutrients are easily accessible only for cultivated plants during their growth, increase mechanization efficiency. To increase production, farmers simplify plant cover structure both within cultivated fields (selection of genetically uniform cultivars and weeds elimination) and within agricultural landscape (elimination of hedges, stretches of meadows and wetlands, small mid-field ponds). Animal communities on cultivated fields are also impoverished (Karg and Ryszkowski 1996). Negative ecological effects of agriculture intensification are connected to:

- impoverishment of plant and animal communities
- decrease of humus resources
- decrease of capacities for water storage
- increase of pollution from non-point sources.

It must be clearly said that although farmers can moderate the intensity of these processes through proper selection of crops and tillage technologies, they are not able to eliminate them entirely. The higher control environmental threats efficiency evoked by agriculture could be achieved by structuring agricultural landscape with various non-productive components like hedges, shelterbelts, stretches of meadows, riparian vegetation, small ponds and so on. Therefore, any activity taken in order to maintain or increase landscape diversity is important not only for aesthetics and recreation reasons, but even more for environment protection, and by the same for the protection of living resources in the countryside. But it has been found that in a mosaic landscape composed of cultivated fields, rich in shelterbelts, stretches of meadows, small ponds and other semi-natural elements of landscape, appear animal communities richer and more diversified plant than in a uniform landscape that is composed only of large fields [34].

In order to maintain and even increase biodiversity we should act in two directions. First is the preservation and restoration of degraded habitats to ensure the existence of many species of plants and animals. Second, enriching crop structure in plants which less important from an economic point of view, but important as refuges for many animal species.

8. Guidelines

There is a growing body of ecological knowledge that management of agricultural landscape for its structural diversity becomes the important pillar of the sustainability of rural areas. Program of environmental protection in rural areas should aim not only at the introduction of environmental friendly technologies of cultivation within farm. They should also be concerned with challenge of how to increase the resistance or resilience of the whole landscape against threats. This could be achieved by stimulating natural processes underpinning the control of diffuse pollution and erosion, increase efficiency of water retention and biodiversity conservancy, which can not be controlled only at the farm level but have to be managed by increasing the landscape structure diversity.

The spatial net rangelands and shelterbelts in landscape of cultivated fields usually allows us to lower the concentration of harmful substances in environment to the level that is not dangerous for people's health and the normal course of natural processes. Understood this way, biogeochemical barriers constitute at the same time a kind of "shelters" for all living organisms. Thanks to this, they can influence, in a very positive way, the shaping and protection of the biodiversity in agroecosystems and agriculture landscape.

This does not mean that environmentally friendly technologies are not important. On the contrary, environment friendly technologies could mitigate negative effects of production but cycling of water or spreading pollution with water and wind operate in much larger scales than farm. It is also true for modification of microclimatic conditions or protection of biodiversity. The larger spatial scale e.g. watershed enables diversification of landscape structures which support higher stability and resistance of total landscape including individual farms to threats induced by production intensification.

The above considerations lead us to conclusion that activities aiming at optimisation of farm production and environment as well as biodiversity protection should be carried out in two different but mutually supportive directions. The first one involves actions within the cultivated areas. Their objective is to maintain possibly high level of the storing capacities of soil and to preserve or improve its physical, chemical and biological properties. They include agrotechnologies, which increase humus resources or counteract soil compaction, and rely on differentiated crop rotations. An important effect of humus resources augmentation would be improved water storage capacity, more intensive processes of ions sorption etc. Integrated methods of pest and pathogen control and proper dosing of mineral fertilisers adapted to crop requirements and to chemical properties of soil allow to diminish to same degree non-point pollution. The effectiveness of so directed activities, which could be called methods of integrated agriculture, depends on good agricultural knowledge.

The second component of the integration programme of farm production and nature protection is the management of landscape diversity. It consists in such differentiation of the rural landscape as to create various kinds of so -called biogeochemical barriers, which restrict dispersion of chemical compounds in the landscape, modify water cycling, improve microclimate conditions and ensure refuge sites for living organisms. In landscapes having mosaic structure higher doses of fertilisers can be applied than in homogenous ones which are composed of arable fields only (Ryszkowski 2002). This is a very important conclusion for the program of sustainable development of the countryside. Implementation of those ecological guidelines into the integrated agriculture policy will help to develop new environment friendly agro-technologies which, at the same time, enable intensive production balanced with ability of natural systems to absorb side effects of agriculture without being damaged.

Long term investigation carried out by Institute for Agricultural and Forest Environment showed **that increasing complexity of agricultural landscape mainly by introduction of non- productive elements like shelterbelts, strips of meadows, bushes and small midfield ponds** is one of the best tool for controlling water cycling and chemical pollution of surface and ground water in agricultural landscape.

Such findings open new frontiers for conservancy of living resources outside protected areas and open new prospects for reconciliation of agriculture with nature protection. Paradigms that are presently in force should be changed as follow:

Present paradigms	Future paradigms
Conservation	Management of protected ecosystems
Protection by isolation	Reconciliation with economic activities
Species and communities	Ecosystems
Inventory of species and recognition of their adaptations to environment	Understanding life supporting processes (energy flows, matter cycling and their control mechanisms, transmission of information)
Rare, endangered or protected species	Guilds or functional groups, key stone species
Homogeneity	Heterogeneity (diversity)

Table 4. Conclusions

Author details

Andrzej Kędziora, Krzysztof Kujawa, Hanna Gołdyn, Jerzy Karg, Zdzisław Bernacki,
Anna Kujawa, Stanisław Bałazy, Maria Oleszczuk, Mariusz Rybacki,
Ewa Arczyńska-Chudy, Rafał Łęcki, Maria Szyszkiewicz-Golis,
Piotr Pińskwar, Dariusz Sobczyk and Joanna Andrusiak
Institute for Agricultural and Forest Environment, Polish Academy of Sciences, Poznań, Poland

Cezary Tkaczuk
Siedlce University of Natural Sciences and Humanities, Siedlce, Poland

Acknowledgement

Authors thank Mrs. M. Sc. Wenesa Synowiec for ensuring linguistic correctness of our elaboration.

9. References

[1] Dalesman R, F (1968) A Different Kind of Country. MacMillan Company, New York.

[2] Soulé ME, Wilcox BA (1980) Conservation Biology: An Evolutionary-Ecological Perspective. Sunderland, Massachusetts: Sinauer Associates Inc. 395 p.

[3] Millennium Ecosystem Assessment (2005) Washington DC: Island Press.

[4] Bourdeau P (2001) Biodiversity. In: Tolba MK, editor. Our Fragile World. Oxford: EOLSS Publishers. pp. 299-308.

[5] Kędziora A (2010) Landscape Management Practices for Maintenance and Enhancement of Ecosystem Services in a Countryside. Int. J. Ecohydrol. Hydrobiol. 10 (2-4), 133-152.

[6] Ryszkowski L, Karg J, Kujawa K, Gołdyn H, Arczyńska-Chudy E (2002) Influence of Landscape Mosaic Structure on Diversity of Wild Plant and Animal Communities in

Agricultural Landscapes of Poland. In: Ryszkowski L, editor. Landscape Ecology in Agroecosystems Management. Boca Raton: CRC Press. pp. 185-217.

[7] Tryjanowski P, Hartel P, Baldi A, Szymański P, Tobółka M, Goławski A, Konvicka M, Hromada M, Jerzak L, Kujawa K, Lenda M, Orłowski G, Panek M, Skórka P, Sparks TH, Tworek S, Wuczyński A, Żmihorski M (2011) Conservation of Farmland Birds Faces Different Challenges in Western and Central-Eastern Europe. Acta Orn. 46: 1-12.

[8] Ryszkowski L (2000) The Coming Change in the Environmental Protection Paradigm. In: Crabbe P, Holland A, Ryszkowski L, Westra L, editors. Implementing Ecological Integrity. NATO Science Series, Vol. 1. Dordrecht: Kluwer Academic Press. pp. 37-56.

[9] Marcinek J (1996) Soil of the Turew Agricultural landscape. In: Ryszkowski L, French NR, Kędziora A, editors. Dynamics of Agricultural Landscape. Poznań: PWRiL. pp. 19-27.

[10] Palmer MW (1991) Estimating Species Richness: The Second-order Jackknife Reconsidered. Ecology 72: 1512-1513.

[11] Gołdyn H (2009b) Effect of Landscape Structure on the Floristic Diversity of Farmland: a Case Study of the General Chłapowski Landscape Park [in Polish]. Rozprawy Naukowe 403. Poznań: Uniwersytet Przyrodniczy w Poznaniu. 197 p.

[12] Matuszkiewicz W (2001) Guide to Estimation of Plant Associations of Poland [in Polish]. Warszawa: PWN Sci. Publ. 537 p.

[13] Pysek P, Richardson DM, Rejmánek M, Webster GL, Williamson M, Kirschner J (2004) Alien Plants in Checklist and Floras: Towards Better Communication Between Taxonomists and Ecologist. Taxon 53: 131-143.

[14] Lambdon PW, Pysek P, Basnou C, Hejda M, Arianoutsou M, Essl F, Jarošík V, Pergl J, Winter M, Anastasiu P, Andriopoulos P, Bazos I, Brundu G, Celesti-Grapow L, Chassot P, Delipetrou P, Josefsson M, Kark S, Klotz S, Kokkoris Y, Kühn I, Marchante H, Perglová I, Pino J, Vila M, Zikos A, Roy D, Hulme P (2008) Alien flora of Europe: Species Diversity, Temporal Trends, Geographical Patterns and Research Needs. Preslia 80: 101–149.

[15] Jackowiak B (1990) Anthropogenic Changes of the Flora of Vascular Plant of Poznań [in Polish]. Poznań: Adam Mickiewicz University Press. 232 p.

[16] Mirek Z, Piękoś-Mirkowa H, Zając A, Zając M (2002) Flowering Plants and Pteridophytes of Poland. A Checklist. Kraków: W Szafer Institute of Botany, PAS. 442 p.

[17] Szyszkiewicz-Golis M (2004) Structure and Functional Evaluation of Midfield Afforestation in Agricultural Landscape of Greater Poland Region - Row Wyskoc Catchment's Case [in Polish]. PhD thesis. Poznań, Agricultural University. 188p.

[18] Gołdyn H, Arczyńska-Chudy E, Pińskwar P, Jezierska-Madziar M (2008a) Natural and Anthropogenic Transformation of Water and Marsh Vegetation in Lake Zbęchy (Wielkopolska Region). Ocean. Hydrob. Studies 37: 77-87.

[19] Gołdyn H (2009a) Changes in Plant Communities of a Drainage Ditch Over the Last 30 Years. EJPAU Seria Environmental Development. Avaiable: www.ejpau.media.pl/. Accessed 2009 Nov 09.

[20] Gołdyn H (2010) Changes in Plant Species Diversity of Aquatic Ecosystems in the Agricultural Landscape in West Poland in the Last 30 Years. Biodiversity Conserv. 19: 61-80.

[21] Gołdyn H, Arczyńska-Chudy E (2010) Changes of Plant Associations and Their Succession Processes Course in Peat Pits as an Effect of Peat Transformations and the

Intensification of Meadow Management. In: Szajdak LW, Karabanov AK, editors. Physical, Chemical and Biological Processes in Soils. Poznan: The Committee on Land Reclamation and Agricultural Environment Engineering, Polish Academy of Sciences. pp. 549-559.

[22] Brzeg A, Wojterska M (2001) Plant Communities of Wielkopolska, State of Knowledge and Threats [in Polish]. In: Wojterska M, editor. Plant Cover of Wielkopolska and South Pomerania Lakeland. Guide of Terrain Sessions of 52 PTB Congress. Poznań: Bogucki Wyd. Nauk. pp. 39-110.

[23] Arczyńska-Chudy E (2008) Small Mid-Field Ponds as Biogeochemical Barriers and Phytocenotic Diversity Richness [in Polish]. PhD thesies 199 p.

[24] Ryszkowski L, Karg J (1977) Variability in Biomass of Epigeic Insect in the Agricultural Landscape. Ecol. Pol. 25 (3): 501 517.

[25] Karg J (1980) Differentiation of Insect Biomass in Agricultural Landscape. Pol. Ecol. Stud. 6 (2): 317 328.

[26] Pollard E (1977) A Method for Assessing Changes in the Abundance of Butterflies. Biol. Cons. 12: 115-134.

[27] Świerzowski A (1973) Influence of Electricity on Fishes and Invertebrates [in Polish]. Roczniki Nauk Rolniczych 95: 141-149.

[28] Rybacki M, Berger L (1997) Amphibians of the General Dezydery Chłapowski Landscape Park [in Polish]. Biuletyn Parków Krajobrazowych Wielkopolski. 2: 22-40.

[29] Rybacki M, Berger L (2003) Contemporary Fauna of Amphibians in Wielkopolska Region in Relation to Disappearance of Their Breeding Sites [in Polish]. In: Banaszak J, editor. Stepowienie Wielkopolski Pół Wieku Później. Bydgoszcz: Wydawnictwo Akademii Bydgoskiej. pp 143-173.

[30] Tomiałojć L (1980) The Combined Version of the Mapping Method. In: Oelke H, editor. Vogelerfassung und Naturschutz. Proc. VI Intern. Conf. Bird Census Work. Göttingen. pp. 92-106.

[31] Bibby CJ, Burgess ND, Hill DA (1992) Bird Census Techniques. London: Academic Press. pp. 257.

[32] Kujawa K (1997) Relationships Between the Structure of Midfield Woods and Their Breeding Bird Communities. Acta Orn. 32: 175-184.

[33] Kujawa K (2006) Effect of Afforestation and Agricultural Landscape Structure on Breeding Bird Communities in Afforestations [in Polish]. Rozprawy Naukowe 381. Poznań: Uniwersytet Przyrodniczy w Poznaniu. 160 pp.

[34] Karg J, Ryszkowski L (1996) Animals in arable land. In: Ryszkowski L, French NR, Kędziora A, editors. Dynamics of an Agricultural Landscape. Poznan: PWRiL, pp. 138–172.

[35] Kujawa A, Kujawa K (2008) Effect of Young Midfield Shelterbelts Development on Species Richness of Macrofungi Communities and Their Functional Structure. Pol. J. Ecol. 56: 45-56.

[36] Bałazy S (2003) On Some Little Known Epizootics in Noxious and Beneficial Arthropod Populations Caused by Entomophthoralean Fungi. IOBC-WPRS Bull. 26,1: 63-68.

[37] Bałazy S (1997) Diversity of Entomopathogenic Fungi in Agricultural Landscape of Poland and France. In: Ryszkowski L, Wicherek S, editors. Ecological Management of Countryside in Poland and France. Poznań: RCAFE PAN. pp. 101-111.

[38] Bałazy S, Miętkiewski R, Tkaczuk C, Wegensteiner R, Wrzosek M (2008) Diversity of Acaropathogenic Fungi in Poland and Other European Countries. Exper. Appl. Acarol. 46,1-4: 53-70.

[39] Bałazy S (in press). 2012. Antagonistic Interactions Between Arthropods and Their Fungous Pathogens in Feeding Sites of Saproxylic Pest Insects [in Polish]. Poznań: Wydawnictwo Uniwersytetu Przyrodniczego.

[40] Zimmermann G (1986) Galleria Bait Method for Detection of Entomopathogenic Fungi in Soil. J. App. Entomol. 2: 213-215.

[41] Strasser H, Forrer A, Schinner F (1996) Development of Media for the Selective Isolation and Maintenance of Virulence of *Beauveria brongniartii* In: Jackson TA, Glare TR, editors. Microbial Control of Soil Dwelling Pests. Lincoln: AgResearch. pp. 125-130.

[42] DAISIE – 100 of the worst [Internet]. Available: http://www.europe-aliens.org/speciesTheWorst.do

[43] Kaźmierczakowa R, Zarzycki K (editors) (2001) Polish Red Book of Plants. Pteridophytes and Flowering Plants [in Polish]. Kraków: PAS, W. Szafer Institute of Botany, Institute of Nature Conservation. 664 p.

[44] Szyszkiewicz-Golis M., 2001, Land use changes and the structure of agricultural landscape in the Wielkopolska Region-western part of the Row Wyskoc catchment. The Problems of Landscape Ecology – volume X: Transformations of the natural environment of Poland and its functioning. Institute of Geography and Spatial Management of Jagiellonian University and Polish Association for Landscape Ecology, Krakow, 92-99 [In Polish]

[45] Ryszkowski L, Karg J, Kujawa K (1999) Protection and Management of Biodiversity in Agricultural Landscape [in Polish]. In: Ryszkowski L, Bałazy S, editors. Uwarunkowania Ochrony Różnorodności Biologicznej i Krajobrazowej. Zakład Badań Środowiska Rolniczego i Leśnego PAN. Poznań: pp. 59-80.

[46] Wasilewska L (1979) The Structure and Function of Soil Nematode Communities in Natural Ecosystems and Agrocenoses. Pol. Ecol. Stud. 5: 97-145.

[47] Kasprzak K, Ryl B (1978) The Influence of Agriculture on the Occurrence of Oligochaeta in Arable Soils [in Polish]. Wiadomości Ekologiczne. 24: 333-366.

[48] Ryl B (1984) Comparison of Communities of Earthworms (Lumbricidae) Occurring in Different Ecosystems of Agricultural Landscape. Ekol. Pol. 32: 155-165.

[49] Wolak M, Karg J (2002) Spiders Wintering in Mid-field Wooded Patches [in Polish]. In: Banaszak J, editor. Wyspy Środowiskowe. Bioróżnorodność i Próby Typologii: Wyd. Akademii Bydgoskiej. pp. 147–158.

[50] Kajak A, Oleszczuk M (2004) Effect of Shelterbelts on Adjoining Cultivated Fields: Patrolling Intensity of Carabid Beetles (Carabidae) and Spiders (Araneae). Pol. J. Ecol. 52 (2): 155–172.

[51] Mielewczyk S. Qualitative and Quantitative Changes in Occurrence of Ephemeroptera, Odonata, Heteroptera and Coleoptera in Young Mid-field Pond [in Polish]. In: Mat. Zjazdowe XVII Zjazdu Hydrobiologów Polskich. 1997 wrzesień 8-11; Poznań: 96.

[52] Banaszak J (1983) Ecology of Bees (Apoidea) of Agricultural Landscape. Pol. Ecol. Stud. 9: 421-505.

[53] Szeflińska D (1997) Thrips (Thysanoptera) of Grassland Habitats in the General Dezydery Chłapowski Landscape Park [in Polish]. Biuletyn Parków Krajobrazowych Wielkopolski. 2 (4): 72-77.

[54] Szeflińska D (2002) Thysanoptera of Cereals in Central Wielkopolska. Bulletin of the Polish Academy of Sciences. 50 (3): 177-181.

[55] Oleszczuk M (2010) Refugial Areas in Farmland as Habitats for Rarely Found and Threatened Species of Spiders (Araneae) in Poland [in Polish]. Chrońmy Przyr. Ojcz. 66: 361–374.

[56] Oleszczuk M (2011) Plant-dwelling Spiders (Araneae) of Selected Habitats in Manor Park in Turew [in Polish]. Biuletyn Parków Krajobrazowych Wielkopolski 19: 98-103.

[57] Berger L, Rybacki M (1999) Composition and Ecology of Water Frog Populations in Agricultural Landscape in Wielkopolska. Biological Bulletin of Poznań. Zoology 35(2): 103-111.

[58] Kujawa K (2000) Avifauna of the General Dezydery Chłapowski Landscape Park [in Polish]. Wielkopolskie Prace Ornitologiczne 9:89-121.

[59] Kujawa K (1996) The Influence of Agricultural Landscape Structure on Breeding Avifauna [in Polish]. PhD Thesis. University of Wrocław.

[60] Kujawa K, Łęcki R (2008) Does Red Fox Vulpes vulpes Affect Bird Species Richness and Abundance in an Agricultural Landscape? Acta Orn. 43:167-178.

[61] Ryszkowski L (1982) Structure and Function of the Mammal Community in an Agricultural Landscape. Acta Zool. Fenn. 169: 45–59.

[62] Kozakiewicz M (1993) Habitat Isolation and Ecological Barriers - the Effect on Small Mammal Populations and Communities. Acta Theriol. 38: 1–30.

[63] Kozakiewicz M, Jurasińska E (1989) The Role of Habitat Barriers in Woodlot Recolonization by Small Mammals. Holarctic Ecol. 12: 106–111.

[64] Hansson L (1987) Dispersal Routes of Small Mammals at an Abandoned Field in Central Sweden. Hol. Ecol. 10: 154-159.

[65] Yahner RH (1983) Small Mammals in Farmstead Shelterbelts: Habitat Correlates of Seasonal Abundance and Community Structure. J. Wildl. Manage. 47: 74–84.

[66] Rozporządzenie Ministra Środowiska z dnia 9 lipca 2004 r. w sprawie gatunków dziko występujących grzybów objętych ochroną. Dz.U. 04.168.1765 z dnia 28 lipca 2004 r.

[67] Wojewoda W, Ławrynowicz M (2006) Red List of the Macrofungi in Poland. In: Mirek Z, Zarzycki K, Wojewoda W, Szeląg Z, editors. Red List of Plants and Fungi in Poland. 3rd ed. Kraków: W. Szafer Institute of Botany, Polish Academy of Sciences. pp. 53-70.

[68] Wojewoda W (2003) Checklist of Polish Larger Basidiomycetes. In: Mirek Z, editor. Biodiversity of Poland. Vol. 7. Kraków: W. Szafer Institute of Botany, Polish Academy of Sciences. pp. 1-812.

[69] Chmiel MA (2006) Checklist of Polish Larger Ascomycetes. In: Mirek Z, editor. Biodiversity of Poland. Vol. 8. Kraków: W. Szafer Institute of Botany, Polish Academy of Sciences. pp. 1-152.

[70] Mułenko W, Majewski T, Ruszkiewicz-Michalska M (2008), editors. A Preliminary Checklist of Micromycetes in Poland. In: Mirek Z, editor. Biodiversity of Poland. Vol. 9. Kraków: W. Szafer Institute of Botany, Polish Academy of Sciences. pp. 1-752.

[71] Karg J, Bałazy S (2009). Effect of Landscape Structure on the Occurrence of Agrophagous Pests and Their Antagonists [in Polish]. Progress in Plant Protection 49, 3: 1015-1034.

[72] Tkaczuk C (2008) Occurrence and Infectious Potential of Entomopathogenic Fungi of Soils in Agrocenoses and Semi-natural Habitats in an Agricultural Landscape [in Polish]. Rozprawy naukowe 94. Siedlce: Wyd. Akademii Podlaskiej. 160 p.

[73] Kujawa A, Gołdyn H, Arczyńska-Chudy E (2005) A New Locality of Daisy Leaf Grape Fern (Botrychium matricariifolium) in Western Poland. Rocz. AR Pozn. 372, Bot-Stec. 8: 129-132.

[74] Zając M, Zając A, Tokarska-Guzik B (2009) Extinct and Endangered Archaeophytes and the Dynamics of their Diversity in Poland. Biodiv. Res. Conserv. 13: 17-24.

[75] Żukowski W, Jackowiak B (1995) List of Endangered and Threatened Vascular Plants in Western Pomerania and Wielkopolska (Great Poland). In: Żukowski W, Jackowiak B, editors. Endangered and Threatened Vascular Plants of Western Pomerania and Wielkopolska. Publications of the Department of Plant Taxonomy of the Adam Mickiewicz University in Poznań. Poznań: Bogucki Wyd. Nauk. pp. 9-96.

[76] Ratyńska H (2003) Plant Cover as a Result of Anthropogenic Changes in Landscape Exemplified by the River Główna Catchment Area (the middle Wielkopolska Province, Poland) [in Polish]. Bydgoszcz: Wyd. Akademii Bydgoskiej im. Kazimierza Wielkiego. 381 p.

[77] Pliński M (2009) Dangerous Water Blooms - Not Only a Hydrobiological Problem.XXI Congress of the Polish Hydrobiologist, Lublin, 9-12 Sep 2009. Session V- Ecotoxicology waters. p. 5.

[78] Mazur-Marzec H (2011) Toxic Blooms of Cyanobacteria in the Baltic Sea and Their Impact on Human Health. WWF Report Poland.

[79] Burchardt L, Pawlik-Skowrońska B (2005) Blue-Green Algal Blooms-Interspecific Competition and Environmental Threat. Wiad. Bot. 49: 39-49.

[80] Wilcox TP, Hornbach DJ (1991) Macrobenthic Community Response to Carp (Cyprinus carpio L.) Foraging. J. Freshwat. Ecol. 6: 170-183.

[81] Gehrke PC, Harris JH (1994) The Role of Fish in Cyanobacterial Blooms in Australia. Aust. J. Mar. Fresk. Res. 45: 905-15.

[82] Breukelaret AW, Lammens EH, Breteler K (1994) Effects of Benthivorous Bream (Aramis brama) and Carp (Cyprinus Carpio) on Sediment Re-Suspension and Concentrations of Nutrients and Chlorophyll a. Freshw. Biol. 32: 113-121.

[83] Williams AE, Moss W, Eaton J (2002) Fish Induced Macrophyte Loss in Shallow Lakes: Top-down and Bottom-up Processes in Mesocosm Experiments. Freshw. Biol. 47: 2216 – 2232.

[84] Karg J (2004) Importance of Midfield Shelterbelts for Over-wintering Entomofauna (Turew area, West Poland). Pol. J. Ecol. 52 (4): 421-431.

[85] Ryszkowski L, Karg J, Glura M (2009) Influence of Agricultural Landscape Structure on Diversity of Insect Communities. Pol. J. Ecol. 57 (4): 697 – 713.

[86] Tryjanowski P, Kuźniak S, Kujawa K, Jerzak L (2009) Ecology of Farmland Birds [in Polish]. Poznań: Bogucki Wyd. Nauk. 390 p.

[87] Pugacewicz E (2000) Breeding Avifauna of an Agricultural Landscape in Bielska Plain [in Polish]. Not. Ornitol. 41:1-28.

[88] Kujawa K, Tryjanowski P (2000) Relationships Between the Abundance of Breeding Birds in Western Poland and the Structure of Agricultural Landscape. Acta Zool. Hung. 46: 103-114.

[89] Kujawa K (1996a) The Influence of Landscape Pattern on Breeding Bird Communities in D. Chlapowski Landscape Park [in Polish]. Biuletyn Parków Krajobrazowych 3: 83-90.

[90] Kujawa K (2002) Population Density and Species Composition Changes for Breeding Bird Species in Farmland Woodlots in Western Poland Between 1964 and 1994. Agric. Ecosyst. Environ. 91: 261-271.

[91] Kujawa K (2008a) Long Term Changes (1964-2006) in Breeding Bird Community of Midfield Afforestations in the Dezydery Chłapowski Landscape Park (Wielkopolska) [in Polish]. In: Kaczmarek S, editor. Krajobraz i bioróżnorodność. Bydgoszcz: Wyd. Uczel. UKW. pp. 151-167.

[92] Sosa-Gomez DR, Delpin KE, Moscardi F, Farias JRB (2001) Natural Occurrence of the Entomopathogenic Fungi *Metarhizium, Beauveria* and *Paecilomyces* in Soybean Under Till and No-till Cultivation Systems. Neotrop. Entomol. 30,1: 407-410.

[93] Hummel RL, Walgenbach JF, Barbercheck ME, Kennedy GG, Hoyt GD, Arellano C (2002) Effects of Production Practices on Soil-Borne Entomopathogens in Western North Carolina Vegetable Systems. Environ. Entomol. 31: 84-91.

[94] Root TL, Price JT, Hall KR, Schneider SH, Rosenzweig C, Pounds JA (2003) Fingerprints of Global Warming on Wild Animals and Plants. Nature 421: 57–60.

[95] Ryszkowski L (1990) Agricultural Landscape of Turew Vicinity [in Polish]. In: Ryszkowski L, Marcinek J, Kędziora A, editors. Obieg Wody i Bariery Biogeochemiczne w Krajobrazie Rolniczym. Poznań: Wydawnictwo UAM. pp. 5-12.

[96] Ryszkowski L, Kędziora A (2007) Sustainability and Multifunctionality of Agricultural Landscapes. In: Plieninger T, editor. Zukunftsorientierte Nutzung Ländlicher Räume – Land Innovation. Materialien 15. Berlin-Brandenburgische Akademie der Wissenschaften, Berlin: pp. 5 23..

[97] Herk CM, Aptroot A, Dobber HF (2002) Long-term Monitoring in the Netherlands Suggests that Lichens Respond to Global Warming. Lichenologist 34: 1–15.

[98] Kreisel H (2006) Global Warming and Mycoflora in the Baltic Region. Acta Mycol. 41:79-94.

[99] Emanuel WR, Shugart HH, Stevenson MP (1985) Climate Change and the Broad Scale Distribution of Terrestial Ecosystem Complexes. Climatic Change 7: 29-43.

[100] Notaro M, Vavrus S, Liu Z (2007) Global Vegetation and Climate Change due to Future Increases in CO2 as Projected by a Fully Coupled Model with Dynamic Vegetation. J. Climate 20: 70-90.

[101] White P, Kerr JT (2006) Contrasting Spatial and Temporal Global Change Impacts on Butterfly Species Richness During the 20th Century. Ecography 29: 908 918.

[102] Kerbiriou C, Le Viol I, Jiguet F, Devictor V (2009) More Species, Fewer Specialists: 100 years of Changes in Community Composition in an Island Biogeographical Study. Diversity Distrib. 15: 641–648.

[103] Benning TL, LaPointe D, Atkinson CT, Vitousek PM (2002) Interactions of Climate Change with Biological Invasions and Land Use in the Hawaiian Islands: Modeling the Fate of Endemic Birds Using a Geographic Information System. Proc. Nat. Acad. Sc. J. 29: 14246–14249.

[104] Isaac JL (2009) Effects of Climate Change on Life History: Implications for Extinction Risk in Mammals. Endang. Species. Res. 7: 115–123.

[105] Malcolm JR, Liu C, Neilson RP, Hansen L, Hannah L (2006) Global Warming and Extinctions of Endemic Species from Biodiversity Hotspots. Conserv. Biol. 20: 538 548.

[106] Davey CM, Chamberlain DE, Newson SE, Noble DG, Johnston A (2011) Rise of the Generalists: Evidence for Climate Driven Homogenization in Avian Communities. Global Ecology and Biogeography 21: 568–578.

[107] Pounds JA, Bustamente MR, Coloma LA, Consuegra JA, Fogden MPL, et al. (2006) Widespread Amphibian Extinctions from Epidemic Disease Driven by Global Warming. Nature 439: 161–67.

[108] Deschka G, Dimic N (1986) Cameraria ohridella n. sp. aus Mazedonien, Jugoslawien (Lepidoptera, Lithocelletidae). Acta Ent. Jugosl. 22: 11-23.

[109] Łabanowski G, Soika G (1998) The Horse Chestnut Leafminer Infesting Chestnut in Poland. Ochrona Roślin. 42(12): 12.

[110] Gifford RM (1977) Growth Pattern, Carbon Dioxide Exchange and Dry Weight Distribution in Wheat Growing Under Differing Photosynthetic Environments. Aust. Jour. Plant Physiol. 4: 99-110.

[111] Sionit N, Morten DA, Strain BR, Helmers H (1981) Growth Response of Wheat to CO_2 Enrichment and Different Levels of Mineral Nutrition. Agr. Jour. 73: 1023-1027.

[112] Rogers HH, Thomas JF, Bingham GE (1983) Response of Agronomic and Forest Species to Elevated Atmospheric Carbon Dioxide. Science 220: 428 429.

[113] Kimball BA, Kobayashi K, Bindi M (2002) Responses of Agricultural Crops to Free-air CO2 Enrichment. Adv. Agron. 77: 293–368.

[114] Emmerich WE (2007) Ecosystem Water Use Efficiency in a Semiarid Shrubland and Grassland Community. Rangeland Ecol. Manage. 60: 464–470.

[115] Chylarecki P, Jawińska D, Kuczyński (2006) Monitoring of Common Breeding Birds - Report from Years 2003-2004. Warszawa: OTOP.

[116] Fog K (1988) Reinvestigation of 1300 Amphibian Localities recorded in the 1940s. Memoranda Soc. Fauna Flora Fennica 64: 134-135.

[117] Fog K (1997) A survey of the results of pond projects for rare Amphibians in Denmark. Memoranda Soc. Fauna Flora Fennica.73: 91-100.

[118] Stasiak P (1991) Disappearance of Small Mid-field Water Bodies in Wielkopolska Lowland in the Light of Cartographic Materials [in Polish]. M.Sc. Thesis. Adam Mickiewicz University, Poznań.

Chapter 3

The Ongoing Shift of Mediterranean Coastal Fish Assemblages and the Spread of Non-Indigenous Species

Stefanos Kalogirou, Ernesto Azzurro and Michel Bariche

Additional information is available at the end of the chapter

1. Introduction

Geological history of life on earth tells that continents have been isolated for long periods. It also reveals that collisions of land masses as well as lower sea levels allowed the spread of fauna and flora (Stachowicz and Tilman, 2005). In today's seas, marine communities are being altered and remodelled at an unprecedented rate, when compared to natural changes which occured over geological times. While many marine species populations are dwindling due to overfishing and habitat destruction (Jackson *et al.*, 2001), other species are invading new areas through anthropogenic vectors (Carlton 1985, Galil 2006, Galil et al. 2007). During the last centuries, human transport has increased the number of non-indigenous species (NIS) introductions. For example, half of the plant species of Hawaii are exotics (Sax *et al.*, 2002) as are about 20% of plants in California bay (Sax, 2002) and about 18% of fish species in the eastern Mediterranean Sea (Golani et al. 2002, Golani et al. 2006, EastMed 2010, Golani 2010).

Understanding invasion ecology requires a good knowledge of ecological processes in the systems under study, prior to invasion. Diversity, structure, and function of natural communities would give insights into fundamental ecological processes which could in turn give a better understanding of potential effects following the introduction of NIS.

From a societal perspective, species invasions might pose serious threats to human economic interests and health (Yang *et al.*, 1996; Sabrah *et al.*, 2006; Katikou *et al.*, 2009). Species invasions have also been considered to have negative impacts on native biodiversity (Reise *et al.*, 2006; Streftaris and Zenetos, 2006; Galil, 2007; Lasram and Mouillot, 2008; Zenetos *et al.*, 2009). Furthermore, invasions interacts with other disturbing factors to the marine ecosystem functioning such as habitat destruction, pollution and climate change

(Rilov and Crooks, 2009). Disturbance caused by habitat destruction may open up space for invaders but space can also be released by the invaders themselves. Consider the example given by Rilov and Galil (2009) where two non-indigenous siganids might have modified the competition between algae and mussels through intensive grazing, thus providing space for the non-indigenous mussel *Brachiodontes pharaonis*. Pollution can make environmental conditions less tolerable for native species, and perhaps provide opportunities for opportunists, among which non-indigenous species could be found (Occhipinti-Ambrogi and Savini, 2003; Wallentinus and Nyberg, 2007). Global warming is causing the shifts and poleward migrations of many taxa that are now extending their biogeographical range (Parmesan and Yohe, 2003; Perry *et al.*, 2005). This tendency is also observed in the Mediterranean Sea (Bianchi, 2007; Raitsos *et al.*, 2010). Some species, typically confined to the warmer parts of the Mediterranean, are currently colonizing the northern sectors. This phenomenon has been termed "meridionalization" (Azzurro, 2008).

The increase of water temperature is also allowing the success of tropical exotic species in the Mediterranean Sea, a phenomenon that has been called 'tropicalization' (Bianchi and Morri, 2003). Conditions facilitating invasions are usually related to the physical and biological attributes of the new colonized habitats. Biological impact studies include mostly those species of economic interests (e.g. fisheries) (Streftaris and Zenetos, 2006), human health (e.g. toxic species) (Yang *et al.*, 1996; Bentur *et al.*, 2008; Katikou *et al.*, 2009) and biodiversity (e.g. competition with indigenous species or habitat modifiers) (Golani, 1993a; Golani, 1994; Bariche *et al.*, 2004; Azzurro *et al.*, 2007a; Kalogirou *et al.*, 2007; Wallentinus and Nyberg, 2007; Bariche *et al.*, 2009). A lot of research has also focused on the factors controlling success or failure of invasive species by considering mechanisms of interactions between indigenous and NIS. There is no universal model explaining the mechanisms controlling the success or failure of an invading species (Stachowicz and Tilman, 2005). As far as the Mediterranean Sea is concerned, important mechanisms include competition for resources or space (Bariche *et al.*, 2004; Kalogirou *et al.*, 2007), top-down forces (Goldschimdt *et al.*, 1993), herbivory (Lundberg and Golani, 1995; Galil, 2007), and parasites (Diamant, 2010).

A widely cited theory in invasion ecology is about the relationship between diversity and invasibility of an ecosystem (i.e. more diverse communities should be more resistant to invasion) (Leppäkoski and Olenin, 2000). The mechanism suggests that as species richness increases the competition intensifies and less food resources remain available for new colonizers (MacArthur, 1955; Levine and D' Antonio, 1999). Less diverse ecosystems possessing fewer species and simpler food-web interactions would therefore provide available niches for the establishment of NIS. This hypothesis is known as the "biotic resistance hypothesis" (Levine and Adler, 2004). As an aid to understand this mechanism, both observational and experimental approaches have been applied with conflicting results (Levine and D' Antonio, 1999). Studies that employ both observational and experimental approaches show that high diverse systems does reduce invasion success (Stachowicz and Tilman, 2005). There is a long history of theoretical discussions about the relationship between species richness and productivity or stability of a system. Threats to global species diversity caused by human activities have raised concern on the consequences of species losses to the functioning of ecosystems. In ecology, this concern has received a lot of attention. During the last 20 years, experimental tests of the relationship between species richness and ecosystem processes such as productivity, stability and invasibility have increased rapidly (Stachowicz and Whitlatch, 1999).

Other theories go back to the work of Darwin. Darwin's "naturalization hypothesis" predicts that NIS are less prone to invade areas where closely related species are present. Those species would compete with their relatives and would encounter predators and pathogens. An opposing view is the "pre-adaptation" hypothesis predicting that NIS should succeed in areas where indigenous closely related species are present because they are more likely to share traits that pre-adapt them to their own environment. So far, these theories have been seldom tested on fish species and no clear pattern has emerged so far for these taxa (Ricciardi and Mottiar (2006). Ricciardi and Mottiar (2006) agreed with Moyle and Light (1996) that success is primarily determined by competitive interactions (e.g. "biotic resistance" hypothesis), propagulae pressure and environmental abiotic factors (i.e. the degree to which NIS physiological tolerances are compatible to local physical conditions). Rapid changes in environmental conditions, caused by human activities, have also been mentioned as to increase invasiveness (Occhipinti-Ambrogi and Savini, 2003). Habitats that lack predators are also suggested to be more prone to introductions of NIS (Moyle and Light, 1996). There is also a higher risk of further establishment of species in habitats that have already been invaded, referred as the "invasional meltdown" (Simberloff and Von Holle, 1999; Ricciardi, 2001). In a study from Great Lakes, Ricciardi (2001) found support for the "invasional meltdown" hypothesis by showing that positive interactions (mutualistic) among NIS are more common than negative (competitive). In further support of the "invasional meltdown" hypothesis, Ricciardi (2001) showed that exploitative interactions (e.g. predator-prey) among NIS are strongly asymmetrical to the benefit of one invading species at a negligible cost to another.

2. Current patterns of change of the Mediterranean biota

In the last century the Mediterranean Sea has been a receptacle of NIS, most of them arrived by mean of direct or indirect mediation of humans. Today, the Mediterranean Sea can be considered as one of the main hotspots of marine bio-invasions on earth (Quignard and Tomasini, 2000), and is by far the major recipient of NIS among European seas including macrophytes, invertebrates and fishes (Streftaris et al., 2005; Zenetos et al., 2010). The Mediterranean is unique because of its connection to the Indo-West Pacific realm via the Suez Canal (Fig. 1), allowing the so called Lessepsian migration (Por, 1978). The rate of this immigration has increased in recent decades and has ecological, social and economic impacts (Zenetos et al., 2008; Bilecenoglu, 2010; Zenetos et al., 2010). The Eastern Mediterranean basin is potentially more prone to introductions of subtropical and tropical NIS than the western basin. This has been mainly attributed to different physical and biological conditions between the two basins. It is to mention that the construction of the

Aswan Dam on the Nile River in 1966 reduced significantly the freshwater flood into the Mediterranean Sea. This led to an increased salinity of 2-3% along the Mediterranean coast of Egypt and to a reduction of the most important sources of nutrients in the eastern Mediterranean Sea (Galil, 2006). The damming of the Nile might have positively favoured the westward dispersion of Lessepsian NIS along the Northern African shores (Ben-Tuvia, 1973).

Figure 1. The Suez Canal

New terms have been recently created to describe current changes of the Mediterranean biodiversity. Due to the tropical nature of most of the exotic species that enter the Mediterranean, various authors have defined the process of entrance and spread of these organisms as 'tropicalization' (Bianchi and Morri, 2004; Bianchi, 2007). Another definition that has been used is "demediterranization"(Quignard and Tomasini, 2000) that put the emphasis on the process of biotic homogenization of the Mediterranean Sea. Instead Massuti et al. (2010) used the term 'meridianization' to indicate the increasing divergence (in terms of composition of the biological communities) between the Eastern and Western sectors of the Mediterranean, due to the continuous influx of Lessepsian and Atlantic biota. This latter term, 'meridianization' should not be confused with 'meridionalization', which instead would indicate the northward expansion of southern ('meridional') species towards the northern sectors of the basin (Azzurro, 2008). Several indigenous species such as *Sparisoma cretense* and *Thalassoma pavo* have been regarded as "meridional" (CIESM, 2008) since they have been recently found to reproduce and have established populations in the coldest part of the Mediterranean Sea (Ligurian Sea) (Guidetti *et al.*, 2002). Additionally, a reduction of temperate species followed the increase of tropical species in the Ligurian Sea (Bianchi and Morri, 2003). These trends of change in the biological diversity of the Mediterranean Sea are part of a general reshuffling of species, that is happening at the global level (Vitousek *et al.*, 1997), and climate warming would contribute to promote the shifts of species distribution, an evidence that is particularly clear among marine fishes (Perry *et al.*, 2005). As fish are particularly sensitive to changes in water temperatures, physiological processes directly alters behaviour, generating active movement and migratory patterns of these organisms (Roessig *et al.*, 2004). Other indirect effects of climate change, such as those related to the change of currents, could affect larval dispersal, retention and recruitment of marine organisms (Bianchi and Morri, 2004).

More than 700 fish species inhabit the Mediterranean Sea with a general decrease in number moving eastwards (Quignard and Tomasini, 2000; Lasram *et al.*, 2009) . Among these, at least 80 are non-indigenous of Indo-West Pacific and Red Sea origin (Cicek and Bilecenoglu, 2009; Bariche, 2010b; EastMed, 2010; Golani, 2010; Bariche, 2011b; Sakinan and Örek, 2011; Salameh, 2011; Bariche and Heemstra, 2012). The abundance of these non-indigenous species is not well documented. The list of non-indigenous fish species with quantitative information from the Mediterranean Sea can be found in Table 1.

Family	Species	Reference
Atherinidae	*Atherinomorus forskalii*	(Bariche *et al.*, 2007; Shakman and Kinzelbach, 2007)
Callionymidae	*Callionymus filamentosus*	(Gucu and Bingel, 1994; Kalogirou *et al.*, 2010)
Carangidae	*Alepes djedaba*	(Shakman and Kinzelbach, 2007)
Clupeidae	*Etrumeus teres*	(Bariche *et al.*, 2006; Bariche *et al.*, 2007; Carpentieri *et al.*, 2009)
Clupeidae	*Herklotsichthys punctatus*	(Bariche *et al.*, 2006; Bariche *et al.*, 2007; Carpentieri *et al.*, 2009)
Dussumieriidae	*Dussumieria elopsoides*	(Goren and Galil, 2005; Bariche *et al.*, 2007)
Fistulariidae	*Fistularia commersonii*	(Shakman and Kinzelbach, 2007; Carpentieri *et al.*, 2009; Kalogirou *et al.*, 2010; Kalogirou *et al.*, 2012b)
Hemiramphidae	*Hemiramphus far*	(Shakman and Kinzelbach, 2007; Carpentieri *et al.*, 2009)
Holocentridae	*Sargocentron rubrum*	(Carpentieri *et al.*, 2009)
Labridae	*Pteragogus pelycus*	(Kalogirou *et al.*, 2010)
Leiognathidae	*Leiognathus klunzingeri*	(Gucu and Bingel, 1994)
Monacanthidae	*Stephanolepis diaspros*	(Gucu and Bingel, 1994; Harmelin-Vivien *et al.*, 2005; Shakman and Kinzelbach, 2007; Carpentieri *et al.*, 2009; Kalogirou *et al.*, 2010; Kalogirou *et al.*, 2012b)

Mullidae	*Upeneus moluccensis*	(Gottlieb, 1960; Oren *et al.*, 1971; Gucu and Bingel, 1994; Golani and Ben-Tuvia, 1995; Sonin *et al.*, 1996; Goren and Galil, 2005; Harmelin-Vivien *et al.*, 2005; Carpentieri *et al.*, 2009; Kalogirou *et al.*, 2010)
Mullidae	*Upeneus pori*	(Gucu and Bingel, 1994; Golani and Ben-Tuvia, 1995; Goren and Galil, 2005; Shakman and Kinzelbach, 2007; Carpentieri *et al.*, 2009; Kalogirou *et al.*, 2010; Kalogirou *et al.*, 2012b)
Nemipteridae	*Nemipterus randalli*	(Carpentieri *et al.*, 2009)
Pempheridae	*Pempheris vanicolensis*	(Carpentieri *et al.*, 2009)
Scaridae	*Scarus ghobban*	(Bariche and Saad, 2008; Carpentieri *et al.*, 2009)
Scombridae	*Scomberomorus commerson*	(Shakman and Kinzelbach, 2007; Carpentieri *et al.*, 2009)
Siganidae	*Siganus luridus*	(Gucu and Bingel, 1994; Bariche *et al.*, 2004; Harmelin-Vivien *et al.*, 2005; Shakman and Kinzelbach, 2007; Carpentieri *et al.*, 2009; Kalogirou *et al.*, 2010; Kalogirou *et al.*, 2012b)
Siganidae	*Siganus rivulatus*	(George and Athanassiou, 1967; Bariche *et al.*, 2004; Bariche, 2005; Harmelin-Vivien *et al.*, 2005; Shakman and Kinzelbach, 2007; Carpentieri *et al.*, 2009; Kalogirou *et al.*, 2010; Kalogirou *et al.*, 2012b)
Sphyraenidae	*Sphyraena chrysotaenia*	(Golani and Ben-Tuvia, 1995; Carpentieri *et al.*, 2009; Kalogirou *et al.*, 2010; Kalogirou *et al.*, 2012a)
Sphyraenidae	*Sphyraena flavicauda*	(Kalogirou *et al.*, 2012b)
Synodontidae	*Saurida undosquamis*	(Oren *et al.*, 1971; Ben-Yami and Glaser, 1974; Gucu and Bingel, 1994; Golani and Ben-Tuvia, 1995; Galil and Zenetos, 2002; Goren and Galil, 2005; Harmelin-Vivien *et al.*, 2005; Shakman and Kinzelbach, 2007; Carpentieri *et al.*, 2009)
Pempheridae	*Pempheris vanicolensis*	(Harmelin-Vivien *et al.*, 2005)
Pomacentridae	*Sarogentrum rubrum*	(Harmelin-Vivien *et al.*, 2005)
Tetraodontidae	*Lagocephalus sceleratus*	(Carpentieri *et al.*, 2009; Kalogirou *et al.*, 2010; Aydin, 2011; Kalogirou *et al.*, 2012b)
Tetraodontidae	*Lagocephalus spadiceus*	(Carpentieri *et al.*, 2009)
Tetraodontidae	*Lagocephalus suezensis*	(Carpentieri *et al.*, 2009; Kalogirou *et al.*, 2010)

Table 1. List of the non-indigenous fish species of Indo-Pacific and Red Sea origin with references on quantitative information in abundance in the Mediterranean Sea

The arrival of these invaders raises plain concern on the ecological and economic impact that such migrants have but the available information is still scarce (Rilov and Galil, 2009) and there is an obvious lack of knowledge. It is at the same time obvious that the ecological effect of some species is significant (Kalogirou *et al.*, 2007; Bariche *et al.*, 2009). Competitive exclusion and displacement of native species are often potential expectations in ecological studies (Bariche *et al.*, 2004; Galil, 2007). A noteworthy example is the presence of the two Lessepsian herbivorous *Siganus rivulatus* and *S. luridus*. The impact of the two species on indigenous species has received some scientific attention and comments (Bariche *et al.*, 2004; Azzurro *et al.*, 2007b; Galil, 2007; Golani, 2010). Since their first record in the Mediterranean Sea, respectively 1924 and 1955, the two species have established significant populations in the eastern basin and have spread westwards as far as Sicily and Tunisia (Rilov and Galil, 2009). They constituted one third of the fish biomass over hard bottoms and their contribution to the guild of herbivorous fish species in shallow coastal areas reached 80% along the Levantine coast (Goren and Galil, 2001; Bariche *et al.*, 2004). The two siganids are commercially important in the eastern Mediterranean Sea. The native herbivorous *Sarpa salpa* (Sparidae) was relatively an abundant species and a possible competition with *S. rivulatus* was already highlighted along the coast of Lebanon (Gruvel, 1931; George and Athanassiou, 1967). Nowadays, records of *S. salpa* from Lebanon are very scarce and it has been suggested that the native species has been outcompeted by the Lessepsian invaders (Bariche *et al.*, 2004; Galil, 2007). In contradiction, Golani (2010) rejected this hypothesis and considered that Gruvel was unable to recognize *S. salpa* from another common sparid *Boops boops*. We consider that this assumption is not accurate as Gruvel described separately *B. boops* (as *Box vulgaris* C.V. = *Box boops* L.) and *S. salpa* (*Box salpa* Cuv.) and made clear reference to their local abundance and flesh palatability, showing that he knew well how to identify them (Gruvel, 1931). The author was assessing the fishery resources of the Levantine coast by mean of experimental trawling. The clear competitive superiority of siganids may be due to a greater adaptability to fluctuating environmental conditions and other biological advantages (Bariche *et al.*, 2004). Moreover, macrophytes are considered to be abundant along the eastern Mediterranean coast (Lipkin and Safriel, 1971; Lundberg and Golani, 1995). Lundberg & Golani (1995) compared the stomach contents of the siganids in relation to food availability in the source (Red Sea) and invaded (Mediterranean Sea) areas. They found a scarcity of food species underwater and abundance on vermetid reefs and beach rocks, which are situated at sea level and are thus rather inaccessible (Lundberg and Golani, 1995; Lundberg *et al.*, 2004). *Siganus rivulatus* and *S. luridus* were shown to be selective in the eastern Mediterranean, when macrophytes are diverse and abundant and will eat what is available when food is scarce (Bariche, 2006). Selectivity being more important in the Mediterranean Sea is probably due to a larger choice in food species (Golani, 1993b; Lundberg *et al.*, 2004). Nevertheless, the success of siganids also shows a larger trophic or eco-physiological flexibility in the Mediterranean Sea (Hassan *et al.*, 2003; Bariche, 2006). This also reveals the lack of available data regarding herbivorous Lessepsian siganids being better competitors than indigenous ones for food until it is proven that trophic resources constitute the most important limiting factor (Golani, 2010). Finally, negative consequences of fish invasions are not only restricted to native fish communities. The intensive grazing of macrophytes by the same siganids might have reduced the competition between algae and mussels, and thus released space for the establishment of a non-indigenous mussel, *Brachiodontes pharaonis*, on rocky shores along the coasts of the Levant (Rilov and Galil, 2009).

3. The case of the invasive pufferfish *Lagocephalus sceleratus*: Ecological consequences, economic impacts and risks for human health

Pufferfishes are marine fish species that are distributed in tropical and subtropical areas of the Atlantic, Indian and Pacific Ocean. Puffers include 121 species within the Tetraodontidae family among which nine (*T. flavimaculosus, L. sceleratus, L. spadiceus, L. suezensis, S. pachygaster, S. spengleri, T. spinosissimus, S. marmoratus, L. lagocephalus*) are found in the Mediterranean Sea. Some puffers contain the strongest paralytic toxin known to date, tetrodotoxin (Sabrah *et al.*, 2006). European legislation (854/2004/EC) states that toxic fish of the Tetraodontidae family should not enter the European markets. In a global perspective, occasional accidental poisonings have lead to numerous human deaths, the majority of which have been documented in southeastern Asia, including Malaysia, Taiwan, Hong Kong, and Korea (Kan *et al.*, 1987; Yang *et al.*, 1996).

Lagocephalus sceleratus received considerable public attention shortly after its first record in 2003 from the Gökova bay in the south-eastern coasts of the Aegean Sea due to the presence of significant amounts of tetrodotoxin (Akyol *et al.*, 2005). The distribution of *L. sceleratus* is currently limited to the eastern Mediterranean Sea but the species is showing a rapid spread westward.

Few dozens of tetrodotoxin poisoning cases occurred along the Levantine coast and in Cyprus (Bentur *et al.*, 2008). *Lagocephalus spadiceus*, which has been present in the Mediterranean for several decades, was rarely marketed but regularly consumed by fishermen in Lebanon and Syria without any noticeable concern. The sudden appearance of the highly toxic *L. sceleratus* had a serious impact on those fishermen, who used to eat *L. spadiceus* (Bariche *pers. obs.*). The large numbers of *L. sceleratus* that have been caught by coastal fishermen in the eastern Mediterranean has initiated major national efforts to alert fishermen and the public about the toxicity of this fish (Fig. 2).

These efforts have included setting up posters warning the public about the lethal effects if consumed, but also that small individuals could easily be misidentified with other small commercial edible species such as *Spicara smaris, Boops boops* and *Atherina hepsetus* (Kalogirou pers. obs.). Studies from the Mediterranean Sea showed that there is a significant positive correlation between toxicity levels and size of fish (Katikou *et al.*, 2009). According to the results of Katikou *et al.* (2009) individuals smaller than 16 cm in length do not possess toxicity levels that could be lethal. This reduces the risks in connection with misidentification since commercial *S. smaris, B. boops* and *A. hepsetus* rarely exceed this size.

Figure 2. A sampling from sandy bottoms on the coast of Rhodes Island in the southeastern Aegean Sea with a large number of Lagocephalus sceleratus individuals

Lagocephalus sceleratus has also been considered an economical pest by fishermen since it is affected the local fish markets in three ways; deterring customers from buying fish, introducing additional work to discard toxic fish and predating on local stocks of commercially important squids and octopuses (Fig. 3). Lebanese fishermen are considerably affected by the damage caused by *L. sceleratus* on their fishing gears and catches. In fact, the puffers damage considerably fishing nets and longlines with their strong fused teeth. This is evident from the numerous complaints done by Lebanese fishermen and the presence of fishing hooks and fishing nets fragments respectively in oral cavities and stomachs (Bariche, pers. obs.).

Lagocephalus sceleratus ranked among the 100 'worst' Invasive Alien Species (IAS) in the Mediterranean Sea with profound social and ecological impacts (Streftaris and Zenetos, 2006). Social impacts are obvious due to toxicity but the lack of quantitative data does not support ecological impacts. Despite the successful establishment in the eastern Mediterranean, little is known concerning the ecology of the fish (Kalogirou *et al.*, 2010; Kalogirou *et al.*, 2012b).

An invading species might sometimes go to a peak of density and then decline, a path often called boom and bust (Williamson and Fitter, 1996). This path followed the NIS bluespotted cornetfish *Fistularia commersonii* in Rhodes Island, SE Aegean Sea (Kalogirou, pers. obs.). When a NIS is established into waters where its preferred food is under-utilized by

Figure 3. Lagocephalus sceleratus individual from the coast of Rhodes Island in the south-eastern Aegean Sea

indigenous species the resulting population explosion is later brought into equilibrium with available resources. Even though this dynamic leads to the significant reduction of the invading species population size, only very few studies have reported subsequent extinction of the NIS. Competition, despite strong advocacy (Moulton, 1993), seems to be the least likely explanation for most of the examples. Decline and extinction from a build-up of enemies (predators and pathogens) and lack of sufficient resources is more likely to be important explanations in failure of invading animals to establish permanent populations (Williamson and Fitter, 1996).

4. Do we need new methodologies to monitor current changes of Mediterranean fish diversity?

Concern has been expressed to the lack of monitoring, coordination, and study in relation to the changing diversity of the Mediterranean Sea. As a matter of fact, exotic fishes spreading in the Mediterranean Sea are usually found by chance as specific procedures for their detection are lacking (Azzurro, 2010). Consequently, the extent of these changes may be underestimated as usually happens in several other marine systems (Witenberg and Cock, 2001). Increasing efforts are being devoted to the survey of marine habitats but one of the major obstacles to research remains the lack of data at large geographical scales. This would be important to perceive temporal and spatial trends and to fill important existing information gaps. New methodologies involving local communities have recently proved to be successful in discovering trends of change in Mediterranean fish diversity (Azzurro *et al.*, 2011). As a matter of fact, collaboration with local communities are increasingly used to approach the study of large scale changes in the natural world. As a matter of facts some countries, such as Australia and the USA (California; Hawaii) have already started monitoring projects which involve community-based actions for the detection of marine invasive species. People are basically asked to 'monitor' the marine environment around them, in the course of their daily activities and to provide reports of invasions and various tools and detection kits have been developed all around the world with the aim to widely disseminate information about potential invaders to target communities. In a pilot study called 'alien fish alert' fishermen and divers of the Sicily Strait were asked to provide reports of all "unusual occurrences" (Azzurro, 2010). Given the familiarity of fishermen with local species, no training on fish taxonomy was considered necessary and no black list was proposed, with the following slogan: *"there is no need of any expertise in identifying alien species – those familiar with our sea will immediately recognize a 'strange' fish that they have never seen before – it is such records that we are after!"*. An awareness campaign was realized by means of media promotion, posters and personal interactions. As a matter of fact, this organization was found to provide researchers with an excellent tool for early detection of newly established NIS. These activities are encouraged by the European Strategy of Invasive Alien Species for the up-building of public awareness and collection and distribution of information. In a parallel way, a set of posters with NIS fish photos were distributed among fishermen and at fish auctions along the coast of Lebanon and at port authorities and local fish markets of SE Aegean Sea, Greece. These posters showed only selected fish species displaying characteristic and recognizable family appearances, such as Apogonidae, Chaetodontidae, Scaridae etc. Additionally, posters in Lebanon also provided a phone number with a promise of payment when a new species is collected and delivered. Many fishermen showed a positive response to the advertisement and several first records were collected as such (Bariche, 2010a; 2011a; Bariche and Heemstra, 2012).

The collaboration with local fishery communities has several advantages when the species to monitor are fishes in respect to other groups of organisms. Fishery landings also provide quantitative data, samples and additional information. In addition, the identification of many fishes is relatively easy and this is an obvious advantage for their detection (Fig. 3). Therefore, members of local fishery communities, with broad geographical distributions and familiarity of natural environments could play a dynamic role for the early detection of environmental changes.

Another significant example of innovative ideas to monitor fish diversity changes in the Mediterranean Sea was "Local Ecological Knowledge" (LEK). In recent years, LEK has emerged as an alternative information source on species presence or qualitative and

Figure 4. Sargocentrum rubrum individuals caught by trammel nets in Lebanon

quantitative indices of species abundance (Rasalato *et al.*, 2010). Local Ecological Knowledge can be defined as the information that a group of people have about local ecosystems. We usually rely on knowledge gained by individuals over their lifetimes, and not on what information has been handed through generations. To extract data and information from individuals' memory, semi-structured or unstructured conversations between the researcher and the participant were used, a practice commonly called "oral history". In a recent study, Azzurro et al. (2011) provided evidence of a trend for thermophilic taxa to increase in the Central Mediterranean Sea on the basis of a set of interviews to local fishermen. The study was based on interviews to local fishermen and divers with more than ten years of experience. Species mentioned in each interview were used to build a presence-absence dataset that provided extremely coherent results about the northward expansion of families such as Carangidae and Sphyraenidae, whose expansion was only previously noted by occasional records in the scientific literature. These new methodologies give us the chance to get information that otherwise cannot be obtained from the efforts of single researchers. Hopefully in the next future their potential will be increasingly exploited for the monitoring and the understanding of the biodiversity changes in the Mediterranean Sea.

Author details

Stefanos Kalogirou
Hellenic Centre for Marine Research, Hydrobiological Station of Rhodes, Greece

Ernesto Azzurro
ISPRA, National Institute for Environmental Protection and Research, Livorno, Italy

Michel Bariche
Department of Biology, American University of Beirut, Beirut, Lebanon

5. References

Akyol, O., Unal, V., Ceyhan, T. & Bilecenoglu, M. (2005). First confirmed record of *Lagocephalus sceleratus* Gmelin, 1789 in the Mediterranean Sea. *Journal of fish biology* 66, 1183-1186.

Aydin, M. (2011). Growth, reproduction and diet of pufferfish (*Lagocephalus sceleratus* Gmelin, 1789) from Turkey's Mediterranean sea coast. *Turkish Journal of Fisheries and Aquatic Sciences* 11, 589-596.

Azzurro, E., ed. (2008). *The advance of thermophilic fishes in the Mediterranean sea: overview and methodological questions.* Monaco: *CIESM Workshop Monographs*

Azzurro, E. (2010). Unusual occurences of fish in the Mediterranean Sea: an insight into early detection. In *Fish invasions of the Mediterranean Sea: Change and Renewal* (Golani, D. & Appelbaum-Golani, B., eds.). Sofia-Moscow: Pensoft.

Azzurro, E., Fanelli, E., Mostarda, E., Catra, M. & Andaloro, F. (2007a). Resource partitioning among early colonizing *Siganus luridus* and native herbivorous fish in the Mediterranean: an integrated study based on gut-content analysis and stable isotope signatures. *Journal of the Marine Biological Association of the UK* 87, 991-998

Azzurro, E., Moschella, P. & Maynou, F. (2011). Tracking Signals of Change in Mediterranean Fish Diversity Based on Local Ecological Knowledge. *PLoS ONE* 6, e24885.

Azzurro, E., Pais, A., Consoli, P. & Andaloro, F. (2007b). Evaluating daynight changes in shallow Mediterranean rocky reef fish assemblages by visual census. *Marine Biology* 151, 2245-2253.

Bariche, M. (2005). Age and growth of Lessepsian rabbitfish from the eastern Mediterranean. *Journal of Applied Ichthyology* 21, 141-145.

Bariche, M. (2006). Diet of the Lessepsian fishes, *Siganus rivulatus* and *S. luridus* (Siganidae) in the eastern Mediterranean: A bibliographic analysis. *Cybium* 30, 41-49.

Bariche, M. (2010a). Champsodon vorax (Teleostei: Champsodontidae), a new alien fish in the Mediterranean. *Aqua, International Journal of Ichthyology* 16, 197-200.

Bariche, M. (2010b). *Champsodon vorax* (Teleostei: Champsodontidae), a new alien fish in the Mediterranean. *Aqua* 16, 197-200.

Bariche, M. (2011a). First record of the cube boxfish *Ostracion cubicus* (Ostraciidae) and additional records of *Champsodon vorax* (Champsodontidae) from the Mediterranean. *Aqua, International Journal of Ichthyology* 17, 181-184.

Bariche, M. (2011b). First record of the cube boxfish Ostracion cubicus (Ostraciidae) and additional records of Champsodon vorax (Champsodontidae) from the Mediterranean. *Aqua, International Journal of Ichthyology* 17, 181-184.

Bariche, M., Alwan, N., El-Assi, H. & Zurayk, R. (2009). Diet composition of the Lessepsian bluespotted cornetfish *Fistularia commersonii* in the eastern Mediterranean. *Journal of Applied Ichthyology* 24, 460-465.

Bariche, M., Alwan, N. & El-Fadel, M. (2006). Structure and biological characteristics of purse seine landings off the Lebanese coast (eastern Mediterranean). *Fisheries Research* 82, 246-252.

Bariche, M. & Heemstra, P. (2012). First record of the blackstrip grouper *Epinephelus fasciatus* (Teleostei: Serranidae) in the Mediterranean Sea. *Marine Biodiversity Records* 5.

Bariche, M., Letourneur, Y. & Harmelin-Vivien, M. (2004). Temporal fluctuations and settlement patterns of native and lessepsian herbivorous fishes on the Lebanese coast (Eastern Mediterranean). *Environmental Biology of Fishes* 70, 81-90.

Bariche, M. & Saad, M. (2008). Settlement of the lessepsian blue-barred parrotfish *Scarus ghobban* (Teleostei: Scaridae) in the eastern Mediterranean. *Marine Biodiversity Records* 1, e5.

Bariche, M., Sadek, R., Al-Zein, M. S. & El-Fadel, M. (2007). Diversity of juvenile fish assemblages in the pelagic waters of Lebanon (eastern Mediterranean). *Hydrobiologia* 580, 109-115.

Ben-Tuvia, A. (1973). Man-made changes in the eastern Mediterranean Sea and their effect on the fishery resources. *Marine Biology* 19, 197-203.

Ben-Yami, M. & Glaser, T. (1974). The invasion of *Saurida undosquamis* (Richardson) into te Levant basin - an example of biological effect on interoceanic canals. *Fishery Bulletin* 72, 359-373.

Bentur, Y., Ashkar, J., Lurie, Y., Levy, Y., Azzam, Z. S., Litmanovich, M., Golik, M., Gurevych, B., Golani, D. & Eisenman, A. (2008). Lessepsian migration and tetrodotoxin poisoning due to *Lagocephalus sceleratus* in the eastern Mediterranean. *Toxicon* 52, 964-968.

Bianchi, C. (2007). Biodiversity issues for the forthcoming tropical Mediterranean Sea. *Hydrobiologia* 580, 7-21.

Bianchi, C. N. & Morri, C. (2003). Global sea warming and "tropicalization" of the Mediterranean Sea: biogeographic and ecological aspects *Biogeographia* 24, 319-328.

Bianchi, C. N. & Morri, C. (2004). Climate change and biological response in Mediterranean Sea ecosystems - a need for broad-scale and long-term research. *Ocean Challenge* 13.

Bilecenoglu, M. (2010). Alien marine fishes of Turkey - an updated review. In *Fish Invasions of the Mediterranean Sea: Change and Renewal* (Golani, D. & Appelbaum-Golani, B., eds.), pp. 189-217. Sofia-Moscow: Pensoft Publishers.

Carpentieri, P., Lelli, S., Colloca, F., Mohanna, C., Bartolino, V., Moubayed, S. & Ardizzone, G. D. (2009). Incidence of lessepsian migrants on landings of the artisanal fishery of south Lebanon. *Marine Biodiversity Records* 2.

Cicek, E. & Bilecenoglu, M. (2009). A new alien fish in the Mediterranean Sea: Champsodon nudivittis (Actinopterygii: Perciformes: Champsodontidae). *Acta Ichtyological Et Piscatoria* 39, 67-69.

CIESM (2008). Climate warming and related changes in the Mediterranean marine biota. In *CIESM Workshop Monographs* (Briand, F., ed.), p. 152. Monaco.

Diamant, A. (2010). Red-Med immigration: a fish parasitology perspective, with special reference to Mxyxosporea. In *Fish Invasions of the Mediterranean Sea: Change and Renewal* (Golani, D. & Appelbaum-Golani, B., eds.), pp. 85-97. Sofia-Moscow: Pensoft Publishers.

EastMed (2010). Report of the sub-regional technical meeting on the lessepsian migration and its impact on eastern Mediterranean fishery. p. 25. GCP/INT/041/EC-GRE-ITA/TD-04.

Galil, B. & Zenetos, A. (2002). A sea change - Exotics in the eastern Mediterranean. In *Invasive Aquatic Species of Europe: Distribution, Impacts and Management* (Leppakoski, E., Olenin, S. & Gollasch, S., eds.), pp. 325-336: Kluwer Academic Publishers.

Galil, B. S. (2006). The marine caravan - the Suez Canal and the Erythrean invasion. In *Bridging divides* (Gollasch, S., Galil, B. S. & Cohen, A., eds.), pp. 207-300. Netherlands: Springer.

Galil, B. S. (2007). Loss or gain? Invasive aliens and biodiversity in the Mediterranean Sea. *Marine Pollution Bulletin* 55, 314-322.

George, C. J. & Athanassiou, V. A. (1967). A two year study on the fishes appearing in the seine fishery of the St. George bay, Lebanon *Ann. Mus. Civ. Stor. Nat. Genova* 76, 237-294.

Golani, D. (1993a). The biology of the Red Sea migrant, *Saurida undosquamis* in the Mediterranean and comparison with the indigenous confamilial *Synodus saurus* (Teleostei: Synodontidae). *Hydrobiologia* 271, 109-117.

Golani, D. (1993b). Trophic adaption of Red Sea fishes to the eastern Mediterranean environment - Review and new data. *Israel Journal of Zoology* 39, 391-402.

Golani, D. (1994). Niche separation between colonizing and indigenous goatfish (Mullidae) along the Mediterranean coast of Israel. *Journal of fish biology* 45, 503-513.

Golani, D. (2010). Colonization of the Mediterranean by Red Sea fishes via the Suez Canal - Lessepsian migration. In *Fish Invasions of the Mediterranean Sea: Change and Renewal* (Golani, D. & Appelbaum-Golani, B., eds.), pp. 145-188. Sofia-Moscow: Pensoft Publishers.

Golani, D. & Ben-Tuvia, A. (1995). Lessepsian migration and the Mediterranean fisheries of Israel. In *World Fisheries Congress* (Armentrout, N. B., ed.), pp. 279-289. New Delhi: Oxford & IBH Publishing Company.

Goldschimdt, T., Witte, F. & Wanink, J. (1993). Cascading effects of the Introduced Nile Perch on the Detrivorous/ Phytoplanktivorous Species in the Sublittoral Areas of Lake Victoria *Conservation Biology* 7, 686-700.

Goren, M. & Galil, B. S. (2001). Fish Biodiversity in the Vermetid Reef Of Shiqmona (Israel). *Marine Ecology* 22, 369-378.

Goren, M. & Galil, B. S. (2005). A review of changes in the fish assemblages of Levantine inland and marine ecosystems following the introduction of non-native fishes. *Journal of Applied Ichthyology* 21, 364-370.

Gottlieb, E. (1960). On the selection of *Upeneus moluccensis* and *Mullus barbatus* by trawl cod-ends in the Israeli fisheries. *General Fisheries Council for the Mediterranean* 8, 1-10 + VI.

Gruvel, A. (1931). Les Etats de Syria. Richness marines et fluviales. Exploitation actuelle, avenir. . In *Societe d'editions Geographiques, Marines et Coloniales*, pp. 72-134. Paris.

Gucu, A. C. & Bingel, F. (1994). Trawlable species assemblages on the continental shelf of the Northeastern Levan Sea (Mediterranean) with an emphasis on Lessepsian migration. *Acta Adriatica* 35, 83-100.

Guidetti, P., Bianchi, C. N., La Mesa, G., Modena, M., Morri, C., Sara, G. & Vacchi, M. (2002). Abundance and size structure of Thalassoma pavo (Pisces: Labridae) in the western Mediterranean Sea: variability at different spatial scales. *Journal of the Marine Biological Association of the United Kingdom* 82, 495-500.

Harmelin-Vivien, M. L., Bitar, G., Harmelin, J.-G. & Monestiez, P. (2005). The littoral fish community of the Lebanese rocky coast (eastern Mediterranean Sea) with emphasis on Red Sea immigrants. *Biological Invasions* 7, 625-637.

Hassan, M., Harmelin-Vivien, M. & Bonhomme, F. (2003). Lessepsian invasion without bottleneck: example of two rabbitfish species (Siganus rivulatus and Siganus luridus). *Journal of Experimental Marine Biology and Ecology* 291, 219-232.

Jackson, J. B. C., Kirby, M. X., Berger, W. H., Bjorndal, K. A., Botsford, L. W., Bourque, B. J., Bradbury, R. H., Cooke, R., Erlandson, J., Estes, J. A., Hughes, T. P., Kidwell, S., Lange, C. B., Lenihan, H. S., Pandolfi, J. M., Peterson, C. H., Steneck, R. S., Tegner, M. J. & Warner, R. R. (2001). Historical Overfishing and the Recent Collapse of Coastal Ecosystems. *Science* 293, 629-637.

Kalogirou, S., Corsini Foka, M., Sioulas, A., Wennhage, H. & Pihl, L. (2010). Diversity, structure and function of fish assemblages associated with *Posidonia oceanica* beds in an area of the eastern Mediterranean Sea and the role of non-indigenous species. *Journal of fish biology* 77, 2338-2357.

Kalogirou, S., Corsini, M., Kondilatos, G. & Wennhage, H. (2007). Diet of the invasive piscivorous fish *Fistularia commersonii* in a recently colonized area of eastern Mediterranean. *Biological Invasions* 9, 887-896.

Kalogirou, S., Mittermayer, F., Pihl, L. & Wennhage, H. (2012a). Feeding ecology of indigenous and non-indigenous fish species within the family Sphyraenidae. *Journal of fish biology* 80, 2528-2548.

Kalogirou, S., Wennhage, H. & Pihl, L. (2012b). Non-indigenous species in Mediterranean fish assemblages: Contrasting feeding guilds of *Posidonia oceanica* meadows and sandy habitats. *Estuarine, Coastal and Shelf Science* 96, 209-218.

Kan, S., Chan, M. K. & David, P. (1987). Nine fatal cases of Puffer fish poisoning in Sabah, Malaysia. *Medical Journal of Malaysia* 42, 199-200.

Katikou, P., Georgantelis, D., Sinouris, N., Petsi, A. & Fotaras, T. (2009). First report on toxicity assessment of the Lessepsian migrant pufferfish *Lagocephalus sceleratus* (Gmelin, 1789) from European waters (Aegean Sea, Greece). *Toxicon* 54, 50-55.

Lasram, F. B. R., Guilhaumon, F. & Mouillot, D. (2009). Fish diversity patterns in the Mediterranean Sea: deviations from a mid-domain model. *Marine Ecology Progress Series* 376, 253-267.

Lasram, F. B. R. & Mouillot, D. (2008). Increasing southern invasion enhances congruence between endemic and exotic Mediterranean fish fauna. *Biological Invasions*.

Leppäkoski, E. & Olenin, S. (2000). Non-native species and rates of spread: lessons from the brackish Baltic Sea. *Biological Invasions* 2, 151-163.

Levine, J. M. & Adler, P. B. (2004). A meta-analysis of biotic resistance to exotic plant invasions. *Ecology Letters* 7, 975-989.

Levine, J. M. & D' Antonio, C. M. (1999). Elton revisited: a review of evidence linking diversity and invasibility. *Oikos* 87, 15-26.

Lipkin, J. & Safriel, U. N. (1971). Intertidal zonation on rocky shores at Mikhmoret (Mediterranean, Israel). *J. Ecol.* 59, 1-30.

Lundberg, B. & Golani, D. (1995). Diet adaptations of Lessepsian migrant rabbitfishes, *Siganus luridus* and *S. rivulatus*, to the algal resources of the Mediterranean coast of Israel. *Marine Ecology* 16, 73-89.

Lundberg, B., Ogorek, R., Galil, B. S. & Goren, M. (2004). Dietary choices of siganid fish in Shiqmona reef, Israel. *Israel Journal of Zoology* 50, 39-53.

MacArthur, R. (1955). Fluctuations of Animal Populations and a Measure of Community Stability. *Ecology* 36, 533-536.

Massuti, E., Valls, M. & Ordines, F. (2010). Changes in the western Mediterranean ichthyofauna: signs of tropicalization and meridianization. In *Fish Invasions of the Mediterranean Sea: Change and Renewal* (Golani, D. & Appelbaum-Golani, B., eds.), pp. 293-312. Sofia-Moscow: Pensoft Publishers.

Moulton, M. P. (1993). The all-or-none pattern in introduced Hawaiian passeriforms: The role of competition sustained. *The American Naturalist* 141, 105-119.

Moyle, P. B. & Light, T. (1996). Biological invasions of fresh water: Empirical rules and assembly theory. *Biological Conservation* 78, 149-161.

Occhipinti-Ambrogi, A. & Savini, D. (2003). Biological invasions as a component of global change in stressed marine ecosystems. *Marine Pollution Bulletin* 46, 542-551.

Oren, O. H., Ben-Yami, M. & Zismann, L. (1971). Explorations of the possible deep-water trawling grounds in the Levant basin. *Proceedings of the General Fisheries Council for the Mediterranean, Studies and Reviews* 49, 41-71.

Parmesan, C. & Yohe, G. (2003). A globally coherent fingerprint of climate change impacts across natural systems. *Nature* 421, 37-42.

Perry, A. L., Low, P. J., Ellis, J. R. & Reynolds, J. D. (2005). Climate Change and Distribution Shifts in Marine Fishes. *Science* 308, 1912-1915.

Por, F. D. (1978). *Lessepsian migration: the influx of Red Sea biota into the Mediterranean by way of the Suez canal*. Berlin-Heidenberg: Springer.

Quignard, J. P. & Tomasini, J. A. (2000). Mediterranean fish biodiversity. *Biologia Marina Mediterranea* 7, 1-66.

Raitsos, D. E., Beaugrand, G., Georgopoulos, D., Zenetos, A., Pancucci-Papadopoulou, A. M., Theocharis, A. & Papathanassiou, E. (2010). Global climate change amplifies the entry of tropical species into the eastern Mediterranean Sea. *Limnology and Oceanograph* 55, 1478-1484.

Rasalato, E., Maginnity, V. & Brunnschweiler, J. M. (2010). Using local ecological knowledge to identify shark river habitats in Fiji (South Pacific). *Environmental Conservation* 37, 90-97.

Reise, K., Olenin, S. & Thieltges, D. (2006). Are aliens threatening European aquatic coastal ecosystems? *Helgoland Marine Research* 60, 77-83.

Ricciardi, A. (2001). Facilitative interactions among aquatic invaders: is an "invasional meltdown" occurring in the Great Lakes? *Canadian Journal of Fisheries and Aquatic Sciences* 58, 2513-2525.

Ricciardi, A. & Mottiar, M. (2006). Does Darwin's naturalization hypothesis explain fish invasions? *Biological Invasions* 8, 1403-1407.

Rilov, G. & Crooks, J. A. (2009). Marine Bioinvasions: Conservation Hazards and Vehicles for Ecological Understanding. In *Biological Invasions in Marine Ecosystems* (Rilov, G. & Crooks, J. A., eds.). Berlin Heidenberg: Springer-Verlag.

Rilov, G. & Galil, B. (2009). Marine bioinvasions in the Mediterranean Sea – History, distribution and ecology. In *Biological Invasions in Marine Ecosystems* (Rilov, G. & Crooks, J. A., eds.), pp. 549-575. Berlin Heidelberg: Springer.

Roessig, J. M., Woodley, C. M., Cech, J. J. & Hansen, L. J. (2004). Effects of global climate change on marine and estuarine fishes and fisheries. *Reviews in Fish Biology and Fisheries* 14, 251-275.

Sabrah, M. M., El-Ganainy, A. A. & Zaky, M. A. (2006). Biology and toxicity of the pufferfish *Lagocephalus sceleratus* (Gmelin, 1789) from the gulf of Suez *Egyptian Journal of Aquatic Research* 32, 283-297.

Sakinan, S. & Örek, Y. A. (2011). First record of Indo-Pacific Indian scad fish, *Decapterus russelli*, on the north-eastern Mediterranean coast of Turkey. *Marine Biodiversity Records* 4.

Salameh, P., Sonin, O., Edelis, D. & Golani, D. (2011). First record of the Red Sea orangeface butterflyfish *Chaetodon larvatus* Cuvier, 1831 in the Mediterranean. *Aquatic Invasions* 6, S53-S55.

Sax, D. F. (2002). Native and naturalized plant diversity are positively correlated in scrub communities of California and Chile. *Diversity and Distributions* 8, 193-210.

Sax, DovB F., Gaines, StevenB D. & Brown, JamesB H. (2002). Species Invasions Exceed Extinctions on Islands Worldwide: A Comparative Study of Plants and Birds. *The American Naturalist* 160, 766-783.

Shakman, E. A. & Kinzelbach, R. (2007). Distribution and characterization of lessepsian migrant fishes along the coast of Libya. *Acta Ichtyological Et Piscatoria* 37, 7-15.

Simberloff, D. & Von Holle, B. (1999). Positive Interactions of Nonindigenous Species: Invasional Meltdown? *Biological Invasions* 1, 21-32.

Sonin, O., Spanier, E. & Pisanty, S. (1996). Undersized fishes in the catch of the Israeli Mediterranean fisheries - are there differences between shallow and deeper water ? In *Preservation of our world in the wake of change* (Steinberger, ed.), pp. 449-454. Jerusalem: ISEEQS Publication.

Stachowicz, J. J. & Tilman, D. (2005). Species invasions and the relationships between species diversity, community saturation, and ecosystem functioning. In *Species Invasions: Insights into ecology, evolution and biogeography* (Sax, D. F., Stachowicz, J. J. & Gaines, S. D., eds.). Sunderland, Massachusetts: Sinauer Associates.

Stachowicz, J. J. & Whitlatch, R. B. (1999). Species Diversity and Invasion Resistance in a Marine Ecosystem. *Science* 286, 1577.

Streftaris, N. & Zenetos, A. (2006). Alien marine species in the Mediterranean - the 100 'worst invasives' and their impact. *Mediterranean Marine Science* 7, 87-118.

Streftaris, N., Zenetos, A. & Papathanassiou, E. (2005). Globalisation in marine ecosystems - The story of non indigenous marine species across European Seas. *Reviews in Oceanography and Marine Biology* 43, 419-454.

Vitousek, P. M., Mooney, H. A., Lubchenco, J. & Melillo, J. M. (1997). Human Domination of Earth's Ecosystems. *Science* 277, 494-499.

Wallentinus, I. & Nyberg, C. D. (2007). Introduced marine organisms as habitat modifiers. *Marine Pollution Bulletin* 55, 323-332.

Williamson, M. H. & Fitter, A. (1996). The characters of successful invaders. *Biological Conservation* 78, 163-170.

Witenberg, R. & Cock, M. J. W., eds. (2001). *Invasive alien species: a toolkit of best prevention and management practices*: GISP.

Yang, C. C., Liao, S. C. & Deng, J. F. (1996). Tetrodotoxin poisoning in Taiwan; an analysis of posion center data. *Veterinary and Human Toxicology* 38, 282-286.

Zenetos, A., Gofas, S., Verlaque, M., Cinar, M. E., Garcia Raso, J. E., Bianchi, C. N., Morri, C., Azzurro, E., Bilecenoglu, M., Froglia, C., Siokou I., Violanti, D., Sfriso, A., San Mart, G., Giangrande, A., Kata An, T., Ballesteros, E., Ramos-Espla, A., Mastrototaro, F., Oca A, O., Zingone, A., Gambi, M. C. & Streftaris, N. (2010). Alien species in the Mediterranean Sea by 2010. A contribution to the application of European Union's Marine Strategy Framework Directive (MSFD). Part I. Spatial distribution. *Mediterranean Marine Science* 11, 381-493.

Zenetos, A., Meric, E., Verlaque, M., Galli, P., Boudouresque, C.-F., Giangrande, A., Cinar, E. & Bilecenoglu, M. (2008). Additions to the annotated list of marine alien biota in the Mediterranean with special emphasis on Foraminifera and Parasites. *Mediterranean Marine Science* 9, 119-165.

Zenetos, A., Pancucci Papadopoulou, M. A., Zogaris, S., Papastergiadou, A., Vardakas, L., Aligizaki, K. & Economou, A. N. (2009). Aquatic alien species in Greece: tracking sources, patterns and effects on the ecosystem. *Journal of Biological Research-Thessaloniki* 12, 135-172.

Socio-Economic Impact

Creation of the New System of Management of Important Transition Zones in the Nature

Jing Tan, Tian Yan, Wang Shang-Wu
and Jie Feng

Additional information is available at the end of the chapter

1. Introduction

The construction of ecology management system in natural ecological reserve (reserve) has been a long-term and complex project led by Chinese government. Its fundamental purpose is to protect natural biodiversity, save endangered species, maintain ecology balance, explore the means and approach of using rationally natural resources and ultimately to promote the harmonious development between natural ecosystems and human socio-economy. International management and protection of reserve is a multi-level ecosystem management and protection including the core zone、experimental zone、buffer transition zone、business district and living area. This idea has great significance to guide the management system construction of natural ecosystem in China.

Chinese western nature reserve system is well known for its characteristics such as large-scale, high-altitude, long-ecological transition zone and system vulnerability, which leading to the arduous task for government in natural ecological reserve protection and management, so the government usually attached more importance to the protection of the core and experimental zone, while relatively ignored the management and operation to transition zone. This situation doesn't meet the basic principles of natural ecological system protection and management of systematization, safety and integrity. Therefore, the most urgent priority is how to change the reserve excessive weak management present situation, save reserve biodiversity crisis, alleviate government management function pressure in future, which has important practical significance and long-term history meaning to ensure the natural ecological system security and integrity.

2. Section I The crisis and management system defects of the important transition zones of western nature ecology reserve

2.1. The crisis and pressure of the reserve transition zones

The transition zone is one of the important components of varieties of living organisms and the genetic variation patterns combination in the reserve biodiversity system. The 1997 statistics shows that the scale and amount of typical biology species transition zone in western region accounts for more than 80% of the nation totals [1], which consists of hundreds of thousands of hectares of transition zones, such as Sichuan Wo Long Nature Reserve and its transition zone with more than 20 hectares; Thousands of hectares of transition zone, such as the crossing transition zone between Sichuan MiYaLuo Natural Ecological Reserve System and Sichuan Cao PO Provincial Nature Ecology Reserve System - ---- Mashan Village transition zone with 1110.2 hectares in Aba Prefecture. Sichuan Jin Fo Mashan reserve transition zone [2] with more than 900 hectares, etc.

However, these transition zones are facing double crisis in the long run, natural ecological security and system integrity crisis, the survival and development crisis from the transition zone community.

a.   The Ecological Crisis Of The Reserve Transition Zone

The ecological crisis is mainly from three aspects which are as follows:

The first is the ecological crisis caused by Local residents' over-reliance on natural and ecological resources. Reserve transition zone residents are dependent on long-term eco-environmental resources for their livelihoods, thus the excessive deforestation and use leads to the fragmentation of biodiversity habitat, obvious reduction of biological species distribution, quick decreasing of herd numbers, species inbreeding and genetic heterozygosity falling; The excessive deforestation and use changed the forest composition and organic ecological relations among species, reduced the ability of resistance to diseases and pests for forest itself, increased the scale and frequency of forest diseases and pests occurrence, which made the transition zone natural forests appear many remnants, degradation even extinction. Such as recently decade, in China, 200 kinds of plants have extinct, 4,000 - 5,000 species of higher plants are facing threat, accounting for 15% -20% of total number of species. In China, the first batch rare and endangered plants published in the international convention have been up to 388 kinds of [3];

The second is ecology crisis resulted from the traditional hunting customs of community residents and the increasing demand for wild animals and plants consumer market. For a long time, because the transition zone has been non-control zone, which resulted in its acquisition and poaching behavior is always rampant. In addition to, lots of treasure herbs and edible fungi are collected, and the 20 kinds of national first level protection animals (giant panda, golden monkey, tibetan antelope and so on) are being endangered. Only in a smuggling raid inspection of 1996 in Sichuan, 146 copies of giant panda skin were seized by them [4].

The third is ecology crisis because of the community environment pollution in the transition zones. In this paper, the author's investigation indicates that the Sichuan Wang Lang national nature ecological reserve important transition zone ----- Ping Wu Guan Ba community and whose surrounding communities in Mian Yang City , and the Sichuan Miyaluo nature ecology reserve important transition zone----- Mashan in village in Aba Prefecture Li County and it's surrounding communities, face the waste and water pollution problems. And others data showed 44% of communities in the nature ecology reserve transition zones exist the garbage pollution diffusion phenomenon, 12% have the water pollution, 22% of the core area are suffering the direct threat [6] [7].

b.   The Community Survival And Development Pressure Of The Reserve Transition Zone

Most of Western Reserve transition zones not only belong to minority areas which are representative of Tibetan, Qiang, Yi and so on, but also locate the poor communities in which the socio-economic conditions are relatively backward. Statistics show there are nearly 300 national standard poor counties surrounding the 926 national reserves [5]. According to the 85 investigations into the reserves and their surrounding communities, the population size is 14000 averagely in protection zone, while the transition zone is about 59000. In terms of population density, the population density of 85 protection areas and transition zones is 5.75 persons / km2, which is 1.8 and 2.4 times compared with developed countries [8]. The survival and development pressure of transition zones is mainly from the following aspects:

The first is the pressure from cumulative increase of population. The long-term lenient family planning policy in these areas results in the obvious growth of population density and size for the three decades. Community residents have to live on subsidies relief and the original means of livelihood resulting from the inability to transfer and placement, when the government faces the pressure from community population cumulative increase.

The second is the dependence on ecological resources, excessive deforestation and hunting, which causes the area expansion of zoology forest, wasteland, mark place, and the frequent natural disasters such as regional climate, soil change and landslide. Therefore, some communities are facing the transfer pressure in the short run.

The third, Survival dilemma has resulted in the intensification of the contradictions between the conservation and community development. In some areas of the communities living in the alpine valley transition zone, farmers' own arable crop fails completely, due to water shortage, drought, soil fertility declination, decline in productivity as well as of human illegal logging and the destruction to wild animals; The lack of projects or technology, backwardness of transportation and educational and other factors make farmers be unable to save themselves; The difference of subsidies standard in different reserves transition zones cause the intensification of the contradictions among communities and governments.

2.2. The management system questions of the reserve transition zones

The geographical and scale characteristics of transition zones increase the protection difficulty of reserves. Statistics show that, by the end of 2008, China has established 2538 of different levels of nature ecology reserves, and the scale of them amounts to about 148.943 million hectares, accounting for 15.5 percent of the land area [6]. Moreover the area of protection transition zone is larger, which has far exceeded the scale and level of developed countries.

The investigations into nature ecology reserve in 1997 and 1999 , as well as "sustainable management policy research report of Chinese natural ecological conservation area" held by the Man and the Biosphere National Committee in 2000, pointed out that the majority of our reserves exist universally some problems such as poor management, irrational use of resources[8] and so on, in the ecosystems protection, promotion the ecological all-round development, scientific research, popular science and other functions. This requires we have to review from the management system of nature ecology protection.

a.   The Management System Defects Of The Nature Ecology Reserve

Facing of the ecological crisis and community survival development pressure in the reserve, government should make efforts mainly from three aspects: First, the government need take leading management system, taking "rescuing protection mode, first program after construction, perfection gradually"[9] management measures to carry out mainly compulsory management and protection on the core areas; Second, the government should take system and legislation to restrict economy development of the community surrounding transition zone; Third, it's important to take community relief and subsidies measures to relieve the conflict between protection area and its surrounding communities development. For example, the protection of rare animals such as giant panda and Chinese alligator core area is a kind of typical crisis rescuing protection.

Under the guidance of "rescuing principle" imposed by the Government, the management system adopted of reserve in China is the government-led and top-down arrangement. Namely, it is the management system combining the comprehensive management with departmental management based on the category management and classification management. The so-called category management is that the area was divided into two levels including core area management and the experimental area management in line with degree of importance to protected objects; The so-called hierarchical management means that the management institution is composed of nation, province, city and county according to the representativeness, importance and crisis degree of protected objects; The so-called management combining the comprehensive management with departmental management, is according to the regulations provisions 1 , paragraph 2 1 in the law of 《People's Republic Of China Ecological Reserves Ordinance》 and the regulations paragraph 8 in the law of 《 Natural Ecological Reserve Ordinance 》: "the state environment protection administration departments shall be responsible for the comprehensive management of nature ecology reserves. Nearly ten administrative departments: State Forestry Administration, State Oceanic Administration, Ministry of Agriculture, Ministry of Construction, Ministry of Land Resources, Ministry of Water Resources and Chinese Academy of Sciences is in charge of the relevant conservation area in their respective responsibility range."

In theory, this management system arrangement is nearly comprehensive and thoughtful. In fact, it will lead to some problems such as overlapping management, cross-management, bull management and poor management in practice. First of all, a reserve can be managed simultaneously by 4 levels management agencies and multiple departments, which will lead to duties fuzzy and duplication and making responsibilities move down. But the corresponding resources and right don't move down, which leading to the separation of powers and responsibilities, management responsibility dislocation, management dislocation or management absence. As a result, this system causes high management costs and poor management effect because of coordination difficulty.

Second, category management generally just takes care of the core and experimental areas, ignoring the management and conservation of the whole natural protection system. Thus transition zone of protection system appears lots of management vacuum.

Third, hierarchical management、bull management and cross-management increase system construction costs and ineffective implementation. Different departments are in charge of the same protected area, thus they will easily make regulations from their own management or interests due to the dislocation of system or management standards introduced by them. That's to say, legislation and regulation has become the way of sharing resources and the means of vested interests distribution among government departments.

b.   The Contradictions Between Ecology Protection And Community Development Intensified By The Mandatory And Exclusive Management Mechanism

The so-called "mandatory and exclusive management mechanism", established under the government-led management mode, has unfavorable effects on the protection activities. On the one hand, it take the core area as the mandatory protection and the scope of subsidies, these compensation policies objectively broke the balance among communities, which will sharpen the conflict among communities as well as the government; On the other hand, the current management system considers government as the management subject, while rules out other social members especially community residents, which will make the government locate the helpless and passive situation in the long run. This likely leads to the opposition sentiment between the reserve and communities, thus many farmers don't care the illegal behavior even develop and utilize maliciously the natural ecological resources, so management and protection effectiveness isn't satisfactory.

c.   The Lack Of Scientific Partition And Systematic Concepts In Reserves

The natural ecology reserve refers to the ecology systems composed of representative, typical and integrated biological communities and the abiotic environment. In line with the biodiversity and systematic features, the protection areas is identified as the ecology system consisting of the core zone, experimental zone, buffer zone, transition zone, business district, tourist area, living service areas and other functional areas.

However, China's reserve only contains core area and experimental zone. According to the policy, only core area is the focus of compulsory protection and resource allocation. While the important ecology transition zone likely become vacuum area, because it has not been identified as the core area and the protected area. This kind of management thoughts isn't helpful for the development and balance of reserve ecology system, especially for the protection of important transition zone in the ecology system, which impacts directly on the integrity of the structure and function of ecology system.

Therefore, how to bring important transition zones into protection system? The author proposed the management system which introduces "Agreement Protection" mechanism based on project and brought farmers and social organizations into the legal management subjects, and they will participate in the protection with government together. This system can stimulate effectively the protection initiative and enthusiasm of community and social, ease the contradictions between the ecology protection and community development and functions pressure of government, ultimately, can resolve effectively the problem of subject vacancy of management system in western natural ecological reserve.

2.3. Creation the new management system of important transition zones in Chinese Western nature ecology reserves

a.   The Basic Idea Of Creating Co-Management System Of The Reserve Transition Zone

Based on reserve transition zone crisis and the existing problems of reserve management system, firstly, the author believes that the reserve transition zone should be brought into the scope of protection and management to ensure the integrity of the ecology protection system; Secondly, community-based organizations and farmers in the transition zone also should be considered as the legal protection subject to make up for management system subject vacancy; Thirdly, on the condition of not changing ownership of state-owned ecological forest , let community-based organizations and farmers participate in the transition zone ecological forest management by the way of signing Agreement Protection based on projects. Simultaneously, in nature ecology Protection projections, entitle them some privileges such as information, participation, supervision and decision-making power to improve their host status and enhance responsibility consciousness in ecological protection; Create the new system (common management) of ecological protection in reserve transition zone (see Figure 1).

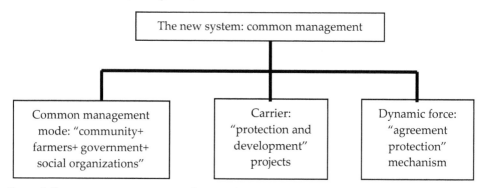

Figure 1. Common management system in natural protection reserve transition zone

b. Creation The New Mode Of Common Management: "Community+ Farmers+ Government+ Social Organizations"

The so-called common management mode: "community + farmers + government + social organizations" is that the government and community members consider the "Agreement Protection" mechanism as driving force to participate in jointly management such as the design, decision making, implementation, supervision and evaluation based on projects. The content of " Agreement Protection" agreed the common management rights, responsibilities and interests among "community and famers"(party a)、"government" (party b) and social organizations (third party), to form the interests joint through the resources integrity way, thus, the protection and construction tasks will be completed more easily. The focus of common management is target consistenc, interest-sharing, joint participation and risk-sharing. Compared with government-led mode, this mode has the following features:

1. Relying on community development projects

Common management mode regards a "Agreement Protection " projects as carrier, project target covers the transition zone protection target and community development goal, in details, including the international protection and development programs, the state development and poverty alleviation projects and new rural construction projects, especially GET (the global environment facility) aid projects (see Table 1).Projects aid agencies may be government and other social welfare organizations at home or abroad.

Projects type	Projects breakthrough point	Remarks
Economic development projects	Economic forest construction and its application; Ecological farming base construction; Ecological farm households tourism economic construction; Alpine ecological flowers construction; Special economic development projects	All kinds of fruit and vegetable planting base, medicinal plants base; Eco-bee breeding base; ecological poultry breeding base; Flower type ornamental and consumer base; Prickly ash and silkworm base, etc
Alternative Resource projects	Promotion and utilization of solar energy Digester construction and use Energy saving and environmental protection of life and production of construction Life and production construction of energy conservation and environmental protection	
Circulation resource industry construction projects	"Three links", human and livestock drinking water projects	"Three links" includes Water, electricity and road

Table 1. The coverage of community development projects

2.   The main functions of "Agreement Protection" projects (see Table 2).

The functions of "Agreement Protection" projects	Specific tasks
The biodiversity management and protection of reserve transition zone	Eliminating of illegal logging and dredging; Eliminating illegal fishing; Eliminating foreign sabotage; Firewood cutting in accordance with provisions; Collecting the understory of grass and fruit, fungi, ferns and other non-wood products in accordance with regulations; Restrictive hunting wild boar and other destructive animals
Communities development	Infrastructure construction: human and livestock drinking project and new energy promotion, etc.; Rural economy construction: courtyard, fruit trees, rubber, walnut and other with local characteristics economic construction; Breeding and its processing; The new rural construction: electric light, TV ,phone, ecological toilet, rural cultural center construction, etc. The rural new practical technical training: planting and breeding industry, comprehensive prevention and control of disease, migrant workers' skills training, women's ability construction, etc.

Table 2. The main functions of "Agreement Protection" projects

3.   The subjects' roles and responsibilities of co-management mode

The subjects of the common management mode include community-based organizations, farmers, government and social organizations. Government and social organizations are advocates and supporters of "Agreement Protection" projects. Community-based organizations and farmers are supporters and executors of "Agreement Protection" projects. First, the main responsibilities of the advocates and supporters: selection and determination of project, setting goals, drawing up jointly tasks with community organizations, providing project-related fund, materials and technical services; Second, community organizations and farmers considered as implementers participate in jointly the construction of project's implementation plan, management mechanism and management system with the advocates and supporters. Third, the advocates and supporters provide services such as guidance, service, monitoring and evaluation on the progress and performance of project of "Agreement Protection" projects (see Figure 2).

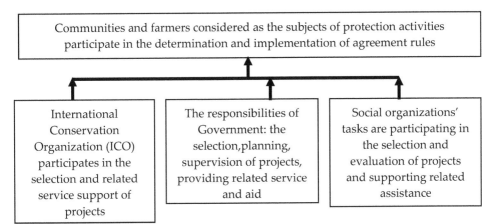

Figure 2. Subjects' responsibilities of co-management mode

c.   Establishment the Related Mechanism Of Co-Management "Agreement Protection"

1.   Establishment the "Agreement Protection" Mechanism Of Co-Management

The so-called "Agreement Protection" mechanism of co-management doesn't change the structure of forest rights, it make ecology forest management rights and protection rights delegated or transferred to the communities and any other organizations or individuals who are willing to assume the protection responsibility to ensure community's development and security and integrity of reserve ecosystem. This kind of mechanism sets responsibilities, rights and benefits of interested parties in the process of management and protection by means of agreement, forms the co-management mechanism based on community organizations and farmers, supplemented by Government and other organizations.

2.   Establishing various forms of incentive and restraint mechanisms based on local conditions

In accordance with economic development regulations, under the condition of avoiding greater risk, to meet effectively personal interests is the direct driving force of promoting economic development. Therefore, the effective compensation and incentive way has positive significance to encourage communities and farmers to participate in the protecting action actively.

Compensation mechanism is a kind of coordinating manner for government and social organizations to support and relieve community livelihoods and development. So how to transfer the compensation way into the incentive and constraint functions is critical to encourage community and farmers to complete initiatively the management and protection task.

d.   Establishment Of Multi-Level Education And Technical Training Mechanism

Education and technical training plays a very important role for the "Agreement Protection" mode. The persistence of farmers' protection consciousness, awareness and conservation actions on the transition zone usually can be reflected after education and training. For example, the provisions of contents and frequency of education publicity and the technical training in Mashan Village "Agreement Protection" project have achieved remarkable effect. Through education and technical training, farmers can really feel the importance of conservation actions on the ecologyl safety and community economic development, what's more, they will truly feel a range of democratic rights and master consciousness in the community ecology protection activities. Therefore, the establishment of education and technical training mechanism provide objectively the most fundamental and powerful guarantee for the continuous process of protection action. In summary, the root of management and protection issues in reserve transition zone is existing defects of management system. Through the establishment of "Agreement Protection" mechanisms and co-management mode, the management and protection of reserve transition zone was transferred to communities and farmers. This kind of new system is not only helpful to transfer resources and role of communities and other social organization, but also useful to relieve the pressure from protection function of government. So it can effectively make up for the poor management and the absence of subjects in the management system of nature ecology reserve. Therefore, it is the new breakthroughs for further reformation of natural protection management system.

3. Section II The "Agreement Protection" mechanism and pattern of Western conservation transition zone

3.1. Background

a.   Overview of the "agreement to protection"

The "agreement to protect" research on Mashan Village of Li County, in Aba State, Sichuan Province(Hereinafter called as " agreement to protect "research) is raised by "Beijing Natural Landscape Protection Center"(Hereinafter referred to as "Landscape").It advocates and encourages local governments to transfer collective woodland protection, and lets the community and the villagers participate in the protection action. The " agreement to protect "based on various incentives such as ecological compensation is aimed to expand the scope of protection, ease the contradiction between protection and economic development, and finally promote community socio-economic development.

Based on the "Mutual agreement protection" signed by the Forestry Bureau of Li County and the Landscape, the Mashan Village is chosen as the pilot village of "agreement to protect". In December 2008, Forestry Bureau of Li County (Party A) and Mashan Village (Party B) signed the "agreement to protect" which continued from December 2008 to December 2010. According to the research requirements and conventions, we made mid-term evaluation in November 2009 and annual assessment in August 2010.

b.   The nature and purpose of the study

Under the agreement, the implementations of the project are local government (Party A - Bureau of Forestry) and community residents (Party B - Horse Village community), and also invite independent experts to Party A, Party B and the third party (the "landscape" donors) group involved in the assessment. The study includes the implementation of the matters specified, the agreement's impact on community social and economic development ,and the agreement's effectiveness of community-based biodiversity protection etc..

There are three aspects of the study. Firstly, the study evaluated the effectiveness of the biodiversity conservation protection and socio-economic development; secondly, the study checked the process of the implementation, and provided an important reference for sustainable planning; thirdly, the study summarized the experiences and lessons of the protocol protected mode, timely detected problems and proposed recommendations for improvement, and finally provided technical support and enlightenment for the mode's demonstration and promotion.

c.   The purpose of the agreement to protection

Ecology, society and economy form a complete organic system, and there are close internal relations and interactions among them. Therefore, the eco-environmental protection and economic and social development of the Mashan Village should promote each other and complement each other. The research on "Agreement to protect" is based on interactions of ecology, society and economy, and the decisions are made for the following reasons.

1.   To regulate and restrain the villagers' behavior we should strengthen the protection awareness of the villagers, and let them recognize the vital and long-term interests they can get from the protection of the ecology; Only the villagers agree with the purpose of protection, will they take the appropriate measures and actions.
2.   To improve the quality of the villagers' living environment by the protection of the natural environment and the using of the new energy, we can improve the living and health conditions of Mashan Village ,which will contribute to improving the quality of the living environment of villagers.
3.   To provide financial compensation and maintain social stability. On the basis of the protection obligations and responsibilities on the villagers, we should provide financial compensation and spiritual encouragement to their protection behaviors. With a variety of compensation, we can resolve the contradiction between the protection and development, and fight for the win-win between community development and ecological protection.
4.   4) To protect and inherit Qiang culture. When it was designed, the agreement focused on the characteristics of the Qing culture and established cultural activity funds for purchasing necessary items, which provided important safeguard and support to the protection and promotion for national culture.

3.2. Natural and social status of Mashan Village

Observed from the perspective of the natural ecosystem's integrity and security, Mashan Village in Li County of Aba autonomous prefecture, Sichuan province, is a transition zone with great conservation value. Mashan Village is located in the arid valleys of the Hengduan

Mountainous and mountain forest Transitional Zone, the community is closed to the Wolong National Nature Ecological Reserve of Sichuan bordering on the Miyaluo Nature Ecological Reserve and Caopo Provincial Nature Ecological Reserve of Sichuan, which belongs to communities of Miyaluo Provincial Nature Ecological Reserve.

The topography in Mashan Village has created more abundant types of natural ecosystems, and gave birth to the rich resources of the species. Survey shows, there are 13 species of state emphatically protected plants and rare and endangered plants, 21 species of state-level protected wild animals and 18 species of state-level protected birds in nature reserve. The species of wild plants and animals under the first-grade state protection were *Fargesia spathacea Franch, Taxus chinensis, Rhinopithecus roxellanae, Ailuopoda melanolcuca, Panthera pardus, Panthera uncial, Moschus berezovskii, Moschus sifanicus, Budorcas taxicolor, Aqulia chrysaetos, Bonasa sewerzowi, Tetraophasis obscures, Lophophorus lhuysii* etc.

Mashan Village is a mixed village with Tibetan, Yi and Qiang where farmers rely on the natural environment, vegetation and animals, all of which are living resources of their lives. The nature ecological environment of community due to long-term deforestation, resulting in serious damage to vegetation, landslides, debris flows and other natural disasters, hillside planted almost reaped nothing. Over the past decade, local farmers mainly rely on government assistance. The farmers have not received ecological protection education and the technical training of the management and protection woodland, their unconscious awareness of participation in the protection of the ecological, turned a blind eye to community poaching, or even involved in illegal logging, all of which plunged the Miyaluo natural ecosystems security into a crisis.

How to change this situation, by whom or in what way to restore these damaged vegetation system? This section from the "agreement protection" mechanism and natural ecosystem co-management pattern of innovative practice, revealed that by optimizing the relations of production and a wider mobilization of social forces to participate in completion of the protection and management of natural ecosystems is the key to institutional innovation of the natural protection system.

3.3. "Agreement Protection" pattern and its characterisitics

"Agreement protection" is a new mechanism for biodiversity conservation. The concept is proposed by the Conservation International (Conservation International CI). Agreement protection is the protection right that aimed at separating ownership from managerial authority of protective area on the premise of unchanged ownership of woodland. And then participation management and protection of subject whose responsibilities, powers and interest are consolidated in the form of "agreement", and protection right and managerial authority are turned over to communities and farmers which promised to undertake management and the responsibility of protection, as well as any organization willing to undertake the responsibility of protection, and ultimately we achieve the goal of effective protection and development of natural ecosystems.

The author participated in the baseline survey and final research assessment of the project which called "Mashan Village in Xue town of Li County wild plants and animals agreement Protection projection" (hereinafter referred to "Protection projection") in Aba autonomous prefecture, Sichuan province in 2008 and 2010, considering that the "Protection projection "is a beneficial exploration of Nature Ecological Reserve transition zone joint management and protection system innovation, and very worthy of studying and summarizing.

a.   "Agreement Protection" Pattern and Its Characteristics

Mashan Village's "Protection projection" operation pattern is based on the four subject containing community, farmers, government and society organizations which participate in operation pattern of joint management and protection (hereinafter referred to as "co-participatory pattern"). The pattern takes "Protection projection" as a support, the "agreement protection" as the ligament, agreed to communities with farmers (Party A) and Government (Party B) whose powers, responsibilities and interest in Protection projections. The basic features are as follows:

The first, the organization of the management of Protection projection is the common Management Committee including village committees and farmers representative of the Mashan Village, county-level Forestry Bureau and International Protection Organization; the second, the main role in the Protection projections: International Protection organizations are the sponsors and supporters of the project, county-level Forestry Bureau is the supporters and service providers of the project, communities and farmers are the specific actors of the protective action. The third, agreement protection content is developed by the members of the four parties and then was submitted to community farmers' general assembly for discussion to pass and practice. Fourth, "Protection" agreement clearly agreed responsibilities, rights and interest in order to make member's responsibilities and division clear. Under the framework of the agreement protection, county-level Forestry Bureau is responsible to assist the village committee of Mashan Village to build a conservation organizations and institutions including "patrolling management and protection group, "construction group of ecological forest and production forest", "oversight group of agreement Protection projection", "advocacy group of the national culture "and correlation protection institutions. International Protection Organizations and backbone of village committees are responsible for the coordination of Protection actions, and mobilize community residents' perceived protection and self-help management action and so on.

b.   The Comparison of the "Agreement Protection" Pattern and the Government-led Pattern

1.   Government-led pattern, in which government does as a dominant implementation of the protection and management of protected areas, and agreement protection pattern, in which the government, supporting and helping of social organizations, communities and farmers do as a dominant implementation of the protection and management of protected areas. Clearly defined communities and farmers are the main body of the natural ecological environment protection of community and the first responsible person status in the "agreement protection" pattern. The farmer should have right

relevant to the protection system to know, to speak, to participate, to make decision and management rights of democratic rights and civil rights, which completely changed the bystanders and saboteurs role of the community subject;

2. Agreement protection pattern changed the government-led functions. Government is no longer the only conservation managers in the reserve, but is important mentors, supporters and service providers of the ecological protection. Government provided adequate guidance, support and services to the protection action of communities and farmers, including construction of protection organizations, technical training, to guide management and provide appropriate compensation and activity funds. Moreover, through the guidance of the agreement Protection projection, the transition zone of protected areas is effectively integrated into the natural protection management system, which plays an important role to promote the integrity of the protected area system.

3. Agreement protection pattern introduce incentive and restraint mechanisms to encourage community farmers to join the protection action. Agreement Protection projection fully mobilize social resources ,establish various forms of incentives and constraints, assist Mashan Village community in establishing a stable monitoring and protection system, reward and punishment mechanisms implemented in the whole process of the protective action, not only have encouragement of material and spiritual dimensions but also incentives of technology, bonuses, and honors level. Moreover, it also provide with severe penalties to those who do not fulfill the agreement and not implement protection in accordance with the terms of the agreement. They can be used in the combination of incentive and restraint mechanisms, both simply and effectively.

In a word, "co-participatory protection pattern", on one hand, effectively gives the farmers a variety of democratic rights, increases their political status and enhances their sense of responsibility; on the other hand, objectively the effective means of rewards and punishments for the protective action which was smoothly carried out provide a mechanism guarantee and improve the farmers' enthusiasm, so this "agreement protection" pattern should receive good results.

3.4. "Agreement Protection "mechanism of Mashan Village

The "agreement protection" mechanism through helping the government to transfer the protection rights of woodland to the community organizations and farmers and by means of ecological compensation or a variety of other incentives promotes biodiversity protection and management of the Contract Area.

a. The Organizational Mechanisms and Decision-making Mechanism of Agreement Protection

Under the framework of agreement protection, Mashan Village firstly established organizational mechanisms and decision-making mechanism. In organization building, with the help of village committee, a organization carrier, it set up the "agreement protection" commission which is composed of governments, international protection organizations, village committees and farmers representative and strengthened the management of community protection organization. In the decision-making on major events, the village committee organizes villagers to discuss and make collective decisions, put the provisions of the agreement protection into system regulations of pacts agreed and implementation. Such as farmers can gather the fallen leaves and withered woods freely and fell the trees moderately in their own lands; through resolutions and announcements of the general assembly of farmers, require the implementation of strict state-owned forests and collective forest protection and management is required, and prohibited the community and non-community farmers into the protected areas for illegal logging.

b.   According to Local Conditions, Establishing Classified Management Mechanism

There are a total area of 1110.2 hectares in the protection zone of Mashan Village, involving 502.8 hectares collective forests land and 607.4 hectares state-owned woodland. In view of woodland species and current situation, the community takes in the charge of classified management and protection mechanism of forest land. First, the community regards the collective forests and state-owned woodlands destroyed or degraded as a "reforestation land", through hands out saplings technical aid replants and protects; second, regards the collective forests and state-owned woodlands in good condition as "protection woodlands" for strict patrol and monitoring; third, the community farmers have deserted its own forest land, through compensates walnut saplings and technical guidance to plant and develop economic forest.

The local Forestry Bureau provide technical training and guidance for the three types of woodland, and international protection organizations and government should give appropriate subsidies for protecting behavior to ensure that replanted forest saplings of the survival rate could be more than 90%.

The classification management and protection mechanisms not only can achieve the goals of protective project, but also help communities to build economic self-help system to solve the contradiction between ecological protection and economic development and inspire the communities to actively participate in the awareness and responsibility of the protective actions.

c.   Incentive and Restraint Mechanisms Ensure the "Protection projection" Carried Out Smoothly

Mashan Village "agreement protection" pattern runs in accordance with the principle of the "who builds, who benefits, who protects, who benefits". The incentive and restraint mechanisms are the key to ensure the protective action continued orderly development.

1.   Incentives mechanisms. " Agreement protection " lay down the incentives methods: First, incentives methods such as free distribute walnut saplings and provide cultivation technology training and guidance to help farmers to develop economic forest and injected "blood transfusion and hematopoietic" mechanism for forest property rights of farmers to stimulate recovery and development of eco-forest; second,

encourage community organizations to carry out publicity education activities of conservation actions by way of bonuses and honors; third, methods such as on-site technical training and passing on management experience are used to promote ecological forest protection and forest management; fourth, reward hard-working people and punish lazy people by way of labor subsidies to strengthen the organization and institutionalization of the patrol work to ensure that regulatory work is normal and orderly development; fifth, encourage the community to carry out national characteristics cultural heritage and conservation activities by way of presenting cultural and promotional materials. Sixth, timely periodic summary assessment in accordance with the requirements of the "Agreement", give a reward and honorary title to what are effective protective behaviors and encourage farmers to develop the initiative, consciousness, honor sense, pride of the community eco-protective behaviors.

2. Constraint mechanisms. Mashan Village has developed the appropriate discipline and penalties focusing on the work of supervision and patrol team: Firstly, lay down the rules of the patrol mission and the punishment method of deforestation behavior to restrain the lazy violations. Secondly, the behavior such as those who fail to complete the task on time and timely report and process when have problem, take measures of withholding and temporary withholding subsidies or bonuses, and formulate the stipulation of reorganizing the deadline. Thirdly, appraisal protection motion regularly according to the agreement, the condition such as woodland management and protection and fulfill the guardianship duty are not in place which temporarily have not reward, pending completion of the rectification and then reissue. At last, make strict rules for collective protection activities about the number of participants (higher than 51% of the village) and the number of participation times (more than 4 times / year), and collective rewards and punishments are linked.

3. The distribution of benefits mechanism. In order to defend the collective economic sustainable development and the sustainability of community conservation actions, protocol protection provisions, the expected return of 20% of walnut grove set-aside for collective provident fund which used to guarantee the healthy and sustainable development of community protection organization in the provision of agreement protection. The expected return distribution mechanism, fully affirmed the value of the existence of collective organization, embodies the concept which the organization led farmers to implement protection actions. Such an interest distribution mechanism, increase farmers' hope and enthusiasm for expected income.

3.5. The effect evaluation of Mashan Village "Agreement Protection " pattern

5-7 August 2010, the authors carried out field research for Mashan Village "agreement protection" pattern. Before the survey they used a random number table, randomly selected 50 samples in the village 94 farmers of the General Assembly on the public, and at the same time took 30 samples as a standby; After three days of the household survey, observation, typical interviews and focusing discussion, we collected 54 valid questionnaires copies. Aiming at the discussion topics of research, the author selected 5 key indicators of data processing to form survey results.

a. "Agreement protection" pattern promote the transformation of the socio-economic status of Mashan Village

1. "Agreement protection" pattern significantly increase the protection awareness of farmers

In the investigation, the author selected two indicators to study the change of protection awareness for the farmers, awareness and value of acceptance. The awareness is measured and assessed by the Protection projection, project scope, protected plants, village regulation and so on (see Figure3). The other is described through protective object, community organization, community culture and organizational activities and so on.

The survey results show that the farmers' protection awareness changes significantly. Implementation of the agreement Protection projection made communities and farmers show the master attitude to protecting responsibility and protective behaviors. Farmers have a clear understanding of the practical significance and long-term significance of the ecological protection behavior for themselves and the environment, farmers are willing to participate in patrol work, and they reached a consensus on the validity of agreement protection at 4 aspects.

The First, the natural disasters such as landslides, mudslides can be decreased by protecting forest land.

The Second, the forest land protection is our responsibility. We should do better on our own beneficial protection actions;

The third, the "agreement protection" pattern play a role in improving the natural ecological environment of the community. After two years of hard-working, community and ecological protection zone should be able to see more wild rare animals, and promote the economic development of tourism;

The fourth, the construction of Walnut economic forest is indeed a good project can benefit for the future generations.

2. The "agreement protection" pattern promoted the construction of community organizations system

Protection organizations is the most foundational and most important basic-level organization which mobilizes community farmers' positivity of protection action, organize publicity education, consolidate community the continuity of protection action, organizes publicity education, consolidates the sustainability of community protection action, and improve the protective effect. Because of the imperfect natural resource management system, as all rural communities throughout the country in the past, Mashan Village did not have protective organizations which allow community to participate. "Agreement protection" pattern promoted the system construction of community organization (see table 3).

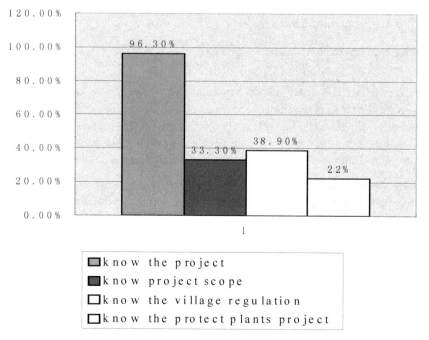

Note: Combined with questionnaire and interview survey comprehensive data, ① 96.3% of the farmers know the "agreement protection" project and its content; ② 54.5% of the farmers think that the woodland should be protected by themselves, 34% of the farmers believe that the woodland should be protected by the government and farmers together, 11.5% is not clear. ③ 61.8% of the farmers think that once found the actions such as deforestation and poaching will resolutely put a stop; Data derived from the obtained first-hand information by the authors' investigation

Figure 3. The awareness for studying protection consciousness

Method of data sources	Indicators: the construction of Organizations and institution
Interviews, Observations, Questionnaires,	The village regulation increased from 3 to 14, including the entire contents of community ecological protection and civilized behavior. Established 4 kinds of protective organizations: "the management and protection group of patrolling mountain", "ecological forest and economic forest construction group", "supervision groups of agreement Protection projection", "advocacy group of national culture". Put forward 4 systems and a require, that is "cultural propaganda system", "management system of ecological forest and economic forest construction", "supervision system", "supervision groups rules and regulations", and "a claim to the management and protection group of patrolling mountain" .

Note: Data derived from the obtained first-hand information by the authors' investigation

Table 3. Organizations and institutions building

The formulation of the organizations and systems effectively not only provide an important safeguard for participating, regulating and supervising protective actions, but also promise the way of making use of natural resources and management oversight, set up a woodland management and protection of technical training platform to improve the protection consciousness, awareness and capacity of farmers.

3.   "Mashan Village" pattern improved the protection ability of farmers

The survey shows that the behaviors such as unauthorized felling bamboo, illegal hunting, forest destruction, digging herbs randomly, picking up mushrooms freely have not happened in experimental communities in 2 years; community farmers are able to moderate harvesting wood according to the provisions of the agreement protection and dependence on forest resources significantly reduced. The "agreement protection" pattern enhance the protection abilities of the community farmers (see table 4).

Method of data sources	Index: the change of the protection abilities
Interviews, Observations, Questionnaires, On-site assessments	The behaviors of random cutting scarcely appear during the term of agreement protection. 15% of the households use their own planting trees as part of live energy resources. More than 95% of the households use electricity to cook food and boil water, even a small number of villagers use solar power. Forest fires did not happen in the village during the term of agreement protection

Note: Data derived from the obtained first-hand information by the authors' investigation

Table 4. The protecting abilities of farmers

4.   The "agreement protection" pattern promoted economic conditions diversification of farmers

Under the joint efforts of governments and international protection organizations ,"agreement protection " takes planting the economical trees as the breakthrough point to help the community farmers learned a variety of techniques such as woodland planting, management, disease prevention and patrolling to help farmers improve the value of knowledge of ecological forestry and the management level. In this paper, the author selected 4 indexes to research the changes of the economic development, that is the disposable income per household, living consumption expenditure per household, number of walnut trees per household and number of livestock per household(table 5).

Method of data sources	Index: the economic changes condition
Observations, Interviews, Questionnaires, On-site assessments	100 walnut trees planted per family. Farmers attach great importance to the walnut forest expected return, after two years, and every single walnut tree can increase the income of ¥200 - ¥500. The average annual income of walnut tree per household can reach about ¥30 000 per household. After Walnut trees mature, the average annual income will also increase more than 10%. Approximately 85% of the rural households are no more engaged in livestock breeding, which caused the annual income will be reduced about ¥500- ¥1000 per household. Reduce farming income has little effect on livelihoods while plays a significant role in protecting ecology. The structure of households' income which changes from "Logging and hunting forest and products income + original planted income + the original aquaculture income" two years ago into "conversion of cropland to forest subsidies + post-disaster reconstruction subsidies + wage income + the original pepper and other forest income". The household disposable income reached ¥15,000- ¥35,000 per year right now. The average household expenditure is mainly education and living expenses, and other expenditure performance was not significant

Note: Data derived from the obtained first-hand information by the authors' investigation

Table 5. The economic development changes of the farmers

The survey shows that the farmers' income structure has changed from aquaculture and migrant workers into the mainly farming gradually after the implementation of the "agreement protection " pattern.

b. The "agreement protection" pattern promoted the protection and management of natural ecosystems

1. Promoted the integrity of natural ecosystems. Transferred the transition zone of protected areas over to community to manage which can make up for the lack of traditional protected areas division and management system and also promote the returning to nature and integrity of protected areas ecosystems.

2. Fill the subject vacancy in the law. Integrating the community farmers into the category of management' main body is the breakthrough point of the management system. This not only solved the problem of government's functions pressure and shortage of human resources, but also more important thing is that the transition zone transformed from the vacuum state to the protection state. Farmers effectively have the democratic status related to protection, which have far-reaching social and political significance.

3. The management functions of government have changed. Government is no longer the protection functions that are monopolized, but is important advocates, supporters and service providers in the natural ecological protection. So they also look forward to the new mechanism to give a guarantee on the allocation of resources.

4. Conclusions

Mashan Village is a transition zone of protection areas which only have nearly 100 families and the transition zone area of 1, 000 hectares. The practical effect of the village in the implementation of the "agreement protection" pattern and mechanism is significant. First, it has expanded the area of the nature ecological reserves to alleviate the contradiction between the ecological protection, community production and economic development; second, it stimulated the protective awareness and consciousness of community organizations and farmers individuals, mobilized the ability of ecological protection and promoted the diversification of community protection organizations and personal relationships, enhanced the cohesion of the community farmers, responsibility and collective honor sense; Third, community and government are propelled to participate protection action together, objectively enhanced the positive interaction development of community ecology, economy, social; fourth, "agreement protection" pattern completely changed a single government-led management model. Similar to the major transition zone of topography and biodiversity of Mashan Village there are still have a lot in western china. Mashan Village "agreement protection" pattern provided creating the management system of nature ecological reserves major transition zone of western parts with strong evidence and basis.

Author details

Tan Jing, Tian Yan, Wang Shang-Wu
Sichuan Agriculture University, Sichuan, China

Feng Jie
Forestry Bureau Li County, Sichuan, China

5. References

[1] Wang Zhi, Jiang Ming-kang. Comparison of Chinese nature reserve classification with IUCN protected area categories. Rural Eco-environment,2004,(4)

[2] Xue Qiao, Fan Hua. Try commenting on the relationship that Conservation Steward Program with Forestry Ecological and environmental protection of Li county.Journal of Sichuan Forestry Science and Technology.2009.Vol.30, (3)

[3] Liu Jian-Guo, M. Linderman. Ecological Degradation of nature reserves: The case of Wolong giant panda Nature Reserve. Science, 2001 (4)

[4] Su Yang. The research and practice of harmonized development of the nature reserves and their ambient communities in West China. Sustainable development in China,2003(4)

[5] Chen Hong-Mei, Research on the Legal System of Nature Reserve's Management in China[D]. HoHai university.2006.(2)

[6] Li Yong-hua. Thoughts of Function and Intensive Management of Nature Protection Zone. Ecological Economy,2001,3,30~32

[7] Zhao Jun. Defects and Improvements of Public Participation s Principle in Chinese Environment Law[J]. Environmental Science and Technology,2005,28(2).

[8] Xia Shao-min, Yan Xian-wei. Management system of China's nature reserve. Journal of Zhejiang Forestry College,2009,26(1)

[9] Han Nian-Yong. A policy study on sustainable management for China nature reserves. Journal of Natural Resources,2000,3,32~37

Chapter 5

Effects of Disturbance on Sandy Coastal Ecosystems of N-Adriatic Coasts (Italy)

Gabriella Buffa, Edy Fantinato and Leonardo Pizzo

Additional information is available at the end of the chapter

1. Introduction

Due to human driving forces, many terrestrial habitats are undergoing striking modifications, destruction and fragmentation at an increasing and historically unprecedented rate (Sala et al., 2000), drawing attention to ecosystems' resilience as a necessary condition for both biodiversity conservation and sustainable development (McLeod et al., 2005).

Among the most endangered and threatened ecosystems worldwide, there are seashore, coastal sand dunes and nearby wet infradunal downs which are facing escalating anthropogenic pressures (Defeo et al., 2009), chiefly from coastal development, direct human use, mainly associated with recreation, and sea level rise.

All coastal European Countries, and particularly those of the Mediterranean Basin (Curr et al., 2000; European Environment Agency [EEA], 1999), suffer from the loss and degradation of sand dune landscape which are leading to a dramatic biodiversity loss, caused by the alteration and disappearance of many habitats and the rarefaction and/or local extinction of the most typical and extremely specialized native species, sometimes replaced with alien species.

Coastal dune systems make up 20% of the area occupied by the world's coastal landscapes (van der Maarel, 2003) and contain diverse and productive habitats important for human settlements, development and local subsistence (Schlacher et al., 2008). According to data reported by the United Nations Conference on Environment and Development [UNCED] (1992), about half of the world's population lives within 60 km of the shoreline, and it is likely to rise to three quarters by the year 2020. Population increase, united with economic progress and development, and the growing demands for spare time opportunities represent the eventual drivers of escalating pressures on sandy beaches (Dugan & Hubbard, 2010). Particularly coast-bound tourism, which became a mass phenomenon after World

War II, is now considered the primary cause of degradation of coastal dunes (Acosta et al., 2000). Among European holidaymakers more than 60% prefer the coast (European Community [EC], 1998) and even more people use sandy beaches which attract the greatest percentage of tourists every year (Davenport & Davenport, 2006; Schlacher et al., 2007). Rapid growth of human populations on the coast is then expected to further influence beaches and coastal sandy ecosystems with effects on biodiversity, community composition and ecological function (Defeo et al., 2009; Dugan & Hubbard, 2010).

As for the Italian coastline, more than 3000 km are represented by sand dune systems, which maintained well preserved morphological, hydrological and naturalistic features until the nineteenth century (Garbari, 1984). From the twentieth century on, they have been suffering from a strong intensification of human activities that mainly include forestry, agriculture, fisheries and aquaculture, transport and tourism, with consequent urbanization, trampling and beach cleaning. During the last century, a loss of about 80% of dune systems has been calculated for the Mediterranean area as a result of increasing urbanization (EEA, 1995).

Similarly, from 1950 onward, in the N-Adriatic region large stretches of coastal seashores, foredunes and infradunal downs have been fragmented by housing and resort development, road construction and coastal armoring, and the remaining sites suffer from increasing erosion, reduction in sand supply, alteration of geomorphic processes and heavy human use in the form of trampling, mechanical grooming and berm building (Nordstrom et al., 2009).

Coastal dune systems are typical transitional ecosystems, linked both to marine environment and terrestrial river basins, usually extending, narrow and long, along the coastline (Acosta et al., 2007), where environmental factors deeply influence their size, shape and boundaries. Like other ecotones, they exhibit a sharp gradient both in biotic and environmental factors, mainly related to substrate coherence and salinity, wind, salt spray and wave regime, which differ with distance from the water and topographic sheltering (Acosta et al., 2007; Carranza et al., 2008; Lortie & Cushman, 2007; Nordstrom et al., 2009; Ranwell, 1972). This steep gradient makes them highly dynamic systems deeply influenced by environmental stressors and drivers (Barbour, 1992), but at the same time, as abiotic patterns change within a short distance, it is responsible for the high level of ecological diversity, environmental heterogeneity and for the coexistence of different communities within a relatively limited space (Frederiksen et al., 2006; Martínez et al. 2004; Wilson & Sykes, 1999). Moreover, coastal dune systems are inhabited by extremely specialized biotic assemblages, rarely shared with adjacent terrestrial ecosystems (Defeo et al., 2009; Kutiel et al., 1999; Schlacher et al., 2008).

Healthy dune ecosystems also provide unique ecological services, such as erosion and salt spray control, storm buffering, water filtration and purification, nutrients mineralisation and recycling, coastal fisheries, functional links between terrestrial and marine environments, provisioning of crucial habitats for endangered species such as birds, as well as cultural services like recreation and education (Barbier et al., 2008; Defeo et al., 2009; Schlacher et al., 2008).

Furthermore, in Europe, coastal dune habitats are listed in the CORINE biotope classification and some of them are regarded as priority habitats or habitats of community interest in Annex I of the EU Habitat Directive (CE 43/92), recognized as a cornerstone of Europe's nature conservation policy (Feola et al., 2011; Mücher et al., 2004). This status implies they deserve special conservation attention and, as asked by the most recent European legislation, a strategic approach to planning and management in order to achieve sustainable development. As the assessment of the state of ecosystems at a given time and place is at the basis of any planning process for the management and conservation of natural resources, the aim of this study was 1) to provide a comprehensive and up-to-date outline of N-Adriatic sandy coastal landscape; 2) to assess the impacts of anthropogenic disturbance on sandy coastal systems by integrating landscape-level and community-level approaches; 3) verify how habitat configuration is important in supporting landscapes', communities' and plant species' diversity and quality.

2. Study site

The investigation was undertaken on the N-Adriatic coast which represents the longest sandy coastal line in Italy. The 100 km long coastline encompassing the study sites correspond to the Venetian portion of this system, isolated by other areas of sandy coastal plain by the estuaries of large rivers: the Tagliamento river northwards and the Brenta-Adige-Po rivers system southwards (Figure 1).

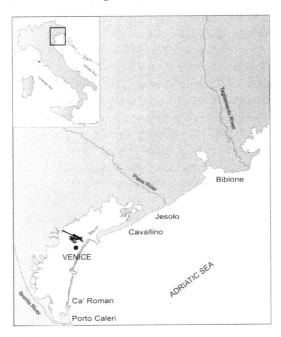

Figure 1. Map of the study site.

Sites consist of narrow, recent dunes (Holocene), that generally occupy a narrow strip along the seashore, bordered by river mouths and tidal inlets, mostly fixed by docks. The natural forces that shape and influence the dynamic of these sandy coastal systems are basically wind, waves and sea level fluctuations, climate and rivers run-off (Nordstrom et al., 2009). Sediments on the backshore and dunes are similar at all sites and are in the range of fine sand (Bezzi et al., 2009). Carbonate clearly rules the mineralogical composition of sands (especially in the northernmost area) due to the lithology of the catchment areas of corresponding rivers; southwards a slight magmatic component arises (Zunica, 1971). Dominant winds blow from the northeast and east (Bezzi et al., 2009). Annual average wave heights are lower than 0.50 m (Dal Cin & Simeoni, 1994); tides are semi-diurnal, with a spring range of about 1.0 m and a neap range of about 0.20 m (Polli, 1970); the combination of spring tides, winds and low atmospheric pressure can raise sea level up to 1.60 m.

Climate deserves a closer examination being one of the most characterizing aspects of the site. From a biogeographic point of view, the N-Adriatic seacoast can be included in the Eurosiberian region, Appennine-Balcanic province and Po-Valley subprovince. The mean annual temperature is about 13°C, with low winter (0.3°C) and high summer (17.7°C) values. Mean annual rainfall is 831.5 mm, with maximum precipitation (89.1 mm) in the spring-autumn season and a minimum (49.3 mm) in summer. Bioclimatic classification (Rivas-Martinez, 2008) shows a Temperate Oceanic type which allow reference to this area as the only sector of the Mediterranean Basin that does not belong to the Mediterranean climatic Region (Buffa et al., 2005; 2007; Sburlino et al., 2008).

Average annual values of both rainfall and temperature have been increasing since 1992 all along the Venetian coast: mean rainfall shows an increase from 688 mm in 1992 to the current 976 mm (p<0.05); mean temperature has grown from 13.04°C to 13.70°C (p<0.10). To evaluate a possible change in climate regime, we compared two time series monitored at the same site (weather station of Cavallino-Venice, located in the central part of the Venetian stretch), comprising the periods between 1961-1990 and 1992-2010 respectively. Although, the two mean annual precipitation do not show a significant difference (P_{61-90}=809.4 mm/P_{92-10}=845.5 mm; df=47; F=1.73; t=0.71), a seasonal rainfall redistribution has occurred with a shift from an oceanic regime to an equinoctial one that shows two maximum (in spring and autumn) and two minimum. Temperature has also changed: thermic regime remains similar, but with a general shift to higher temperature values; the mean annual temperature increased from 12.7°C to the current 13.5°C (p<0.01). Ombrothermic diagrams (Figure 2) allow us to highlight a change of climatic phase which shifted from Low Supratemperate upper subhumid climate to Upper Mesotemperate upper subhumid climate.

Besides the bioclimatic diagnosis, the variation of some bioclimatic indices (reported in Table 1) emphasises an increase of summer aridity, due to the simultaneous increase in temperature and decrease in precipitation, giving evidence to a process of "mediterraneisation" (Fernández-González et al., 2005). Moreover, the general warming lengthened the Period of Plant Activity (i.e., months with mean temperature >3.5°C, Rivas-Martinez, 2008) from 10 to 11 months.

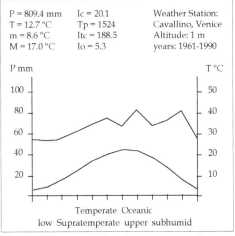

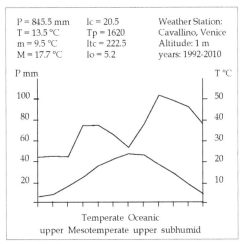

Figure 2. Bioclimatic classification of Cavallino weather station (central part of Venetian coastal stretch), for the period 1961-1990 (left) and 1992-2010 (right). For a detailed description of indices and general theoretical principles see Rivas-Martinez (2008).

variable	1961-1990	1992-2010
P mm (annual)	809.4	845.5
T °C (annual)	12.7	13.5
m °C (annual average of minimum values)	8.6	9.5
M °C (annual average of maximum values)	17.0	17.7
Tp °C (positive annual T, with $T_{i1-12} > 0°C$)	1524	1620
Ts °C (average T of summer three months)	651	686
Ps mm (average P of summer three months)	227.7	195.5
Ios3 (ombrothermic index (Pp/Tp) of summer three months)	3. 5	2.8

Table 1. Climatic parameters and bioclimatic indices. For a detailed description of indices and general theoretical principles see Rivas-Martinez (2008).

Many other studies identify the decade 1981–1990 as the onset of climate change in Europe, with a trend starting in the beginning of the 1970s (Werner et al., 2000). In the Veneto Region, this breakpoint is particularly evident for temperature and evapotranspiration: after the change point, in fact, temperatures show a significant increase (+1.5 and + 0.9 °C for yearly averages of maximum and minimum temperatures respectively relative to the previous phase) in all seasons and particularly clear in spring, summer and winter for maximum temperatures and in summer for the minimum ones (Chiaudani, 2008).

Human pressures are very high along the entire coastline. Until the 1950s, the Veneto coast was almost entirely fronted by dunes up to 10 m high (Bezzi & Fontolan, 2003; Pignatti, 2009), but few of them still survive and beaches generally suffer from the decrease of sediment supplies delivered by rivers, subsidence, and reduction in longshore sediment transport due to interruption by shore-perpendicular structures (Nordstrom et al., 2009).

To defend shorefront buildings and to provide space for leisure use, from 1950, large beach sectors have been protected, and still are, by different structures, such as groynes (shore-perpendicular constructions that catch sand moving alongshore), jetties and revetments, locally called "murazzi"; some of them, like in Pellestrina, were built by the Venetians in the 18th century and later rebuilt to larger dimensions (Nordstrom et al., 2009).

Most sites are managed by private corporations, through beach concessions by national government, and land behind beaches mostly developed as campsites, resorts, towns and villages; only very few sites (Ca' Roman, Porto Caleri and partly the "Laguna del Morto" near Eraclea) are still near-natural, undeveloped and underutilized. In 2011, beach summer tourism (from May to September) in the Veneto region numbered more than 25 million visitors. Summer beach tourism is one of the main resources of the region (Bezzi & Fontolan 2003) and facilities for accommodating people show an average density of about 76.3/100 sqKm. Private beaches management encompasses machine litter cleaning after major storm events and before the tourist season to clean and flatten the sand surface for tourist facilities and during the summer, plant litter and shell fragments are regularly removed and left outside the beaches (Nordstrom et al., 2009).

Therefore, the encroachment of human facilities has severely restricted the space available for natural landforms and vegetation, and environmental gradients have been largely truncated, fragmented or compressed. Scattered here and there, some stretches with high natural value still remain and host high levels of biodiversity. All these well preserved sites have been incorporated in Natura 2000 European network as Sites of Community Importance (SCI) and/or Special Protection Areas (SPAs) (Buffa & Lasen, 2010).

3. Landscape classification

3.1. Background

Modern ecology relies on the concept of ecosystem as a fundamental concept to consistent environmental policy making. As provided in Article 2 of the Convention on Biological Diversity (1992), ecosystem means a *"dynamic complex of plant, animal and micro-organism communities and their non-living environment interacting as a functional unit"*.

From a nature conservation point of view, main concerns should be focused on spatial extension of ecosystems and their quality, as well as on their adaptability and recovery potential. As different ecosystems types are not equally valuable or equally susceptible to human-induced environmental change, to ensure ecologically sound management aimed at sustainability and preservation of biodiversity, ecosystems need to be described, characterised and spatially located (Blasi et al., 2000; Rowe, 1996; Sims et al. 1996).

The increasing attention to biodiversity conservation and natural resources management has reawakened the interest in ecosystem classification and mapping, mostly focusing on flexible and vertical, namely hierarchical, methods (Acosta et al., 2005; Klijn & Udo de Haes 1994; Matson & Power, 1996; Zonneveld, 1995), which can help dealing with the complex and dynamic nature of ecosystems providing instruments to refer to any functional unit at any scale depending on the problem being addressed.

The basic idea is that different ecosystems are detectable as a function of their homogeneity, which depends on the scale of observation. The importance of scale is universally recognized as scale concerns all types of ecological data and it is a fundamental facet of ecological heterogeneity, whose interpretation depends on the level of observation established when studying an ecological system (Levin 1992; Rescia et al., 1997).

Hierarchical structuring of communities and ecosystems has been long recognized (Allen & Starr, 1982; O'Neill et al., 1986) and it basically means that, at a given scale, any biological system is composed of lower-level interacting components and, at the same time, is itself an element of a greater system (O'Neill et al., 1989). The recognition of spatially defined landscape units can thus be useful for stratifying landscapes into ecologically homogeneous units, whose patterns and functions, at each level, depend on both the potentiality of lower levels and the restraints imposed by higher levels (O'Neill et al., 1989). Ecosystem classification set up on the hierarchical concept can thus provide interconnected spatial units with different potential purposes, depending on the scale of the problems under investigation and the requisite precision of the results (Carranza et al., 2008).

Blasi et al. (2000; 2005) have recently proposed a hierarchical framework designed for describing and mapping Italian landscapes at different levels. It is a deductive and spatial explicit method based on the homogeneity of the physical environment aiming at defining land units with different vegetation potential (Blasi & Frondoni, 2011). According to the method, land attributes used for classifying landscapes are those widely recognized (Forman & Godron 1986; Zonneveld 1995) as ecologically relevant: climate, lithology, geomorphology, human activities, soil and vegetation, ordered from the most stable factors controlling processes occurring at larger ecological scales to more dynamic ones working at local levels, so reflecting their hierarchy both in time and space.

All these landscape attributes are integrated with the concept of potential natural vegetation (PNV). This concept is one of the most important concepts developed within plant ecology since it allows the provisioning of predictive models of plant communities dynamics (Biondi, 2011). The concept was firstly formalized by Tüxen (1956) and it can be defined as the plant community that would develop in a given habitat if all human influences would stop (Westhoff & van der Maarel, 1973); in other words, it delineates the spontaneous natural development of landscape within a homogeneous land unit. The concept of PNV correlates to that of vegetation series which is composed of dynamically linked plant communities developing into the same type of mature vegetation, i.e. the PNV (Biondi, 2011; Blasi et al., 2005).

The top-down, deductive approach supports the identification of homogeneous units which can be then classified according to the inner dynamic vegetation pattern.

From higher to lower level, Blasi's framework includes (Blasi et al., 2000; 2005):

- Land Regions (LR): represent the broadest level (scale >250.000); identified by means of macroclimatic features;
- Land Systems (LS): delimited according to significant lithological differences;

- Land Facets (LF): delimited through lithomorphological features (altitude, slope and aspect) and local bioclimatic types (rainfall and temperature regimes);
- Land/Environmental Units (LU): defined at a smaller scale (1:10.000/1:50.000), they are determined by a major vegetation series that evolves in one consequent PNV type.
- Land Elements (LE): correspond to the different plant communities.

Although successfully applied to different ecosystems (Acosta et al., 2003a; 2005; Blasi et al., 2004; 2005; Stanisci et al., 2005), the method appears less workable in sandy coastal landscape, where the identification of single land elements is quite impossible at the scale proposed. For this reason, according to the proposal of Carranza et al. (2008), in this study LU have been described through coastal dune geosigmeta which can be defined as a mosaic of adjoining vegetation series within a geomorphologic, biogeographical and bioclimatic unit (Biondi 1994; Blasi et al., 2005).

3.2. Application to N-Adriatic coastal dunes

Land Regions, Systems, Facets and Environmental Units of N-Adriatic coastal dunes were obtained by overlaying physical information digitalized as GIS data vector layers in ArcGIS 9.3 (Environmental Systems Research Institute [ESRI], 2008). As the entire studied area is included in the same macroclimate zone, starting thematic maps were a geological map (Bondesan et al., 2008), a geomorphological map (Bondesan & Meneghel, 2004) and a pedological map (Bini et al., 2002; Ragazzi et al., 2005; Ragazzi & Zamarchi, 2008). Bioclimatic characterization was set up according to Rivas-Martinez (2008) using data (1992-2010) from seven weather stations located all along the Venetian coastline.

Information concerning land elements distribution (potential coastal vegetation–zonation) was obtained using the phytosociological approach (Braun-Blanquet, 1964). Both up-to-date original data (41 surveys) and data from previous phytosociological studies (204 surveys) were used (Gamper, 2002; Gamper et al., 2008; Géhu et al., 1984; Poldini et al., 1999; Sburlino et al., 2008). Vegetation surveys (phytosociological relevés) were combined in a single matrix and classified by cluster analyses to be assigned to syntaxonomic taxa according to their floristic, structural and coenological features (Westhoff & van der Maarel, 1973).

3.3. Results

The entire study area belongs to the same Land Region and Land System (Table 2).

Rainfall and temperature regimes differentiated an upper Mesotemperate upper sub-humid climate northwards and a lower Supratemperate lower sub-humid climate, with a steppic variant, southwards allowing the identification of two different Land Facets (LF) which are separated by the Piave River that seems to act as a bioclimatic divide. Northern Land Facet (NLF) extends from the Tagliamento River to the Piave River; Southern Land Facet (SLF) from the Piave River to the Brenta-Adige-Po Rivers system in the south (see map in Figure 1).

Main land elements are summarized in Table 3 and grouped according to PNV classes.

LR	LS	LF	EU	LE
1. Temperate	1.1. Coastal sand dune	1.1.1. recent coastal dunes under upper Mesotemperate upper sub-humid climate	1.1.1.1. Coastal dune geosigmetum	1.1.1.1.1. Beach and mobile dunes
				1.1.1.1.2. Fixed dunes
				1.1.1.1.3. Dune slack transition to alluvial deposits
		1.1.2. recent coastal dunes under lower Supratemperate lower sub-humid climate, steppic variant	1.1.1.2. Coastal dune geosigmetum	1.1.1.2.1. Beach and mobile dunes
				1.1.1.2.2. Fixed dunes
				1.1.1.2.3. Dune slack transition to alluvial deposits

Table 2. Land Region (LR), System (LS), Facets (LF), Environmental Units (EU) and Elements (LE) of the Venetian coastal dune systems.

Each Environmental Unit is set up by three systems in contact with each other (Figure 3). The variation of structural types (namely, different life forms that dominate and characterize different plant communities) along the zonation is that typically found in sandy coastal systems worldwide (Biondi, 1999; Carboni et al., 2009; Frederiksen et al., 2006; Sykes & Wilson, 1991). The zonation develops moving inland from the sea edge along the steep environmental gradient, with the most pioneering annual communities on the beach and the woods in the inland sheltered zone.

Apart from the fixed zonation determined by their ecological requirements, N-Adriatic sandy coastal plant communities show a certain degree of uniqueness within the Mediterranean Basin (Pignatti, 1959; Géhu et al., 1984; Gamper et al., 2008; Sburlino et al., 2008), which is supported by various factors ranging from the present geographical and physical characteristics to the past climatic events that drove wide floristic migrations in Northern Italy. The synergy of all these features makes possible the presence of a wide range of species with different geographical distribution encompassing temperate, Mediterranean, western, eastern and mountain species (the latter mainly in the Northernmost part of the coast, carried downwards by torrential rivers like the Tagliamento and the Piave). This singular phytogeographic blend, also recognized for other north-eastern Italian ecosystems (Buffa & Villani, 2012), greatly increases the floristic value of this area, contributing to define plant communities and systems not found elsewhere (Lorenzoni, 1983; Buffa et al., 2007).

	Beaches and mobile dunes		Fixed dunes		Dune slacks transition	
Land facets	1.1.1	1.1.2	1.1.1	1.1.2	1.1.1	1.1.2
Land Elements	1.1.1.1.1	1.1.2.1.1	1.1.1.1.2	1.1.2.1.2	1.1.1.1.3	1.1.2.1.3
PNV classes						
Salsolo-Cakiletum maritimae	*	*				
Sporobolo-Agropyretum juncei	*	*				
Echinophoro-Ammophiletum australis	*	*				
Tortulo-Scabiosetum (e)			*	*		
Sileno coloratae-Vulpietum membranaceae			*	*		
Teucrio-Chrysopogonetum grilli (e) (1)					(*)	
Erico-Osyridetum albae (e)			*	*		
Viburno-Phillyreetum (e)			*	*		
Junipero-Hippophaetum (e)				*		
Vincetoxico-Quercetum ilicis (e)					*	*
Eriantho-Schoenetum nigricantis			*	*		
Juncetalia maritimae communities			*	*		
Phragmito-Magnocaricetea communities					*	*
Erucastro-Schoenetum nigricantis (e)			*			
Plantagini-Molinietum coeruleae (e)			*			
Carici elongatae-Alnetum glutinosae					*	
Alluvial deposits					*	*
Lacustrine deposits with coastal lagoons					*	*
Ancient dune					*	*

Left-side vertical groupings:
- *Annual and perennial grass, chamephytic, nanophanerophytic and phanerophytic vegetation*
  - *Annual and perennial grass vegetation* (Salsolo-Cakiletum maritimae; Sporobolo-Agropyretum juncei; Echinophoro-Ammophiletum australis)
  - *Edapho-xerophilous series* (Tortulo-Scabiosetum – Vincetoxico-Quercetum ilicis)
  - *Edapho-higrophilous series* (Eriantho-Schoenetum nigricantis – Carici elongatae-Alnetum glutinosae)
- *Contact Land Systems* (Alluvial deposits; Lacustrine deposits with coastal lagoons; Ancient dune)

Table 3. Potential natural vegetation types that characterize coastal dune ecosystem zonation on the N-Adriatic coast. Asterisks point to the presence of the different plant communities; (*e*) indicates endemic plant communities; (1) *Teucrio-Chrysopogonetum grilli* dry grasslands are present only on ancient dunes located out of the surveyed area.

Northern Land Facet

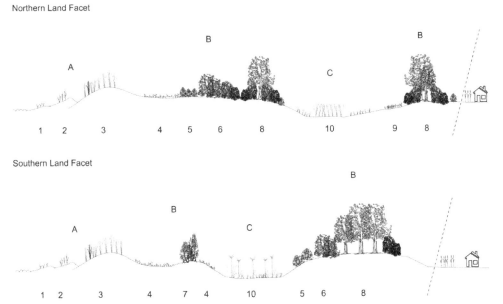

Southern Land Facet

Figure 3. Simplified representation of the Potential Natural Vegetation (PNV) along the N-Adriatic coast. A=Beach and mobile dunes; B=Edapho-xerophilous series; C=Edapho-higrophilous series. 1=Salsolo-Cakiletum maritimae; 2=Sporobolo-Agropyretum juncei; 3=Echinophoro-Ammophiletum australis; 4=Tortulo-Scabiosetum; 5= Erico-Osyridetum albae; 6= Viburno-Phillyreetum angustifoliae; 7= Junipero-Hippophaetum fluviatilis; 8= Vincetoxico-Quercetum ilicis; 9=Teucrio-Chrysopogonetum grilli; 10=higrophilous mosaic.

The sandy coastal system begins with the pioneer, nitrophilous community dominated by annuals of the strandline zone (*Salsolo kali-Cakiletum maritimae* plant community); being exposed to wave inundation, salt spray and wind stress, the community is often patchy and fragmented. Beach land elements are species-poor communities, since few species can survive the stress and disturbance of sand mobility and salt spray (Nordstrom et al., 2009). Pioneer plants of *Salsolo-Cakiletum* are tolerant of salt spray and sand blasting and contribute to the formation of embryo dunes on the backshore while grasses form foredune ridges (Seabloom & Wiedemann, 1994). The *Cakile maritima* plant community is then followed by that of embryo dunes dominated by dune-forming plants such as *Elymus farctus* (*Sporobolo-Agropyretum juncei*). On mobile dunes (white dunes) the *Ammophila arenaria* community (*Echinophoro spinosae-Ammophiletum australis*) establishes; *Ammophila arenaria* is the dominant species and is responsible for stabilizing and building up the foredune by capturing blown sand and binding it together with its tough, fibrous rhizome system (Chapman, 1976). Beaches and mobile dunes soils are shallow, sandy, calcareous, mesic *Typic Xeropsamments* or *Typic Udipsamments* (Bini et al., 2002), depending on local topographical conditions, and have very low organic carbon content and scarce horizons differentiation.

Landward of the foredune crest, increased protection from physical stresses allows the development of woody shrubs in the seaward slopes of the dune and trees and upland species in the landward portions. The inner, more stable dunes host more developed soils: they are thicker and have better horizons differentiation with blocky structure, higher organic carbon content and higher Available Water Capacity (AWC). Depending on the soil moisture regime, they can be classified as *Typic Haploxerepts, Arenic* or *Typic Eutrudepts* (Bini et al., 2002). Consequently, vegetation evolves towards more structured forms, such as medium and high shrubs and ends in the forest dominated by holm-oak. Then, moving inland, mobile dunes systems are spatially replaced by the edapho-xerophilous series of fixed dunes, entirely composed of N-Adriatic endemic communities (Gamper et al., 2008). The series begins with a perennial dry microprairie (*Tortulo-Scabiosetum*) dominated by dwarf shrubs, perennial herbaceous species, mosses and lichens, that covers fixed dunes (grey dunes). On inland ancient dunes, it is replaced by an endemic dry grassland (*Teucrio capitati-Chrysopogonetum grylli*), whose structure is mainly characterized by perennial herbaceous species and lower dwarf shrubs cover. The coastal zonation ends in the dune slack transition with *Quercus ilex* woods (*Vincetoxico-Quercetum ilicis*). This community is currently present although scattered and fragmented as a consequence of agricultural claim, but it shows a good recovery potential under the canopy of senescent pine forests.

Intermediate communities (*Viburno-Phillyreetum angustifoliae* and *Erico-Osyridetum albae*) and the pseudo-macchia (*Junipero-Hippophaetum fluviatilis*) of the seaward side of the semi-consolidated dunes and fixed dunes, exposed to wind action, complete the zonation. Woody communities contribute to outline the biocenotic uniqueness of N-Adriatic coasts relative to Mediterranean coastal sand dune systems: particularly the widespread presence of *Juniperus communis* ssp. *communis* suggests links towards Atlantic coasts rather than Mediterranean ones (Gamper et al., 2008).

Interdunal depressions are colonized by wet grass communities (the edapho-hygrophilous series and the sub-halophilous mosaic of wet infradunal downs). The most interesting are the hygrophilous natural fens (*Erucastro-Schoenetum nigricantis*) and the semi-natural meadows *(Plantagini altissimae-Molinietum caeruleae)*, both endemic, growing on neutral to subalkaline soils enriched in organic matter. *Molinia caerulea* grasslands are a very rare semi-natural community, found where the water table is close to the surface, and its long term conservation needs constant agricultural management such as regular mowing. Managing slowdown determines littering, auto-manuring and development of common shrubs and woody vegetation. Soils of interdunal lowlands somehow resemble those of shallow dunes (coarse texture, subalkaline reaction, scarce horizons differentiation), but with an aquic soil moisture regime, reducing conditions and a slightly saline water-table, at least close to the coastline. Under these conditions, the *Schoenus nigricans* and *Erianthus ravennae* community *(Eriantho-Schoenetum nigricantis)* is the most widespread, although patchy and fragmented.

Inland wetland areas present counteracting aspects, since they are of interest both as regulators of hydrological conditions and for biodiversity conservation. They are characterized by soils with coarse-loamy texture, subalkaline reaction, organic matter accumulation, and a fresh water-table close to the surface (Bini et al., 2002). Plant communities change according to the water-table level and the nutrient content, from the hydro-hygrophilous cane brake (*Phragmition communis* communities) to fens and wet meadows (*Caricion davallianae, Magnocaricion elatae* and *Molinietalia caeruleae*). Further development would foresee the marshy willow shrub (*Frangulo-Salicetum cinereae)* and the black alder wood (*Carici elongatae-Alnetum glutinosae*), nevertheless still found only in a small patchy area southward.

Coastal dune vegetation has been often described as azonal (Acosta et al., 2003a; Buffa et al., 2007); this is mostly true for communities of beaches and mobile dunes which actually have been proved to have a wide geographical range (Biondi, 1999; Carranza et al., 2008). On the contrary, foredune systems, and particularly the edapho-xerophilous series, are delineated by floristic and coenological features which are more related to local conditions such as climate, morphology, lithology and history. In the study area, all these aspects give rise to an outstanding environment so explaining the high level of cenological endemism.

Compared to the xerophilous series, wetland systems are structurally simplified and host a lower number of plant species, usually with wide distributional range (Buffa et al., 2007), but their presence, though extremely patchy and scattered, contributes to increase the N-Adriatic landscape richness and diversity.

Although the two Land Facets show the same systems sequence (beach and mobile dunes-fixed dunes-dune slacks transition) they slightly differentiate in terms of number of land elements and landscape diversity. It is worth noting that the simplified landscape representation drawn in Figure 3 and the series described in Table 3 only represent the potential natural landscape of N-Adriatic sandy costal system but almost nowhere does it actually express from beach to dune slacks transition completely.

Particularly the most complex communities, i.e. woody communities, have almost disappeared and persist only scattered and patchy. The pseudo-macchia (*Junipero-Hippophaetum*), which only 30 years ago was widespread (Géhu et al., 1984), currently exists just in one site (Porto Caleri), thanks to the establishment of a protected area in 1991.

Despite anthropogenic pressure and changes in the coastal dune environment of the N-Adriatic coastline, 20 EU habitats of interest (EU 43/92) have been surveyed. Therefore, regardless of human pressures, the Venetian coastal ecosystem could be regarded as a biodiversity hotspot within the Mediterranean Basin and still conserves many valuable elements to be maintained and emphasized (Buffa & Lasen, 2010).

4. Assessing conservation status

4.1. Background

The increased rate of habitat change and natural resource utilization since the 1950s, and the consequent threats to biodiversity have led to increased concern for monitoring and protecting remaining natural areas.

N-Adriatic coastal dune system suffers from a severe and complex human utilization; meanwhile it holds high landscape, faunal and floral values. These characteristics make it an ideal site to test an analytical approach to conservation status assessment to provide management policy which takes into account ecological values, landscape complexity and driving processes.

In the late 1980', Franklin et al. (1981) recognized three primary attributes of ecosystems: composition, structure, and function, which frame and make up the biodiversity of an area. Composition concerns the identity and variety of elements and includes measures of species diversity; structure represents the physical pattern of a system, from habitat complexity to the landscape level; finally, function relates to ecological and evolutionary processes, such as gene flow, disturbances and nutrient cycling (Noss, 1990).

Recently, progress has been made in developing methods for monitoring compositional diversity and for assessing threats of individual species, mostly in support of the IUCN Red List (Akcakaya et al., 2000; IUCN, 2006). This approach has become so overriding, that ecosystems evaluation is commonly based on the proportion of the threatened or endemic species pool that they encompass (Bonn & Gaston, 2005). Much less advancement has been made in building up an adequate insight into structural and functional diversity and in developing sound methods for assessing threats to habitats and ecosystems. Consequently, structural simplification of ecosystems which leads to the disruption of fundamental ecological processes can remain unappreciated.

The well known complexity of biodiversity and its hierarchical organization make clear that composition, structure and function of ecosystems are interdependent, nested and bounded (Noss, 1990; Margules & Pressey, 2000). These two concepts lead to two main consequences: first, because of the complexity of biodiversity, conservation status assessment can be defined using surrogates or partial measures such as sub-sets of species, species assemblages and habitat types (Hermy & Cornelis, 2000; Margules & Pressey, 2000). In particular, plant communities and vegetation, owing to their specific nature, can be regarded as good indicators of overall biodiversity and specifically of ecosystem integrity of coastal dune ecosystems (Araújo et al., 2002; Carboni et al., 2009; Géhu & Biondi, 1994; Géhu & Géhu, 1980; Lopez & Fennessy, 2002). Second, hierarchy theory suggests that monitoring and assessment can not be limited to one single level of organization and that different levels of resolution are proper for different questions (Noss, 1990; Bonn & Gaston, 2005). Hence, a thorough analysis of the conservation status of landscapes should take into account multiple levels of biodiversity, from community level to the entire landscape, considering compositional and structural features as well. This can be performed choosing a consistent and multiscale set of key indicators, based on field data, embodying the entire complexity of the vegetation system (Carboni et al., 2009; Cingolani et al., 2010; Hermy & Cornelis, 2000).

4.2. Methods

According to a previous research (Buffa et al., 2005) bound to a small coastal area, the conservation status of N-Adriatic coastal landscape has been estimated at two different interconnected scales: at landscape and plant community level. The method has been slightly redefined following the recent proposal by Carboni et al. (2009) and Grunewald & Schubert (2007).

Assessment method grounds on the basic idea that the severe and stressing dunal environment is the dominant factor leading not only to the presence of highly adapted and specialized species and communities (which refers to the identity and variety of elements) but also to a typical and worldwide sea-inland spatial zonation of plant communities (i.e., the physical pattern of the system), which in absence of disturbance events tends to be fixed.

We limited our evaluation only to those PNV with a xeric soil moisture regime, namely "beaches and mobile dunes" and the edapho-xerophilous series of fixed dunes and dune slack transition (see Table 3); as for the edapho-higrophilous series, it was included in "natural surface" category.

4.2.1. Landscape level

Analysis were carried out on the basis of a digital map of the area (1:10.000) encompassing the major natural land cover types as well as artificial surfaces. Natural land elements belonging to coastal zonation were expressed in terms of EU habitats (following EU Habitat Directive).

The map was derived from panchromatic digital aerial ortho-photographs (dated 2010) with a resolution of 6000 x 5600 pixels, covering 1500 m wide stretch from the coastline inward. Land cover was manually interpreted on video, by means of a Geographic Information System (ArcGIS software 9.3), and field survey. The legend follows CORINE land cover expanded, where possible, to a fourth level of detail. If a CORINE land cover category embodied more than one EU habitat (i.e., 3.2.2.2 Termophilous shrubs), it has been accordingly split up.

GIS analysis tools were used to calculate some common landscape metrics (Table 4) regarding both composition and structure. Landscape metrics were figured out for the entire studied coastal stretch to define an overall conservation status; at the same time they allowed a comparative evaluation of the status of PNV land elements and landscape inside the two Land Facets recognized in the area.

Composition	Structure
Total LC type richness	Total number of patches
Natural LC type richness	Number of natural patches
Natural coastal LC type richness	Number of natural coastal patches
Total surface (ha)	Average natural patch size (ha)
Percent natural surface	Mean shape index for natural patches (MSI)
Percent natural coastal vegetation	
Percent urban surface	
Percent agricultural surface	
Shannon Diversity Index (H)	
Eveness (J)	

Table 4. Indices of landscape composition and structure used for assessment.

Landscape composition was evaluated in terms of richness and abundance of the main land cover categories. Particularly urban surface abundance, which causing physical changes strongly influence natural habitat, has been deepened as an indicator of human pressure to natural landscape (Margules & Pressey, 2000; McKinney, 2002). Finally, richness and abundance parameters of each single category have been utilized to compare Shannon diversity index (H) and evenness (J) of the two Land Facets.

Landscape structure was analyzed by means of statistics such as number of patches of each land cover category, mean patch size and their mean shape index. Patch shape and size are important structural features of the landscape mainly related to the concept of "edges", which have been recognized as functional components of the landscape (Cadenasso & Pickett, 2001; Forman, 1995), influencing fluxes of organisms, material, and energy between two adjacent habitats.

Patches spatial arrangement allow the quantitative measurement of heterogeneity of a landscape and the comparison of landscapes, while shape index evaluates landscape configuration in terms of the complexity of patch shape (McGarigal & Marks, 1995). Patton's shape index (Patton 1975) was chosen, among others, because it measures the complexity of patch shape compared to a standard shape (a circular standard) the same size, thus alleviating the size dependency problem. As natural communities of coastal ecosystems usually run stretched out parallel to the coastline, then high MSI values will testify for natural patterns, while less natural typologies or fragmented patches, with more isodiametric or round forms, will show lower MSI values (Carboni et al., 2009).

4.2.2. Community level

Number of categories, their proportions and diversity represent non-spatial system properties (Gustafson, 1998) mostly linked to compositional features, but composition being equal a landscape may exhibit many different patch arrangement, that is many different spatial configuration. When undisturbed, plant communities of sandy coastal ecosystems show a typical distribution pattern along the sea-inland gradient which can thus represent a reference model for evaluating actual spatial configurations.

Spatial configuration integrity was measured by means of richness of boundaries, n (Rescia et al., 1997) and the gamma connectivity, γ (Acosta et al., 2000; 2003b; Forman & Godron, 1986). According to Acosta et al. (2003b), the two indices were calculated through a one-dimensional approach which involves the projection of the plant communities found along sea-inland transects perpendicular to the coastline. Therefore, each transect presents a specific spatial sequence of plant community patches, which can be compared with the reference model.

The number of different types of boundaries (i.e. contacts between plant community types) along each transect represents n, the richness of boundaries, while gamma connectivity index refers to the position of a patch type in relation to other patch types and determines the boundary between patches and the links among them. Gamma connectivity index ranges from 0 (no links among patches) to 1 (every patch is linked to every possible patch), and higher values are normally considered an index of better environmental quality

(Forman, 1996). On the contrary, in coastal dune systems, where plant communities tend to have a strong linear distribution, the best structural quality reflects the natural and undisturbed sequence which has a low number of fixed links. Therefore, higher connectivity values are usually associated to disturbance events which destroy or modify the natural sequence. The two indices have been calculated along 30 transects arranged along the coastline at 3500 m interval, 10 in NLF and 20 in SLF.

Besides, for each plant communities we computed some indicators of species diversity and vegetation quality. We estimated the diversity of plant communities by calculating the diversity index and evenness, following the proposal by Grunewald & Schubert (2007) who adapted Shannon diversity and evenness index specifically for coastal dunes, H_{dune} and E_{dune}, incorporating the parameter "species density" (plant cover relative to the plot-size).

As aliens and ruderal species are predicted to increase as a result of increasing human disturbance (Richardson et al., 2000; Sax & Gaines, 2003), an effect in overall diversity may be expected. Recently, Carboni et al. (2009) proposed relating the number of species of a broad distribution type or of exotic origin, which are generally introduced and/or favoured by human disturbance, to the number of species of a chorological type characteristic for the examined area and strictly dependent on the studied region. As one of the main characteristics of N-Adriatic region is just the co-occurrence of many different chorological types, choosing the most typical was not possible. Therefore, to provide information about quality, or inversely, about the level of "anthropogenization" we calculated a natural diversity index (N) (Grunewald & Schubert, 2007). According to a common procedure in ecological and conservational studies (e.g., Grime, 2002; Martinez et al., 2004; Rodgers, 2002; Rodgers & Parker, 2003), the degree of natural diversity (N) was calculated by classifying species into typical native dune species and untypical dune species often associated with non-dune habitats and disturbed, nutrient-rich sites, including truly alien species as well. Being the ratio between diversity index H_{dune} computed with all species and that without "alien" species, the index N can be read as a sound evaluation of natural diversity; only a maximum value of one can be reached (if no species are excluded = complete natural diversity) (Grunewald & Schubert, 2007).

The significance of overall difference among communities in terms of diversity index, evenness and index N was assessed by one-way ANOVA on transformed data, in order to determine differences at the critical significance level $p<0.01$. One-way ANOVA on transformed data was also used to check differences in quality (H_{dune}, E_{dune}, N_{dune}) between the two coastal LF at the critical significance level $p=0.05$. Finally, we tested differences in the spatial patterns (connectivity and richness of boundaries) and quality (H_{dune}, E_{dune}, N_{dune}) of the two Land Facets comparing values through independent groups t-tests (Sokal & Rohlf, 1995).

4.3. Results

4.3.1. Landscape level

Land cover classification allowed the identification of 30 different CORINE Land Cover Types (Table 5), but as they could embody more than one habitats, total number of land categories summed up to 40.

Level 1	Level 2	Level 3	Level 4
1. Artificial surfaces	1.1. Urban fabric	1.1.1. Continuous urban fabric	
		1.1.2. Discontinuous urban fabric	
	1.2. Industrial commercial and transport units	1.2.1 Industrial or commercial units	
		1.2.2. Road and rail networks and associated lands	
		1.2.3. Port areas	
		1.2.4 Airports	
	1.3. Mine, dump and construction sites	1.3.2 Dump sites	
		1.3.3. Construction sites	
	1.4. Artificial non-agricultural vegetated areas	1.4.1 Green urban areas	
		1.4.2. Sport and leisure facilities	
2. Agricultural areas	2.1. Arable land	2.1.1. Non-irrigated arable land	
	2.2. Permanent crops	2.2.4 Other permanent crops	
	2.4. Heterogeneous agricultural areas	2.4.3. Land principally occupied by agriculture, with significant areas of natural vegetation	
3. Forests and semi-natural areas	3.1. Forests	3.1.1 Broad-leaved forests	3.1.1.1 Quercus ilex woods (9340)
			3.1.1.6 Higrophilous woods
		3.1.2. Coniferous forests	3.1.2.1. Pine woods (2270)
	3.2. Shrub and/or herbaceous vegetation association	3.2.2. Moors and heathland	3.2.2.2 Termophilous shrubs (2250/2160) (three different plant communities)
		3.2.4. Transitional woodland/shrub	
	3.3. Open spaces with little or no vegetation	3.3.1. Beaches, dunes, and sand plains	3.3.1.1. Open sand (1210)
			3.3.1.2. Partially vegetated dunes (2110)
			3.3.1.3. Densely vegetated dunes (2120)
			3.3.1.4. Moderately vegetated slacks (2130)
			3.3.1.5 Interdune annual grasslands (2230)
4. Wetlands	4.1. Inland wetlands	4.1.1. Inland marshes	
		4.1.2. Peatbogs	
	4.2 Coastal wetlands	4.2.1. Salt marshes	
		4.2.3. Intertidal flats	
5. Water bodies	5.1. Inland waters	5.1.1. Water courses	
		5.1.2. Water bodies	
	5.2. Marine waters	5.2.1 Coastal lagoons	

Table 5. CORINE land cover categories surveyed in the study area. For Level 4 correspondence with sandy coastal EU habitats is reported; numbers in brackets correspond to Natura 2000 codes. Note that some Natura 2000 habitats can be represented by more than one plant communities.

Total surveyed area resulted in about 15.800 ha; 30.3% is covered by urbanized surface (mainly represented by towns and villages, roads and tourist facilities) and 22.2% by agricultural areas (mostly arable lands); only 47.5% is included in natural or semi-natural categories, of which only 4.5% represented by dune systems (Figure 4 and Table 6).

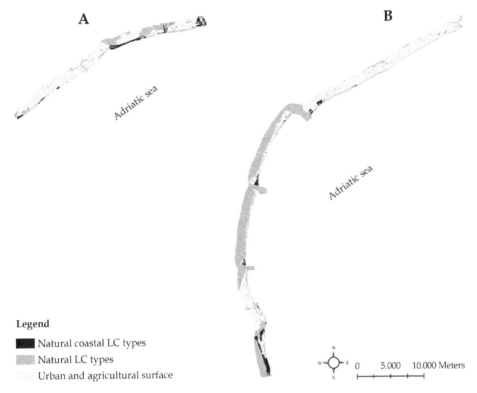

Figure 4. Land cover maps of the two Land Facets showing the main three categories as derived from the CORINE land cover map. A: Northern Land Facet, from Tagliamento River to Piave River; B: Southern Land Facet, from Piave River southwards to Brenta River.

Northern Land Facet (NLF) covers a surface of 4848 ha and extends from the Tagliamento River, in the north, to the Piave River; Southern Land Facet (SLF) covers a larger surface of 10942 ha, from the Piave River to the Brenta-Adige-Po Rivers system in the south.

As for composition, the two Land Facets resulted very similar. In NLF, the natural typologies were 25, 8 of which were coastal ones; the SLF showed 25 natural typologies as well, 9 of which of natural coastal LC types. Abundance of natural LC types discriminated between the two Land Facets: of the total surface surveyed, only 36% was included in natural or semi-natural categories in the Northern part, while in the Southern more than 50% of the total surface fell in this category. Interestingly enough, while urban surface is quite similar (32% in the north, 29% in the south), urbanization pattern differentiates; large and compact cities (such as Bibione, Caorle and Eraclea) predominate in the NLF, while in the SLF urban settlements permeate the landscape with some larger cities (like Jesolo) along with a continuous presence of houses, little villages and tourist facilities, whose spread is favoured by a more developed roads network. The situation is much impressive in the central part of SLF, corresponding to Venice barriers, where urbanization and widespread leisure facilities join a limited barriers width, leaving few sites for undisturbed natural vegetation. Therefore, although percentage of natural surfaces is lower in the Northern LF, they suffer from lower human pressure. Moreover, it is worth noting that the higher average size of natural patches in the SLF (Table 6) is mostly due to the presence of some categories such as coastal lagoons (i.e., particularly the Venice lagoon, in the central part) and salt marshes which usually cover broad areas.

variable	total	NLF	SLF
Composition			
Total LC type richness	40	36	37
Natural LC type richness	27	25	25
Natural coastal LC type richness	9	8	9
Total surface (ha)	15790.96	4848.79	10942.17
Percent natural surface	47.5	35.7	52.7
Percent natural coastal vegetation	4.5	5.8	3.9
Percent on natural surface		16.1	7.4
Percent urban surface	30.3	31.8	29.6
Percent agricultural surface	22.2	32.5	17.6
Shannon Diversity Index (H) (total surface)		2.52	2.34
Eveness (J) (total surface)		0.70	0.65
Shannon Diversity Index (H) (natural surface)		2.56	1.17
Eveness (J) (natural surface)		0.80	0.36
Structure			
Total number of patches	8100	2790	5310
Number of urban patches	4397	1652	2745
Number of agricultural patches	507	119	388
Number of natural patches	3196	1019	2177
Number of natural coastal patches	458	214	244
Average natural patch size (ha)	2.35	1.697	2.651
Average natural coastal patch size	1.54	1.303	1.743
Mean shape index for natural patches (MSI)	2.32	2.82	2.08
Mean shape index for natural coastal patches (MSI)	2.11	2.18	2.05

Table 6. Landscape composition and structure indices for total area and for each Land Facet examined (NLF=Northern Land Facet; SLF=Southern Land Facet).

Total Shannon diversity index, calculated taking into account the proportions of each land cover category, reflects the situation, with a slightly higher diversity northwards. The dominance of coastal lagoons and correlated salt marshes also affects evenness, which is slightly lower for the SLF than for the NLF. Shannon diversity index and evenness, calculated taking into account only the proportions of natural land cover category, still strengthen differences (Table 6) between the LF.

The analysis of the entire area resulted in the identification of a total of 8100 patches, 3654 of which were natural ones, 458 of which represented by coastal vegetation patches (Table 6). The NLF coastal landscape was much more structured and heterogeneous, especially considering the natural coastal patches. In this area number of patches is lower, but they sum to 16% of the total natural patches cover against only 7% of natural coastal patches in the SLF.

The higher heterogeneity reflects on the average size of natural coastal patches, which is little more than a hectare for the former and nearly two for the latter (*t*-test, p<0.05). Mean shape index does highlight significant differences in the shape of the natural coastal patches of the two areas (*t*-test, p<0.01), giving evidences, in the north, of a more natural landscape, where heterogeneity depicts typical features of coastal environments rather than a fragmented landscape, fruit of human disturbance.

Zooming inside the two land elements (beaches and mobile dunes and the edapho-xerophilous series of fixed dunes) helps to further clarify differences between Land Facets. Beaches and mobile dunes are much more abundant in the SLF, both in number of patches (123 against 35) and cover (25% against 15%), with a remarkable 10% covered by *Ammophila arenaria* community, which in NLF is almost replaced by an annual, nitrophilous community (*Sileno-Vulpietum*). Conversely, edapho-xerophilous series is much more widespread in the northern part and its importance is further underlined by the percentage cover of *Tortulo-Scabiosetum* community, one of the rarest and most vulnerable coastal plant association of the Veneto Region, whose abundance is nearly three time than in the south.

4.3.2. Community level

The coastal zonation comprises ten plant communities grouped in nine different EU habitats, as code 2250 embodies both *Erico-Osyridetum* and *Viburno-Phyllireetum* communities. They range from the pioneer beach communities to the holm-oak wood in the inland part.

Plant species richness typically increased following the sea-inland gradient and ranged between 19 species in 29 samples from the upper beach (the *Cakile maritima* community) to 84 species in 38 samples from the transition dune vegetation (*Tortulo-Scabiosetum* community), with a slight decrease in the stabilized dunes (in particular *Quercus ilex* wood).

As expected, H_{dune} index followed the same trend (Table 7), evidencing highly significant differences between communities of beaches and mobile dunes and those of the edapho-xerophilous series, with higher values that concentrated in the intermediate communities of the latter (ANOVA, p<0.0001, d.f. 195, F=27.09). On fixed dunes, plant succession leads to more mature and complex plant communities than on mobile dunes, and more species are characteristically present. Evenness also changed from young successional stages to more mature stages. In the species poor communities of mobile dunes, dominance by one or two species is usually high. Once plant succession has led to more complex plant communities with higher mean coverage, interspecific competition for limited resources becomes the dominant factor limiting dominance and favouring equidistribution of species. Index of natural diversity (N) showed an inverse trend and beaches and mobile dunes land elements were much more subject to invasion by human-favored ruderal or alien species. On the contrary, apart from *Tortulo-Scabiosetum*, intermediate and mature communities of the edapho-xerophilous series highlighted a very low degree of anthropogenization.

Quality of the vegetation	H $_{dune}$	E $_{dune}$	N $_{dune}$
Salsolo-Cakiletum	0.583a	0.743a	0.702a
Sporobolo-Agropyretum	0.906b	0.779ab	0.673ab
Echinophoro-Ammophiletum	0.804abc	0.459c	0.619abc
Sileno-Vulpietum	0.91bcf	0.79ab	0.373g
Tortulo-Scabiosetum	1.540d	0.516cd	0.621abc
Erico-Osyridetum	1.252efg	0.575cdef	0.984de
Viburno-Phyllireetum	1.806deh	0.627ef	0.986def
Juniperum-Hippophaetum	1.519de	0.527cde	0.964d
Vincetoxico-Quercetum	1.426degh	0.402ce	0.931def

Table 7. Indices of overall quality of plant communities of coastal zonation (paired *t*-tests; where letters differ, values are significantly different at p < 0.01).

As for the analysis of the spatial arrangement of communities along the zonation no significant differences emerged between the two Land Facets neither in the spatial connectivity index, nor in richness of boundaries (Table 8).

variable	NLF	SLF	sig.
Spatial structure			
Richness of boundaries (n)	5.429	4.455	0.030
γ sea-inland	0.537	0.603	0.325

Table 8. Indices of spatial structure of the vegetation of the two Land Facets analysed (significance values refer to *t*-tests for independent groups; df = 28).

Richness of boundaries was hardly higher for the NLF as a consequence of the higher heterogeneity of this area, but the difference is only slightly significant. Moreover, the relatively high connectivity average values gave evidence for disturbance for both Land Facets, causing fragmentation of the communities and their repeated presence along the transects.

In Figure 5, some explanatory transects have been reported as examples. The comparison to reference model depicted in Figure 3 described two different disturbance patterns which confirmed the spatial and structural analysis. Transects of NLF mostly lacked first terms of zonation and the sequence started with open sand immediately followed by the edapho-xerophilous series and the edapho-higrophilous mosaic (both fresh and sub-halophilous communities where water-table was slightly saline). Conversely, in SLF edapho-xerophilous series is well conserved only in a few sites (see transect G); first terms of the sequence are much more widespread and sometimes well developed (see transect F), but in general their width is compressed by impending human settlements (see transect E).

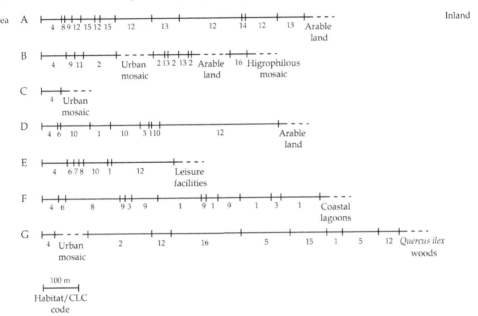

Figure 5. Schematic transects reporting sequence of natural and artificial land cover categories. Transects A and B are representative of NLF; in particular, transect A comes from the northernmost part corresponding to the Tagliamento mouth. Transects C-G come from SLF; F and G represent the sequence in Porto Caleri, at the southern boundary. 1= 3.1.1. Broad-leaved forests; 2=3.1.2. Coniferous forests; 3=3.2.4. Transitional woodland/shrub; 4=3.3.1. Beaches, dunes, and sand plains; 5=4.2.1. Salt marshes; 6=3.3.1.1. Open sand (1210); 7=3.3.1.2. Partially vegetated dunes (2110); 8=3.3.1.3. Densely vegetated dunes (2120); 9=3.3.1.4. Moderately vegetated slacks (2130); 10=3.3.1.5. Interdune annual grasslands (2230); 11=3.2.2.2. Termophilous shrubs (2250/2160) (three different plant communities); 12=3.1.2.1. Pine woods (2270); 13=4.1.1. Inland marshes; 14=4.1.2. Peatbogs; 15=3.1.1.6. Higrophilous woods; 16=3.1.1.1. Quercus ilex woods (9340).

Diversity and evenness indices, calculated taking into account only the proportions of coastal communities found in the two Land Facets, and particularly N_{dune}, the index of natural diversity, confirmed differences (Table 9). Apart from evenness, which never differentiated (E and E_{dune}), other indices were significantly higher for the dune system of NLF than for that of SLF indicating a richer and more diverse vegetation for the former. The better conservation status of the edapho-xerophilous series, which have been mostly eliminated in the southern area as a consequence of agricultural activities and leisure facilities development, is probably responsible for such higher diversity values.

The better conservation status in NLF has been also reinforced by the significant difference emerged for the N index. In this case, N represented the ratio between H calculated taking into account the cover of all communities and that without cover of "alien" communities. Values were significantly different and in the NLF, N index was nearly 1 (that is, almost no communities have been excluded), clearly highlighting how human pressures facilitate natural communities replacement by ruderal, not typical communities such as *Sileno-Vulpietum*, in the seaward slopes of the dunes, or by dynamic stages characterized by transitional woods and shrub communities, landwards.

variable	NLF	SLF	F	p
Shannon Diversity Index (H)	1.540	1.137	18.13	p<0,0001
Eveness (J)	0.582	0.550	n.s.	n.s.
H_{dune}	1.521	1.041	24.24	p<0,0001
E_{dune}	0.553	0.604	n.s.	n.s.
N	0.927	0.677	34.14	p<0,0001

Table 9. Differences in quality between NLF and SLF (one-way ANOVA; df=195).

5. Discussion and conclusion

Coastal areas are responsive systems affected by natural as well as anthropogenic pressures. Specifically, coastal sand dune dynamics, not only that linked to human disturbance, but even the natural cyclic dynamics (i.e. wave and tidal regime, sediment budget with the recurrence of regression and accretion phases), is associated with visible modifications in both plant communities and landscape, so that this sound relationship can be used as a monitoring tool in coastal areas (Araùjo et al., 2002). Particularly plant communities, that represent well-identifiable land elements with a relatively stable composition, structure and mutual relationships, all related to specific environmental conditions, can provide reliable monitoring activities (Loidi, 1994; Lomba et al., 2008).

At present, plant communities zonation along the N-Adriatic coast is complete only at few sites, mostly located at some distance from areas of urban development, where tourism is limited by legislation (Porto Caleri) or simply because of the difficulty in reaching them (i.e., Ca' Roman and the Tagliamento mouth). In almost all the rest of the Veneto coast, actual vegetation zonation is noticeably far from the potential one. Where disturbance events were very strong, some PNV communities completely disappeared and available space has been occupied by replacement ruderal communities.

The worst conservation status has been found on the central part of Venetian coastline, affected by urbanization, agriculture as well as heavy tourist presence, and here the natural landscape appears much more endangered in terms of trampling and alien invasions linked to human settlements and in some portion it has completely disappeared.

Factors causing the disturbance are of various types and act at multiple temporal and spatial scales, translating into effects that manifest themselves differently in space and time.

Moreover, it can be very difficult to distinguish actual sources of impact and to separate their individual effects (e.g., human trampling vs. erosion or embankment removing foredunes) (Defeo et al., 2009), or to find unaltered and natural beaches that could act as good control sites.

In Veneto region, most of the coastline is increasingly deprived of sand since groynes, jetties and revetments trap sediments that would otherwise supply beaches; activities such as land reclamation, urbanisation, afforestation and agricultural use further interfere with the sediment budget (Nordstrom, 2000). As a result, nearly 50% of the Venetian beaches are experiencing erosion (Bondesan & Meneghel, 2004), which mostly impacts beaches and embryonic dunes. Human's response to beach erosion and retreat has historically consisted in placing armouring structures (Charlier et al., 2005), that alter the natural hydrodynamic system of waves and currents, modifying sand transport rates, which in turn affect the erosion/accretion dynamics (Miles et al., 2001), possibly causing further deep habitat changes.

Particularly embryonic dunes are landfoms strongly related to beach dynamics and can thus be regarded as effective geo-indicators of coastal evolution. Absence of embryonic dunes is typical of coastal areas suffering from erosion phenomena which can be recognized through geomorphologic features, but as eroding beaches become narrower, the reduced surface directly reduce diversity of ecosystems, particularly in the upper intertidal zone (Dugan et al., 2008). Coastal erosion can thus lead to structural modifications in terms of denudation of some sites, thus truncating the coastal sequence completely removing the first terms (i.e., *Cakile maritima* and *Elymus farctus* communities) or drastically reducing their space thus causing communities merge together. Geomorphic events in mobile dunes even induce severe changes on inland areas as the defensive physical barrier provided by mobile dunes weakens. Disruptions of mobile dunes promote erosional gaps on the fixed dunes and a decline in vigour of *Ammophila arenaria*. As its resilience declines, marram is joined by more species, first by other specialised dune plants, then by less specialised grasses, drought-tolerant annuals such as those of *Sileno-Vulpietum*, a community rich in ruderal taxa.

Erosion is probably responsible for the small number of beaches and mobile dunes patches in NLF, but human pressures, mainly through trampling and beach grooming, can also promote the disruption of embryonic and foredunes, thus acting in synergy with erosion. As already pointed out by several studies (Brown & McLachlan, 2002; Carboni et al., 2009; Grunewald & Schubert, 2007; Kutiel et al., 1999; but see also Bonte & Hoffmann, 2005 for further references), trampling and other recreation-bound human activities, such as beach cleaning are among the most severe factors impacting sandy shores resulting in fragmentation, communities merging and/or replacement, alien invasion and in an overall lowering of diversity values. In SLF, where beaches are mainly in accretion phase, human disturbance is particularly intense and beach communities have almost completely disappeared from many sites, substituted by ruderal replacement communities.

Strong accumulation of mobile sand in interior areas has severe consequences also on plant communities of grey dunes, which in connection with trampling, causes the occurrence of plant species that are normally typical of embryonic dunes and a decrease in number of character species. Within the edapho-xerophilous series, grey dunes (*Tortulo-Scabiosetum*) show the highest amount of alien species, both exotic and human-favoured ruderal species. Particularly dwarf shrubs, which represent the most typical component, seem to be the least tolerant plants relative to trampling (Cole, 1995).

While pioneer stages mainly suffer from coastal erosion and tourism, at the other extreme, fixed dune communities are affected mostly by urbanization. As erosion and tourism pressures truncate the first elements of the zonation, towns and villages, coastal roads, pines plantations and agriculture truncate the last stages of the typical zonation. Human disturbance on intermediate shrub communities is chiefly represented by urban development, campsites, leisure facilities, roads and afforestation, while urban development and cultivated land have drastically reduced the area covered by holm-oak woods and at present well-preserved woods survive at only very few coastal sites. The effects of disturbance on coastal dune ecosystem vary according to the severity of the disturbance, but on fixed dunes and dune slacks transition disturbance usually drives to the complete disappearance of natural communities.

Therefore, the primary long-term threat coastal sandy ecosystems are facing is a "coastal squeeze" (Defeo et al., 2009; Schlacher et al., 2007), which causes sandy systems to be trapped between erosion on the sea side and human settlements inlands, thus leaving no space for natural sediment dynamics.

The two nested levels of analysis, considering higher organization level, the landscape level (with both composition and structure evaluation) as well as lower level (community level), such as species diversity and vegetation quality, provided a mutual description and evaluation of the naturalness of coastal landscape in the two coastal sectors studied and a sound assessment of their conservation status.

Coastal sand dune landscapes hold habitats of high economic, social and ecological value on a worldwide scale. At the same time, they are among the most threatened ecosystems on a national and European scale, facing escalating anthropogenic pressures. Sand dune habitat loss and degradation is leading to a remarkable biodiversity loss, which in turn can result in irreversible damage to ecosystem functions and ecological services. While some existing geo-physical models can be applied to predict climate-related changes for coastal areas (Zhang et al., 2004), no equivalents exist for the ecological effects of global change which could lead to a significant net loss of dune areas over the next century. Because of the scale of the problem, interdisciplinary and innovative approaches are required and the continued existence of sandy coastal areas as functional ecosystems is likely to depend on direct conservation efforts, which will have to progressively incorporate ecological aspects of coastal landscapes (Schlacher et al., 2007).

The most recent European legislation, in particular the Marine Strategy Framework Directive, which encompasses and reinforces other previous EU Directives such as WFD 2000/60 and Habitat Directive 92/43, calls for a strategic approach to coastal zone management providing sustainable development. The integrated and ecosystem-based approach used in this paper fulfils most EU requirements for policy making and by dealing with the problem both through a community-oriented and a landscape pattern-oriented approach it can provide a comprehensive framework for sustainable coastal management and development and for the improvement of projects or actions supporting biodiversity and ecosystem services. This approach could thus represent an innovative tool for the sustainable management assessment as it provides clear and easy applicable monitoring instruments allowing planners and stakeholders to evaluate the effectiveness of different action plans. Provided a solid classification, plant community types can be considered highly reliable indicators of environmental status in coastal areas. Moreover, the hierarchical landscape classification, coupled with the PNV concept, results in a reference model for environmental monitoring of anthropogenic pressures on coastal areas, providing interconnected spatial units which help dealing with the complex and dynamic nature of ecosystems.

Author details

Gabriella Buffa, Edy Fantinato and Leonardo Pizzo
DAIS - Dept. of Environmental Sciences, Informatics and Statistics, Ca' Foscari University, Italy

6. References

Acosta, A.; Blasi, C. & Stanisci, A. (2000). Spatial connectivity and boundary patterns in coastal dune vegetation in the Circeo National Park, Central Italy. *Journal of Vegetation Science*, Vol. 11, pp. 149-154.

Acosta, A.; Blasi, C.; Carranza, M.L.; Ricotta, C. & Stanisci, A. (2003b). Quantifying ecological mosaic connectivity and hemeroby with a new topoecological index. *Phytocoenologia*, Vol. 33, pp. 623–631.

Acosta, A.; Stanisci, A.; Ercole, S. & Blasi, C. (2003a). Sandy coastal landscape of the Lazio region (Central Italy). *Phytocoenologia*, Vol. 33, pp. 715-726.

Acosta, A.; Carranza, M. L. & Giancola, M. (2005). Landscape change and ecosystem classification in a municipal district of a small city (Isernia, Central Italy). *Environmental Monitoring and Assessment*, Vol. 108, pp. 323-335.

Acosta, A.; Ercole, S.; Stanisci, A.; De Patta Pillar, V. & Blasi, C. (2007). Coastal Vegetation Zonation and Dune Morphology in Some Mediterranean Ecosystems. *Journal of Coastal Research*, Vol. 23, pp. 1518–1524.

Akcakaya, H.R.; Ferson, S.; Burgman, M.A.; Keith, D.A.; Mace, G.M. & Todd, C. (2000). Making consistent IUCN classifications under uncertainty. *Conservation Biology*, Vol. 14, pp. 1001–1013.

Allen, T.F.H. & Starr, T.B. (1982). *Hierarchy: perspectives for ecological complexity*, University of Chicago Press, Chicago.

Araújo, R.; Honrado, J.; Granja, H.M.; De Pinho, S.N. & Caldas, F.B. (2002). Vegetation complexes of coastal sand dunes as an evaluation instrument of geomorphologic changes in the coastline, In: *Littoral 2002. The Changing Coast.* EUROCOAST/EUCC, pp. 337–339, Porto, Portugal.

Barbier, E.B.; Koch, E.W.; Silliman, B.R.; Hacker, S.D.; Wolanski, E.; Primavera, J.; Granek, E.F.; Polasky, S; Aswani, S.; Cramer, L.A.; Stoms, D.M.; Kennedy, C.J.; Bael, D.; Kappel, C.V.; Perillo, G.M.E.; Reed, D.J. (2008). Coastal Ecosystem–Based Management with Nonlinear Ecological Functions and Values. *Science*, Vol. 319, pp. 321-323.

Barbour, M.G. (1992). Life at the leading edge: The beach plant syndrome, In: *Coastal plant communities of Latin America*, Seeliger U. (Ed.), pp. 291–307, Academic, San Diego.

Bezzi, A. & Fontolan, G. (2003). Foredune classification and morphodynamic processes along the Veneto coast (N. Adriatic, Italy), In: *MEDCOAST '03*, Őzhan, E. (Ed.), pp 1425–1434, MEDCOAST Secretariat, Ankara.

Bezzi, A.; Fontolan, G.; Nordstrom, K.F.; Carrer, D. & Jackson, N.L. (2009). Beach nourishment and foredune restoration: practices and constraints along the Venetian shoreline, Italy. *Journal of Coastal Research*, Vol. 56, pp. 287–291.

Bini, C.; Buffa, G.; Gamper, U.; Sburlino, G. & Zilocchi, L. (2002). Soils and vegetation of coastal and wetland areas in Northern Adriatic (NE Italy). 7th Int. Meet. Soils with Mediterranean Type Climate (Selected Papers), *Options Méditerranéènnes*, Ser. A, Vol. 50, pp. 31-36.

Biondi, E. (1994). The phytosociological approach to landscape study. *Ann. Bot.*, Vol. 52, pp. 135–141.

Biondi E. (1999). Diversità fitocenotica degli ambienti costieri italiani, In: *Aspetti ecologici e naturalistici dei sistemi lagunari e costieri*, Bon, M.; Sburlino, G. & Zuccarello, V. (Eds.), pp. 39-105, Arsenale Editrice, Venezia.

Biondi, E. (2011). Phytosociology today: Methodological and conceptual evolution. *Plant Biosystems*, Vol. 145, no. 1, pp. 19-29.

Blasi, C.; Carranza, M.L.; Frondoni, R. & Rosati, L. (2000). Ecosystem classification and mapping: A proposal for Italian landscapes. *Applied Vegetation Science*, Vol. 3, pp. 233–242.

Blasi, C.; Filibeck, G.; Frondoni, R.; Rosati, L. & Smiraglia, D. (2004). The map of the vegetation series of Italy. *Fitosociologia*, Vol. 41, no. 1 Suppl. 1, pp. 21-25.

Blasi, C.; Capotorti, G.; Frondoni, R. (2005). Defining and mapping typological models at the landscape scale. *Plant Biosystems*, Vol. 139, pp. 155-163.

Blasi, C. & Frondoni, R. (2011). Modern perspectives for plant sociology: The case of ecological land classification and the ecoregions of Italy. *Plant Biosystems*, Vol. 145, pp. 30-37.

Bondesan, A. & Meneghel, M. (Eds.). (2004). *Geomorfologia della provincia di Venezia*, Esedra Editrice, Padova.

Bondesan, A.; Primon, S.; Bassan, V. & Vitturi, A. (Eds.). (2008). *Le unità geologiche della provincia di Venezia*, Provincia di Venezia, Università di Padova, Venezia.

Bonn, A. & Gaston, K.J. (2005). Capturing biodiversity: selecting high priority areas for conservation using different criteria. *Biodivers. Conserv.*, Vol. 14, pp. 1083–1100.

Bonte, D. & Hoffmann, M. (2005). Are coastal dune management actions for biodiversity restoration and conservation underpinned by internationally published scientific research?, In: *Dunes and Estuaries 2005*, Herrier, J.-L.; Mees, J.; Salman, A.; Seys, J.; Van Nieuwenhuyse H. & Dobbelaere I. (Eds.), pp. 165-178, Proceedings International

Conference on Nature Restoration Practices in European Coastal Habitats, Koksijde, Belgium, 19-23 September 2005.

Braun-Blanquet, J. (1964). *Pflanzensoziologie*. 3rd edn. Springer, Wien.

Brown, A.C. & McLachlan, A. (2002). Sandy shore ecosystems and the threats facing them: some predictions for the year 2025. *Environ. Conserv.*, Vol. 29, pp. 62-77.

Buffa, G.; Mion, D.; Gamper, U.; Ghirelli, L. & Sburlino, G. (2005). Valutazione della qualità e dello stato di conservazione degli ambienti litoranei: l'esempio del S.I.C. "Penisola del Cavallino: biotopi litoranei" (Venezia, NE-Italia). *Fitosociologia*, Vol. 42, pp. 3-13.

Buffa, G.; Filesi, L.; Gamper, U. & Sburlino, G. (2007). Qualità e grado di conservazione del paesaggio vegetale del litorale sabbioso del Veneto (Italia settentrionale). *Fitosociologia*, Vol. 44, pp. 49-58.

Buffa, G. & Lasen, C. (2010). *Atlante dei siti Natura 2000 del Veneto*, Regione del Veneto - Direzione Pianificazione Territoriale e Parchi, Venezia.

Buffa, G. & Villani, C. (2012). Are the ancient forests of the Eastern Po plain large enough for a long-term conservation oh herbaceous nemoral species? *Plant Biosystems*, in press.

Cadenasso, M.L. & Pickett, S.T.A. (2001). Effect of edge structure on the flux of species into forest interiors. *Conservation Biology*, Vol. 15, pp. 91-97.

Carboni, M.; Carranza, M.L. & Acosta, A. (2009). Assessing conservation status on coastal dunes: A multiscale approach. *Landscape and Urban Planning*, Vol. 91, pp. 17–25

Carranza, M.L.; Acosta, A.T.R.; Stanisci, A.; Pirone, G. & Ciaschetti, G. (2008). Ecosystem classification for EU habitat distribution assessment in sandy coastal environments: An application in central Italy. *Environ. Monit. Assess.*, Vol. 140, pp. 99–107.

Chapman, V. J. (1976). *Coastal Vegetation* (2nd Ed.), Pergamon Press, Oxford.

Charlier, R.H.; Chaineux, M.C.P. & Morcos, S. (2005). Panorama of the history of coastal protection. *Journal of Coastal Research*, Vol. 21, pp. 79–111.

Chiaudani, A. (2008). *Agroclimatologia statica e dinamica del Veneto. Analisi del periodo 1956-2004*, Tesi di Dottorato di Ricerca in Scienze delle Produzioni Vegetali, indirizzo Agronomia Ambientale, Università degli Studi di Padova, Padova.

Cingolani, A.M.; Vaieretti, M.V.; Gurvich, D.E.; Giorgis, M.A. & Cabido, M. (2010). Predicting alpha, beta and gamma plant diversity from physiognomic and physical indicators as a tool for ecosystem monitoring. *Biological Conservation*, Vol. 143, pp. 2570–2577.

Cole, D.N. (1995). Experimental trampling of vegetation. II. Predictors of resistance and resilience. *Journal of Applied Ecology*, Vol. 32, pp. 215-224.

Curr, R.H.F.; Koh, A.; Edwards, E.; Williams, A.T. & Daves, P. (2000). Assessing anthropogenic impact on Mediterranean sand dunes from aerial digital photography. *Journal of Coastal Conservation*, Vol. 6, pp. 15–22.

Dal Cin, R. & Simeoni, U. (1994). A model for determining the classification, vulnerability and risk in the Southern coastal zone of the Marche (Italy). *Journal of Coastal Research*, Vol. 10, pp. 18–29.

Davenport, J. & Davenport, J.L. (2006). The impact of tourism and personal leisure transport on coastal environments: a review. *Estuarine, Coastal and Shelf Science*, Vol. 67, pp. 280–292.

Defeo, O.; McLachlan, A.; Schoeman, D.S.; Schlacher, T.A.; Dugan, J.; Jones, A.; Lastra, M. & Scapini, F. (2009). Threats to sandy beach ecosystems: A review. *Estuarine, Coastal and Shelf Science*, Vol. 81, pp. 1–12.

Dugan, J.E. & Hubbard, D.M. (2010). Loss of Coastal Strand Habitat in Southern California: The Role of Beach Grooming. *Estuaries and Coasts*, Vol. 33, pp. 67–77.

Dugan, J.E.; Hubbard, D.M.; Rodil, I.; Revell, D.L. & Schroeter, S. (2008). Ecological effects of coastal armoring on sandy beaches. *Marine Ecology*, Vol. 29, pp. 160–170.

EEA (1995). *Europe's environment: the Dobříš assessment*, EEA, Copenhagen.

EEA (1999). *State and pressures of the marine and coastal Mediterranean environment*, E. Papathanassiou & G. P. Gabrielidis (Eds.), European Environment Agency, Environmental assessment series No. 5, Available from http://reports.eea.eu.int/ENVSERIES05/en/envissue05.pdf

E.C., DG XXIII (1998). *Fact and figures on the Europeans on holiday 1997-98*, Eurobarometer, no. 48, Brussels.

ESRI (2008). *ArcGIS 9.3*, Environmental Systems Research Institute, Redlands, California.

Feola, S.; Carranza, M.L.; Schaminée, J.H.J.; Janssen, J.A.M. & Acosta, A.T.R. (2011). EU habitats of interest: an insight into Atlantic and Mediterranean beach and foredunes. *Biodivers. Conserv.*, Vol. 20, pp. 1457–1468.

Fernández-González, F.; Loidi, J. & Moreno Sainz, J.C. (2005). Impacts on plant biodiversity, In: *Evaluación preliminar de los impactos en España por efecto del cambio climático*, Moreno J.M. (ed.), pp. 183-247, Ministerio de Medio Ambiente, Madrid.

Forman, R.T.T. (1995). *Land mosaics: the ecology of landscapes and regions*, Cambridge University Press, New York.

Forman, R.T.T. (1996). *Land Mosaic*, Cambridge University Press, Cambridge.

Forman, R. & Godron, M. (1986). *Landscape ecology*, Wiley & Sons, New York.

Franklin, J.F.; Cromack, K.; Denison, W.; McKee, A.; Maser, C.; Sedell, J.; Swanson, F. & Juday, G. (1981). *Ecological characteristics of old-growth Douglas-fir forests*, USDA Forest Service General Technical Report PNW-1, 18, Pacific Northwest Forest and Range Experiment Station, Portland, Oregon.

Frederiksen, L.; Kollmann, J.; Vestergaard, P. & Bruun, H.H. (2006). A multivariate approach to plant community distribution in the coastal dune zonation of NW Denmark. *Phytocoenologia*, Vol. 36, pp. 321-342.

Gamper, U. (2002). *Caratteristiche ecologiche della vegetazione a carattere mediterraneo presente sul litorale sedimentario nord-adriatico (Veneto–NE-Italia), con particolare riguardo alle problematiche di conservazione della biodiversità fitocenotica*, Tesi di Dottorato in "Biologia ed ecologia vegetale in ambiente mediterraneo", Università di Catania.

Gamper, U.; Filesi, L.; Buffa, G. & Sburlino, G. (2008). Diversità fitocenotica delle dune costiere nord-adriatiche 1 – le comunità fanerofitiche. *Fitosociologia*, Vol. 45, pp. 3-21.

Garbari, F. (1984). Aspetti della flora e della vegetazione delle nostre coste marine. *Agricoltura Ambiente*, Vol. 23, pp. 45-48.

Géhu, J.M. & Géhu, J. (1980). A methodology for the objective evaluation of natural environments. Coastal examples, *Third Seminary of Applied Phytosociology "Biological Evaluation of the Territory through Biocenotic Indices"*, Institute Européen d'Ecologie, Metz, pp. 75–94.

Géhu , J.M.; Scoppola, A.; Caniglia, G.; Marchiori, S. & Gehu-Franck, J. (1984). Les systèmes végétaux de la còte nord-adriatique italienne. Leur originalité à l'échelle européenne. *Doc. Phytosoc.*, N.S. Vol. 8, pp. 486-558.

Géhu, J.M. & Biondi, E. (1994). Anthropization of Mediterranean dunes, In: *Alterazioni Ambientali ed Effetti sulle Piante*, Ferrari, C.; Manes, F. & Biondi, E. (Eds.), pp. 160–176, Edagricole, Bologna.

Grime, J.P. (2002). *Plant Strategies, Vegetation Processes, and Ecosystem Properties*, Wiley & Sons, Chichester.

Grunewald, R. & Schubert, H. (2007). The definition of a new plant diversity index "H' dune" for assessing human damage on coastal dunes—Derived from the Shannon index of entropy H'. *Ecological Indicators*, Vol. 7, pp. 1–21.

Gustafson, E. J. (1998). Quantifying Landscape Spatial Pattern: What Is the State of the Art? *Ecosystems*, Vol. 1, no. 2, pp. 143-156.

Hermy, M. & Cornelis, J. (2000). Towards a monitoring method and a number of multifaceted and hierarchical biodiversity indicators for urban and suburban parks. *Landsc. Urban Plan.*, Vol. 49, pp. 149–162.

IUCN, (2006). *Guidelines for Using the IUCN Red List Categories and Criteria. Version 6.2.* Prepared by the Standards and Petitions Working Group of the IUCN SSC Biodiversity Assessments Sub-Committee in December 2006. Available from http://app.iucn.org/webfiles/doc/SSC/RedList/RedListGuidelines.pdf

Klijn, F., & Udo de Haes, H.A. (1994). A hierarchical approach to ecosystems and its implications for ecological land classification. *Landscape Ecology*, Vol. 9, pp. 89–104.

Kutiel, P.; Peled, Y. & Geffen, E. (2000). The effect of removing scrub cover on annual plants and small mammals in a coastal sand dune ecosystem. *Biological Conservation*, Vol. 94, pp. 235–242.

Levin, S.A. (1992). The problem of pattern and scale in ecology. *Ecology*, Vol. 73, pp. 1943-1967.

Loidi, J. (1994). Phytosociology applied to nature conservation and land management. In: Song, Y.; Dierschke, H. & Wang, X. (Eds.), Proceed. 36th IAVS Symp. in Shanghai, East China Normal Univ. Press.

Lomba, A.; Alves, P. & Honrado, J. (2008). Endemic Sand Dune Vegetation of the Northwest Iberian Peninsula: Diversity, Dynamics, and Significance for Bioindication and Monitoring of Coastal Landscapes. *Journal of Coastal Research*, Vol. 24, no. 213 (Supplement 20), pp. 113-121.

Lopez, R.D. & Fennessy, M.S. (2002). Testing the floristic quality assessment index as an indicator of wetland condition. *Ecol. Appl.*, Vol. 12, pp. 487–497.

Lorenzoni, G.G. (1983). Il paesaggio vegetale nord Adriatico. *Atti Mus. civ. St. nat. Trieste*, Vol. 35, pp. 1-34.

Lortie, C.J. & Cushman, J.H. (2007). Effects of a directional abiotic gradient on plant community dynamics and invasion in a coastal dune system. *Journal of Ecology*, Vol. 95, pp. 468–481.

Margules, C.R. & Pressey, R.L. (2000). Systematic conservation planning. *Nature*, Vol. 405, pp. 243–252.

Martínez, M.L.; Psuty, N.P. & Lubke, R.A. (2004). A perspective on coastal dunes, In: *Coastal dunes. Ecology and Conservation*, Martínez, M.L. & Psuty, N.P. (Eds.), pp. 3-10, Ecological Studies, Vol. 171, Springer , Berlin.

Matson, B.E., & Power, R.G. (1996). Developing an ecological land classification for the Fundy Model Forest, south-eastern New Brunswick, Canada. *Environmental Monitoring and Assessment*, Vol. 39, pp. 149–172.

McGarigal, K. & Marks, B.J. (1995). *FRAGSTATS: spatial analysis program for quantifying landscape structure*. USDA Forest Service General Technical Report PNW-GTR-351, USDA, Washington DC.

McKinney, M.L. (2002). Urbanization, biodiversity, and conservation. *BioScience*, Vol. 52, pp. 883–890.

McLeod, K.; Lubchenco, J.; Palumbi, S. & Rosenberg, A. (2005). *Scientific Consensus Statement on Marine Ecosystem-Based Management*, Communication Partnership for Science and the Sea,   2005;   Available   from http://compassonline.org/marinescience/solutions_ecosystem.asp

Miles, J.R.; Russell, P.E. & Huntley, D.A. (2001). Field measurements of sediment dynamics in front of a seawall. *Journal of Coastal Research*, Vol. 17, pp. 195–206.

Mücher, C.A.; Bunce, R.G.H.; Hennekens, S.M. & Shaminée, J.H.J. (2004). *Mapping European habitat to support the design and implementation of a pan-European network: the PEENHAB project*, Alterra report 952, Wageningen.

Nordstrom, K.F. (2000). *Beaches and Dunes on Developed Coasts*. Cambridge University Press, Cambridge, pp. 338.

Nordstrom, K. F.; Gamper, U.; Fontolan, G.; Bezzi, A. & Jackson, N.L. (2009). Characteristics of Coastal Dune Topography and Vegetation in Environments Recently Modified Using Beach Fill and Vegetation Plantings, Veneto, Italy. *Environmental Management*, Vol. 44, pp. 1121–1135.

Noss, R.F. 1990. Indicators for monitoring biodiversity: a hierarchical approach. *Conservation Biology*, Vol. 4, pp. 355–364.

O'Neill, R.V.; DeAngelis, D.L.; Allen, T.F.H. & Waide, J.B. (1986). A hierarchical concept of ecosystems. *Monographs in Population Biology*, 23, Princeton University Press, Princeton.

O'Neill, R.V.; Johnson, A. R. & King, A.W. (1989). A hierarchical framework for the analysis of scale. *Landscape Ecology*, 3: 193–205.

Patton, D.R. (1975). A diversity index for quantifying habitat edge. *Wildl. Soc. Bull.*, Vol. 3, pp. 171-173.

Pignatti, S. (1959). Ricerche sull'ecologia e sul popolamento delle dune del litorale di Venezia. Il popolamento vegetale. *Boll. Mus. Civ. Venezia*, Vol. 12, pp. 61-142.

Pignatti, S. (2009). Com'è triste Venezia, soltanto mezzo secolo dopo. *Parchi*, Vol. 58, pp. 59-70.

Poldini, L.; Vidali, M. & Fabiani, M.L. (1999). La vegetazione del litorale sedimentario del Friuli-Venezia Giulia (NE Italia) con riferimenti alla regione alto-Adriatica. *Studia Geobotanica*, Vol. 17, pp. 3-68.

Polli, S. (1970). *Tabelle di previsione delle maree per Trieste e l'Adriatico Settentrionale per l'anno 1971*. Istituto Talassografico Sperimentale Trieste, Trieste, pp. 20.

Ragazzi, F.; Vinci, I.; Garlato, A.; Giandon, P. & Mozzi P. (Eds.) (2005). *Carta dei suoli del bacino scolante in laguna di Venezia*. ARPAV Osservatorio Regionale Suolo, Castelfranco Veneto, pp. 399.

Ragazzi, F. & Zamarchi, P. (Eds.) (2008). *I suoli della provincia di Venezia*. Servizio Geologico e Difesa del Suolo della Provincia di Venezia, ARPAV Osservatorio Regionale Suolo, Castelfranco Veneto, pp. 268.

Ranwell, D. (1972). *Ecology of Salt Marshes and Sand Dunes*. Chapman and Hall, London, pp.258.

Rescia, A.J.; Schmitz, M.F.; Martín de Agar, P.; de Pablo, C.L. & Pineda, F.D. (1997). A fragmented landscape in northern Spain analyzed at different spatial scales: implications for management. *J. Veg. Sci.*, Vol. 8, pp. 343–352.

Richardson, D.M.; Pyšek, P.; Rejmánek, M.; Barbour, M.G.; Panetta, D.F. & West C.J. (2000). Naturalization and invasion of alien plants: concepts and definitions. *Diversity and Distributions*, Vol. 6, pp. 93-107.

Rivas-Martinez, S. (2008). Global bioclimatics (Clasificación bioclimatica de la Tierra). Version 1/12/2008. Available from http://www.globalbioclimatics.org

Rodgers, J.C. (2002). Effects of human disturbance on the dune vegetation of the Georgia Sea Islands. *Phys. Geogr.*, Vol. 23, no. 1, pp. 79–94.

Rodgers, J.C. & Parker, K.C. (2003). Distribution of alien plant species in relation to human disturbance on the Georgia Sea Islands. *Diversity and Distributions*, Vol. 9, no. 5, pp. 385–398.

Rowe, J.S. (1996). Land classification and ecosystem classification. *Environmental Monitoring and Assessment*, Vol. 39, pp. 11-20.

Sala, O.E.; Chapin, F.S. III; Armesto, J.J.; Berlow, E.; Bloomfield, J.; Dirzo, R.; Huber-Sanwald, E.; Huenneke, L.F.; Jackson, R.B.; Kinzig, A.; Leemans, R.; Lodge, D.M.; Mooney, H.A.; Oesterheld, M.; Poff, N.L.; Sykes, M.T.; Walker, B.H.; Walker, M. & Wall D.H. (2000). Global biodiversity scenarios for the year 2100. *Science*, Vol. 287, pp. 1770-1774.

Sax, D.F. & Gaines, S.D. (2003). Species diversity: from global decrease to local increase. *Trends in Ecology and Evolution*, Vol. 18, pp. 561-566.

Sburlino, G.; Buffa, G.; Filesi, L. & Gamper U. (2008). Phytocoenotic originality of the N-Adriatic coastal sand dunes (Northern Italy) in the European context: The Stipa veneta-rich communities. *Plant Biosystems*, Vol. 142, pp. 533-539.

Schlacher, T.A.; Dugan, J.; Schoeman, D.S.; Lastra, M.; Jones, A.; Scapini, F.; McLachlan, A. & Defeo, O. (2007). Sandy beaches at the brink. *Diversity and Distributions*, Vol. 13, pp. 556–560.

Schlacher, T.A.; Schoeman, D.S.; Dugan, J.; Lastra, M.; Jones, A.; Scapini, F. & McLachlan A. (2008). Sandy beach ecosystems: Key features, management challenges, climate change impacts, and sampling issues. *Marine Ecology*, Vol. 29, pp. 70–90.

Seabloom, E.W. & Wiedemann, A.M. (1994). Distribution and effects of Ammophila breviligulata Fern. (American beachgrass) on the foredunes of the Washington coast. *Journal of Coastal Research*, Vol. 10, pp. 178–188.

Sims, R.A.; Corns, I.G.W. & Klinka, K. (1996). Global to local: Ecological land classification - Introduction. *Environ. Monitor. Assessment*, Vol. 39, pp. 1-10.

Sokal, R.R. & Rohlf, F.J. (1995). *Biometry*.W.H. Freeman, New York.

Stanisci, A.; Feola, S. & Blasi, C. (2005). Map of vegetation series of Ponza island (central Italy). *Lazaroa*, Vol. 26, pp. 93-113.

Sykes, M.T. & Wilson, J.B. (1991). Vegetation of a coastal sand dune system in southern New Zealand. *J. Veg. Sci.*, Vol. 2, pp. 531–538.

Tüxen, R. (1956). Die heutige potentielle natürliche Vegetation als Gegenstand der Vegetationskartierung. *Angewandte Pflanzensoziologie* (Stolzenau), Vol. 13, pp. 5–42.

UNCED (1992). *United Nations Conference on Environment and Development, Agenda 21, Chapter 17: Protection of the Oceans, All Kinds of Seas, Including Enclosed and Semi-Enclosed Seas, and Coastal Areas and the Protection, Rational Use and Development of Their Living Resources.* United Nations Divison for Sustainable Development, New York, pp. 42.

van der Maarel, E. (2003). Some remarks on the functions of European coastal ecosystems. *Phytocoenologia*, Vol. 33, pp. 187-202.

Werner, P.C.; Gerstengarbe, F.W.; Friedrich, K. & Oesterle, H. (2000). Recent climate change in the North Atlantic/European Sector. *Int. Journal of Climatology*, Vol. 20, no. 5, pp. 463-471.

Westhoff, V. & van der Maarel, E. (1973). The Braun–Blanquet approach. In: *Ordination and classification of communities*, Whittaker, R.H. (Ed.), pp. 617–726, Handbook of vegetation science, Vol. 5, Junk, The Hague.

Wilson, B. & Sykes, M.T. (1999). Is zonation on coastal sand dunes determined primarily by sand burial or salt spray? A test in New Zealand dunes. *Ecol. Lett.*, Vol. 2, pp. 233–236.

Zhang, K.Q.; Douglas, B.C. & Leatherman, S.P. (2004). Global warming and coastal erosion. *Climate Change*, Vol. 64, pp. 41–58.

Zonneveld, I. S. (1995). *Land ecology: An introduction to landscape ecology as a base for land evaluation, land management and conservation*, SPB Academic Publishing, Amsterdam.

Zunica, M. (Ed.) (1971). *Evoluzione dei litorali dal Tagliamento all'Adige con particolare riguardo ai lidi della Laguna di Venezia (Relazione definitiva)*. Min. Lav. Pubbl. Com. St. Provv. Venezia, Padova.

Chapter 6

Implication of Alien Species Introduction to Loss of Fish Biodiversity and Livelihoods on Issyk-Kul Lake in Kyrgyzstan

Heimo Mikkola

Additional information is available at the end of the chapter

1. Introduction

Lake Issyk-Kul in Kyrgyzstan is the second largest mountain lake in the world (after Lake Titicaca in South America). It is situated in a basin surrounded by high mountains. While its water level is at 1608 m altitude, the mountain ranges of Kungei Ala-Too in the north reach 4711 meters, and those of Terskei Ala-Too in the south 5216 m. These mountains represent the major part of the Issyk-Kul catchment of 22,080 km² and provide most of the water to the lake.

Issyk-Kul Lake is 180 km long and 60 km wide, its surface area is 6240 km², and the shoreline 670 km (Figure 1.). The mean depth is 280 m, and the maximum 702 m, making it the fifth deepest lake in the world. The area covered by a depth of 0-100 m represents 38% of the total area. This is the major production zone of the lake [1].

Native fish stocks in this high-altitude saline lake have been subjected to predatory pressure from large number of introduced alien fish species. Previous papers and fishers are convinced that these predators are the most destructive to fish biodiversity in the lake, but this study wants to raise also other reasons which could explain at least part of the loss of fish stocks. The rapid growth in human activities with the development of tourism industry; irrigation; water eutrophication and pollution, and climate change impacts are alternative factors this presentation focuses on. This chapter also reviews the fish stocks and fishery management measures to increase the fish yields at the Lake.

Measures taken in order to protect the decreased fish stocks and endemic fish species include a Moratorium for Artisanal and Commercial Fish catching for a period of 10 years (2003-2013). Despite of the Moratorium at least 500 people continue their activity as illegal fishermen. Impacts of illegal and over-fishing are evaluated as anthropogenic activities. It is also noted that the disintegration of the Soviet Union had profound economic and social effects on many of the newly independent transition economies, like Kyrgyzstan.

Figure 1. Lake Issyk-Kul is large like a sea. Opposite shore is often not visible. Photo: Azat Alamanov

2. Description of study region

2.1. Physical and chemical environment

Although located at a high altitude, Issyk-Kul Lake never freezes over. The water temperature does not fall below the temperature of maximum density (2.75 °C at the mineral concentration of 6‰) except in shallow Rybachinsky and Tyup bays. The climate is continental with a short hot summer and a cold long winter. In summer the surface water temperature in the central part reaches 18°C, in winter it is seldom above 4°C [2]. The temperature may drop by 12-14°C down to 50 m depth and a further 1.5-2°C to the depth of 100 m. The water layer at 100-200 m depth maintains almost a stable temperature with changes kept within 0.1-0.3°C [3].

The chemical composition of the lake is as follows in mg/l[-1]: Calcium – 121, Magnesium – 287, Sodium+Potassium – 1544, Chloride – 1596, Bicarbonate+Carbon trioxide – 318, Sulfate – 2102. Total cations: 1952, total anions 4016 [4]. So its mineral content is chloride/sulfate/sodium/magnesium based. With the drop in water level also comes a certain increase in salinity. Data from 1932 shows that the salinity measured 5.8‰, and by 1984 it had increased to 6.0‰. Over this period water level dropped by 2.5 m. Current measurements show that the salinity between October 2008 and November 2009 varied in Bosteri between 2 and 9‰ and the average was 5.9‰; which indicates that it is going down (Mikkola, unpublished). Since 1986, the decline in water level has stopped and the lake level has started to rise again. Low salinity (less than 20% that of seawater) indicates that in historical terms Issyk-Kul has only relatively recently become a closed lake. Hydrologists have suggested that, deep underground, the lake water filters into the Chui River. It looks as though the river Chui never "found" its way to the lake as the river bends a mere 4 km of the lake to the west, disappearing into the desert of Kazakhstan. During the very high water cycles the lake's water may overflow to the river through a natural depression – the Kutemaldu channel.

Currently the pH ranges between 7.7 and 7.9. The waters of Issyk-Kul are rich in oxygen, as a result of aeration and movement of lake waters. First of all, water is well oxygenated because it is regularly mixed by strong winds. During the warm period of the year, the surface water moves from the central part of the lake towards the shores and it is replaced by deeper cool water. In the middle of the lake the water is stratified down to 5-10 m whereas near the shores the thermal discontinuity is at a depth of 20-30 m owing to the warm water inflow. Apart from the central upwelling there is also lateral upwelling that is, caused by the wind driving surface water from the shore to the open parts of the lake [5]. Two major currents, driven by two wind streams, can almost always be observed: one follows the northern shore in a westerly direction, and the other flows east along the southern shore. The transparency of Issyk-Kul waters approaches that of seawater, and in the open part of the lake Secchi disc readings range from 30-47 m, but are reduced even down to 50 cm at the mouths of the inflowing rivers.

2.2. Biological features

Lake Issyk-Kul flora contains emerged macrophytes, like *Phragmites australis, Typha latifolia* and *Scirpus tabermaemontani* until the depth of 1.5 m. Submerged macrophytes like *Potamogeton pectinatus, Myriophyllum spicatum* and *Najas marina* and attached algae can go down to 30-40 m. Mean annual macrophyte production is about 277 g/m² [6]. *Characeae* are the most common macrophytes, representing 96% of the total annual macrophyte production, and are present in almost all plant associations. Four species of *Chara* grow in shallow water, and three benthic species exist further down. Dense growth of *Charophyte* green algae extends to 40 m depth. Issyk-Kul Lake water is rich in phytoplankton, with 299 identified species. Blue-green algae (*Cyanophyceae*) dominate, but their standing crop is low [7]. Phytoplankton production is at the level of 49 g/m³ [6].

Zooplankton includes 117 species and is dominated by *rotifers* (84%), followed by *cladocerans* (9%) and *copepods* (7%). Zooplankton production is 91g/m³ [6]. Zooplankton and phytoplankton distribution in the lake is uneven, with bays and shallows being richer than open water. *Arctodiaptomus salinus* is present in all parts of the lake and over the year it may represent 75-95% of the total number of zooplankton and 95-99% of the biomass. This species migrates during the night into the surface water where its concentration reaches up to 35,000 ind/m³ [8], thus representing an important food source for all plankton-eating fish like Issyk-Kul Dace *Leuciscus bergi*.

Zoobenthos comprises 224 species. Most benthos occurs between the shoreline and 40 m depth, which comprises the *Charophyte* zone. According to [6] the mean annual production of zoobenthos is 10 g/m³. It has been calculated that the average biomass of zoobenthos in the gulfs with open zones is 93.6 kg/ha [9]. *Chironomids*, mollusks, *gammarids* and *Mysis* comprise 75-80 per cent of the total. In the deeper zone beyond the zone of *Characeae* and down to about 70 m, the biomass is 2.5-3.5 g/m² and is dominated by *chironomids* and the mollusk *Radix auricularia*. Three *Mysis* species introduced into Issyk-Kul from Lake Balkhash, Kazakhstan, in 1965-1968 are now permanently established in shallows, mostly in 1.5-1.8 m depth, but reaching down to 10 m. Their mean biomass in such waters has been measured to range from 1.5 to 2.5 g/m² [10].

2.3. Distribution and abundance of fish fauna

The original fish fauna comprised twelve indigenous species and two subspecies particular to this lake (Table 1). The long historical and geographical isolation of the lake favoured the formation of endemic forms. This fauna is a typical example of the local Central Asian fish complex originating from Central Asian Mountain fauna (a term used by Berg, 1949), which is characterized by the presence of the *loaches* and *cyprinids*, with a small addition of *leuciscins* of Siberian origin. In the native fish fauna of the lake there were no predators although large Naked Osmans *Gymnodiptychus dybowskii* are said to feed partly on small fish [11].

Strictly endemic fish Schmidt's Dace *Leuciscus schmidti* is present throughout the shallow littoral zone but goes during winter down to 35-40 m. It appears in two forms, a common fast-growing lake form and a slow-growing bay form. The fast-growing form reaches 31 cm, a weight of 650 g, and age of 11 years. It spawns on stony beds at depths of 0.5-10 m between the end of March (water temperature 5ºC) and mid-May. Fecundity is 6,000-65,000 eggs per year. It feeds largely on *Characeae*, but also on mollusks. The slow-growing form is present throughout the shallows. It reaches a length of 23 cm, a weight of 220 g, and a maximum age of 13 years. It has a similar fecundity and it spawns on the same substrate as the other form, but later.

Issyk-Kul Dace was the dominant fish until 1997, when Schmidt's Dace became for the first time the most numerous commercial fish in the lake. Issyk-Kul Dace inhabits the whole littoral zone, but is more pelagic than Schmidt's Dace. During the winter it is found down to depths 120-150 m, and reaches a maximum body-length of 17.5 cm and weight of 60 g. It spawns in shallow waters at depths between 1-8 m, and feeds mostly on plankton. During recent years the number and distribution of this species have sharply declined.

There are two endemic species distributed in mountain waters of Middle and Central Asia. The Scaly Osman *Diptychus maculatus* inhabits high-mountain streams, but enters also into Lake Issyk-Kul. It can grow 50 cm long and weighs up to 1 kg. It feeds on vegetation and invertebrates. The fish spawns in the spring and summer. It has a dwarf form, which lives mainly in the incoming small rivers, and may not live in the lake. It does not exceed 25 cm in length and weighs less than 200 g.

The Naked Osman is found in mountain rivers and lakes (Figure 2.). It also appears in two forms: one inhabits in rivers and the other in lakes. Lake living fish appear to have two ecological morphs: a winter lake morph and a summer migratory morph which spawns in rivers with a sandy bottom. Forms and eco-morphs would indicate that taxonomic studies are needed. The winter morph spawns from February to April and its fecundity is 13,000-14,500 per year. The summer morph is smaller, has a lower fecundity of 5,500-12,000, and spawns from April until September [12,13]. Both morphs are omnivorous. In the lake it feeds mostly on mollusks over muddy and loamy bottoms at 15-30 m deep. The largest Naked Osmans in the lake attain the age of 20 years and can grow up to 60 cm long and 3 kg of weight. It was once important commercial fish in the Issyk-Kul Lake, but there are indications that it is close to extinction [14].

Figure 2. The first Naked Osman captured alive 2009 in the UNDP/GEF Project. Photo:Azat Alamanov

Scientific name	Common name	Indigenous	Introduced
Onchorhynchus mykiss	Rainbow Trout		+
Salmo ischchan	Sevan Trout		+
Coregonus lavaretus	Common Whitefish		+
Coregonus widegreni	Valaam Whitefish		+
Coregonus autumnalis	Baikal Omul		+
Leuciscus schmidti	Schmidt's Dace	e	
Leuciscus bergi	Issyk-Kul Dace	e	
Phoxinus issykkulensis	Issyk-Kul Minnov	e	
Tinca tinca	Tench		+
Gobio gobio latus	Issyk-Kul Gudgeon	e	
Schizothorax pseudoaksaiensis issykkuli	Issyk-Kul Marinka	e	
Diptychus maculatus	Scaly Osman	e	
Gymnodiptychus dybowskii	Naked Osman	e	
Alburnoides taeniatus	Striped Bystranka		+
Abramis brama orientalis	Oriental Bream		+
Cyprinus carpio	Common Carp	o	
Ctenopharyngodon idella	Grass Carp		+
Hypophtalmichtys molitrix	Silver Carp		+
Carassius auratus auratus	Goldfish		+
Pseudoraspora parva	Stone Moroko		+
Capoeta capoeta capoeta	Transcaucasian Barb		+?
Triplophysa stoliczkai	Tibetan Stone Loach	e	
Triplophysa stoliczkai elegans	Tyanschan Loach	e	
Triplophysa dorsalis	Grey Loach	e	
Triplophysa strauchii strauchii	Spotted Thicklip Loach	e	
Triplophysa labiata	Plain Thicklip Loach		+
Triplophysa ulacholicus, including T.u. dorsaloides	Issyk-Kul Naked Loach	e	
Sander lucioperca	Pike-perch		+
Micropercops cinctus	Eleotris or Odontobutid		+
Glyptosternum reticulatum	Turkestan Catfish	e	
Aspius aspius	Asp		+?
Coregonus albula	Vendace (Ryapushka)		+?
Coregonus peled	Peled		+?

e= Indigenous, += Introduced, o= not known if indigenous, and +?=not known if the introduction failed

Table 1. List of fish species in the Issyk-Kul Lake [15].

Issyk-Kul Lake and incoming rivers have five indigenous and one alien loach species which are common in littoral underwater meadows, but are also found down to 100 m depth. They feed on benthos, plankton and eggs of other fish [13]. They have never been recorded by name in the catch of the lake except maybe in the "others" component. Subspecies would urgently require taxonomic revision, especially *Triplophysa ulacholicus versus Triplophysa u. dorsaloides* which are here synonymized.

The Issyk-Kul Gudgeon *Gobio gobio latus* spawns in June-July in shallows and feeds on benthos, detritus and fish eggs. It is preyed upon by Spotted Thicklip Loach *Triplophysa s. strauchii*, Sevan Trout *Salmo ischchan* and Pike-perch *Sander lucioperca* [13]. Again this fish has no commercial value and falls into "others" category in fish statistics.

Issyk-Kul Minnov *Phoxinus issykkulensis* is one of the strictly endemic fish species of Issyk-Kul Lake, but unfortunately there is no data on biology or abundance as it has never been important in commercial fishery.

Common Carp *Cyprinus carpio* is a widespread freshwater fish which has been introduced from Asia to every part of the world and it is included in the list of the world's worst invasive species. Many people in Kyrgyzstan, however, see Common Carp as indigenous calling it 'wild form' of Common Carp (Sazan). Most likely it was also introduced into the lake, but probably during the ancient times. It was known to be cultured in Kyrgyzstan at least since 1852 [15]. If accepting the 'wild origin' then the Issyk-Kul populations can be considered vulnerable to extinction.

Issyk-Kul Marinka *Schizothorax pseudoaksaiensis issykkuli* is an endemic species, which reaches 70 cm and a weight of 8 kg, and spawns from May until mid-July on rocky substratum in shallows near aquatic plants (Figure 3.). Its fecundity is 25,000 per year. It is omnivorous. Between 1985 and 1989 it formed 6% of the fish catch but after 1992 it disappeared completely (Table 2).

3. Background and historical overview

3.1. Introductions of alien fish species

At least 19 species have been introduced to the lake by humans, either on purpose or accidentally. Introduction of Vendace *Coregonus albula* and Peled *Coregonus peled* failed, and also survival of Asp *Aspius aspius* and Transcaucasian Barb *Capoeta c. capoeta* is doubtful as these species have not been reported recently. Formerly, the small Issyk-Kul Dace was the major item in fish catches, where it represented about 90% of total biomass. It was, however, considered to have a low value and this led to a proposal to introduce new fish species into the lake [16]. The introduction of the Sevan Trout from Armenia was recommended and, in 1930, 755,000 fertilised eggs were released, followed in 1936 by a further 800,000. Until 1964 Sevan Trout remained rare in the lake due to the shortage of suitable spawning grounds (Figure 4.). At its best, 1976 and1979, 51,6 and 53,8 tonnes of Sevan Trout were captured from the Issyk-Kul Lake. This was mainly due to state owned hatcheries which released 79 million fry into the lake during the 1970s [17].

Figure 3. One of the few Issyk-Kul Marinka captured alive during the study 2008-2011. Photo: Azat Alamanov

Fish species	1955-1964	1965-1969	1970-1974	1975-1979	1980-1984	1985-1989	1990-1994	1995-1999	2000-2003
Schmidt's Dace	482	225	544	496	292	241	223	105	44
Issyk-Kul Dace	9586	10741	9147	5736	1123	1064	790	94	12
Issyk-Kul Marinka	263	39	16	3	34	138	1	-	-
Osmans*	114	10	13	17	10	10	19	-	-
Common Carp	75	85	29	5	7	32	22	-	-
Whitefish	-	-	1	35	106	248	163	57	11
Sevan Trout	0.3	30	123	457	244	206	91	29	1
Pike-perch	-	287	1364	895	340	320	227	25	1
Others	46	51	51	15	15	48	98	74	12
Total	**10566**	**11468**	**11288**	**7659**	**2171**	**2307**	**1634**	**384**	**81**

Original data from Fisheries Department/Mairam Sarieva. Note that the data is in centners of Soviet Union. One centner is 100 kg not one ton as so often misquoted in previous publications. *Osmans = Scaly- and Naked Osman

Table 2. Fish catch from the Issyk-Kul Lake in five year averages from 1955-2003.

Figure 4. Sevan Trout is a colourful fish which grows well in the Issyk-Kul Lake. Photo: Azat Alamanov

Following its introduction, Sevan Trout became an active predator of other fish in the lake and developed several special features. Its growth rate was 4 to 6 times that in Sevan Lake in Armenia. In Issyk-Kul it grows to a bigger size. It matures earlier, and its fecundity has increased five-fold to 3,300-17,300 eggs per fish. The limiting factors for this species in Issyk-Kul are food resources and habitat for reproduction.

In the 1950s, there were further introductions of fish species in order to establish diverse stocks of piscivorous fish, introduce species feeding on phytoplankton and aquatic *macrophytes*, and increase the number of benthos- and plankton-feeding fish [18]. Pike-perch, Tench *Tinca tinca* and Oriental Bream *Abramis brama orientalis* were introduced in 1954-1956 [19]. They became established predominantly in the eastern part of the lake but started soon spreading all over the lake. The introduction of Grass Carp *Ctenopharyngodon idella* and Silver Carp *Hypophtalmichtys molitrix*, and with them inadvertent introductions of Goldfish *Carassius a. auratus*, Stone Moroko *Pseudoraspora parva* and Eleotris *Micropercops cinctus*, were successful but caused a disaster to 'wild carps'. Grass and Silver Carp brought infectious *ascites* of carps into the lake, and the numbers of 'wild carps' started to decrease due to disease.

In the early 1970s, a decision was taken to transform the lake into a trout-whitefish water body at the expense of the local Issyk-Kul Dace population. However, Whitefish *Coregonus lavaretus* is mainly a plankton and benthos feeder, but large individuals can occasionally take other fish. Common Whitefish was introduced from Lake Sevan, Valaam Whitefish or Ludoga *Coregonus widegreni* from Lake Ladoga, and from Lake Baikal came Arctic Cisco or Baikal Omul *Coregonus autumnalis* and Peled [20]. Eggs of Whitefish from Sevan Lake were transferred to the Ton hatchery from which four-day-old fry were released into Lake Issyk-Kul. During 1966-1973 87 million fry were released. In 1974, the first 500 kg of Whitefish were harvested from Tyup Bay.

There were also proposals to replace the Issyk-Kul Dace with the Peled and Vendace (Ryapushka), more nutritious food fish species. However, Peled and Vendace soon disappeared from the lake most likely due to lack of suitable reproduction conditions. Baikal Omul was still observed in the lake as of late 1970s. After that only Whitefish established itself as commercial fish and the highest catch recorded was 35.3 tonnes in 1989. After that the catch started to go down mainly due to reproduction problems and hatchery failure (Table 4).

Most harmful introduction took place accidentally from the cage culture of Rainbow Trout *Onchorhynchus mykiss* (Figure 5.). Since 1980s a lot of small and some large fish started to escape from the culture operations, and now Rainbow Trout is very common all over the lake, but especially near eight cage culture farms. Rainbow Trout moves to fish diet at the size of 35-40 cm [21]. It is not clear if Rainbow Trout would be able reproduce in some incoming rivers, as nowadays the lake seems to have all aged and sized Rainbow Trout.

3.2. Introduction of alien food species for fish

The fish food base was successfully enriched by the introduction of *mysids*, which became targeted by the introduced *coregonids*. In the Issyk-Kul Lake the introduced Whitefish benefitted most of mollusks and *mysids*. Their growth rate was faster than that of the original stock in Lake Sevan, and they also started maturing at an earlier age. However, there was a high mortality of *coregonid* fry which found insufficient food in Issyk-Kul Lake and were heavily preyed on by the endemic fish. Also, it was believed that the higher salinity of Issyk-Kul as compared to Sevan and Baikal could have had a negative impact on fry [20]. The decision was made to stock advanced fry and fingerlings, of which millions were stocked in the following years (at least 10 million from 1977to 1988).

3.3. Recreational and commercial fishing activities

Amateur fishing in the lake started in the 1870s. At first it was unorganized and no statistics were collected. During the 1890s fish catch ranged between 17 – 105 tonnes [2]. At that time, and for many years after, the major commercial fish were Naked Osman, Common Carp, Issyk-Kul Marinka, Issyk-Kul Dace and Schmidt's Dace. According to available information in 1941-1945 harvests of Issyk-Kul Marinka, Naked Osman and Common Carp reached 61 tonnes per year. During the same period Issyk-Kul Dace catch varied from 551 to 900 tonnes per year. It is important to note that during this period lake had already one alien predator, Sevan Trout, but catch of these had not started. It is also interesting to note that 'native' species did not start to go down after the introduction of Sevan Trout [11].

More detailed catch statistics are available from 1955 until the moratorium in 2003 (Table 2). As Table clearly illustrates, fish landings shrunk sharply after the Issyk-Kul Dace population collapsed. Increased fishing activity could be one explanation, but by the end of the Soviet period the state fishing industry was at its peak and encompassed only 300 fishers, 122 boats and 8640 nets [11]. This number of fishers is not much for that size of lake, and over fishing hardly explains alone the decline of fish catches. However, by 1990 Issyk-Kul landings were barely twenty percent of the levels recorded a quarter of a century earlier.

Introduction of one more predator, Pike-perch, into the lake 1954-56 had no immediate and dramatic impact on both lake landings and the species composition of landings. It seems that the production of the lake went down clearly only 1975 onwards, as in the transition from one feeding level (plankton and benthos feeders) to another (predatory fish); the lost feed coefficient against the productivity is approximately ten times. The Pike-perch production reached its peak in 1974, and then the production of Issyk-Kul Dace started to go down very fast as shown in Table 2. The role of Sevan Trout is not as clear as it achieved the highest population only in 1979 after Issyk-Kul Dace had already diminished to one third from the starting level.

Whitefish is often listed as predatory fish but its possible bad habits are not showing clearly in this material. Whitefish achieved the peak population in 1989 when all other species had started to go down very rapidly. Interesting is that the peak populations of Osmans occurred already 1961-62, when all predatory species recorded zero catch. Issyk-Kul Marinka had the peak population in 1988. Total disappearance of Osmans from the catch data happened 1988 and that of Issyk-Kul Marinka 1993. Disappearance of Common Carp took place during the same year (1993) as that of Issyk-Kul Marinka. The predators alone cannot explain these losses, and it is difficult to see any direct relation with fishing either.

Issyk-Kul Dace made 94 per cent of the catch in 1965-69 and less than 15 per cent in 2000-2003, same time the total catch of the lake went down from 1147 to 8 tonnes. Issyk-Kul Marinka, Naked Osman and Sazan Carp are those commercial and indigenous species which are most seriously endangered.

3.4. Fishing as livelihood

The contribution of fishing to annual average income of Issyk-Kul district families is from 5 to 10 per cent and only for some small groups up to 30%. The monthly income of fishermen is not more than 40 USD and that of women processing the fish 54 USD. Women's income is little higher than men's [22]. Although income from fish is small, it allows the families to have cash on daily basis and facilitates implementation of other cash requiring activities (purchase of seeds, forage for animal-breeding etc.). For more details see [15].

3.5. Water level

In historical terms the water level of Issyk-Kul Lake has obviously fluctuated. Some changes are gradual, others sudden and disastrous since they were caused by earthquakes and torrents of water rush from mountains. Large ancient city has been located at depths of 5 to 10 meters near the north coast of the lake, but it was destroyed maybe some 2500 years ago by one of the many local floods which are known to occur every 500 to 700 years [23]. Between years 600 and 1200 AD Issyk-Kul shoreline was again some 500 m lower and after that in the fifteenth century the water level of the lake was more than 10 m higher than it is today.

On Issyk-Kul basin118 rivers and streams flow toward the lake, but only 42% of them actually drain into it and 25% have discharges less than 1 m³ s⁻¹. Only 9% of these are rivers with catchment areas of more than 300 km². Rivers are fed predominantly by melt water from glaciers and snow above 3300 m. The river system reflects also the distribution of rainfall in the basin with low precipitation in the west, where the river system is poorly developed. In the east, where the precipitation is heavier, the hydro-network is denser and the rivers fuller. The greatest volume of flow comes through rivers on the basin's eastern side. Water from most of the rivers has been completely diverted for irrigation before it enters the lake. Therefore, bays in the northern and western coast suffer from increased mineralization. The rivers supply the lake with 3720 million m³ of water per year [24], but they are not the only water supply the lake is receiving. The annual surface water discharge, precipitation and groundwater discharge to the lake are 21, 29, and 33cm respectively; the evaporation from the lake is 82cm. For more details see [7].

Until 1985 water level in Issyk-Kul was falling. Between 1876 and 1972 the decrease was 9 m [25]. During 1960-1979 when the fish catches started to decrease clearly the total decline of water level was 140 cm, at an average rate of seven cm annually. That loss of water level has been one important factor affecting the fish stocks and fisheries.

3.6. Irrigation

Uptake of water for irrigation is one of the factors seen to be responsible for the present changes in the water level. Irrigation also hinders the river spawning of many species, as it prevents small rivers and streams from reaching the lake. Irrigation has led to drying and silting up of spawning grounds and death of the fry themselves as they are poured out with the river water to irrigated fields. During 1960-1979 irreversible uptakes of water from rivers for irrigation reduced the volume of river water entering the lake by an estimated 23 per cent [6]. While in 1930 there were 50,000 ha of irrigated area in the Lake Issyk-Kul catchment, by 1980 the irrigated area reached 154,000 ha [26]. However, even without this irrigation loss, the lake would still have declined at the rate of five cm annually between 1960 and 1979 [27]. This would indicate that climatic factors have also been involved in the fall of the water level.

4. Material and methods

This chapter is based on my field work and data collection in Kyrgyzstan when working in the UNDP/GEF Project: "Strengthening policy and regulatory framework for mainstreaming biodiversity into fisheries sector" as International Fishery Policy Adviser 2008 to 2009 and in the FAO Project GCP/KYR/003/FIN: "Support to Fishery and Aquaculture Management in the Kyrgyz Republic" 2009-2010 as Technical Advisor. A lot of generally unknown 'grey literature' in Russian has been translated and used in this text. More information about interviews and field experimentation is documented in [15].

5. Results and discussion

5.1. Species introduced

It is obvious that Sevan Trout and Pike-perch introductions can be blamed for the reduction in catch. There is clear positive correlation between Sevan Trout and Pike-perch and Schmidt's Dace catch (Table 3). Interestingly Sevan Trout has correlation also with Pike-perch catch. There is positive correlation between prey species, like between Marinka and Osmans as well as between Issyk-Kul and Schmidt's Dace and Sazan Carp, but only Whitefish has strong negative correlation with Issyk-Kul Dace.

Species	Pike-perch	Sevan Trout	White-fish	Marinka	Osmans	Issyk-Kul Dace	Schmidt's Dace	Sazan Carp
Pike-perch	1	,535	-,056	-,176	,088	,268	,559	-,012
Sevan Trout	,535	1	,311	,013	,036	-,136	,339	-,193
White-fish	-,056	,311	1	,143	-,316	-,574	-,209	-,149
Marinka	-,176	,013	,143	1	,492	,116	,124	-,288
Osmans	,088	,036	-,316	,492	1	,480	,424	,261
Issyk-Kul Dace	,268	-,136	-,574	,116	,480	1	,622	,468
Schmidt's Dace	,559	,339	-.209	,124	,424	,622	1	,065
Sazan Carp	-,012	-,193	-,149	,288	,261	,468	, 065	1

Red marked numbers correlation is significant at the 0.01 level (2-tailed). Green marked have correlation at the 0.05 level.

Table 3. Correlations of main fish species in the catch between 1955 and 2003.

But correlation does not necessarily mean causation. This far the introduction of predatory fish species has been seen as the major if not the only reason why native fish stocks collapsed [11,15,28]. This same conclusion was also made in Africa, where introduction of the Nile Perch *Lates niloticus* was believed to have caused the greatest vertebrate mass extinction in recorded history [29,30]. Approximately 150 different species of *Haplocromis* chicklids became extinct in recent times in Lake Victoria. Now, however, reevaluated data shows that Nile Perch did not really succeed until, and after, its prey (the *haplochromines*) had disappeared. The increased eutrophication of the lake and oxygen problems may explain more the diversity changes than the single species predation or fisheries exploitation [31,32].

This does not mean that one should support the alien introductions, and precautionary principles are necessary. Precautionary principle states that one has to expect that new introduced species, although in closures or in the cage, tend to escape for one reason or another into nature [33]. Any new voluntary alien introduction should be understood with that background and the rule should be clear: that new human introduced alien species are not allowed to enter into the country. Still it remains inevitable that some invasive species will arrive without any help of humans.

As shown in Table 4 there are seven clear reasons which could explain the loss of fish stocks and biodiversity and another five reasons which have had at least some negative impact. The reviews and field work highlights that all these twelve negative factors have been present more or less at the same time, so it is not possible to single out any one of them as the most important. Surely they have rather caused the loss of fish stocks and biodiversity together. Some of these factors have been listed already before, like over fishing, disintegration of the Soviet Union, irrigation and water level. Over fishing of Issyk-Kul Dace stocks by the Soviet fleet based at Issyk-Kul Lake was presented in [11]. The disintegration of the Soviet Union had profound economic and social effects, especially in the fisheries sector of the newly independent transition economies [28]. Nowhere were these production shortfalls bigger than in Kyrgyzstan where Lake Issyk-Kul fish landings were in 2003 less than 7 per cent of the catch level recorded in 1989. The major consequence for the fisheries sector was the spectra of uncertainty which included the uncertainty of, how the sector was to be managed, how access to water bodies was to be regulated, how to maintain the backward and forwards supply chains which underpinned pond aquaculture, and livelihoods – as the Soviet guarantee of job security was rescinded. Many experts and professional fishers left the sector to find employment in other sectors in Kyrgyzstan or abroad. Intensive irrigation led to reduced water levels in the lake and more importantly heavy water abstraction caused drying of many incoming streams that the endemic fish species previously used for feeding and/or spawning [34,35]. It has been shown that biological productivity of Lake Issyk-Kul decreased from 1973 to 1981 when the water level was declining at a rate of 7-10 cm per year [36].

5.2. Hatchery failure

Reproduction of many alien and endemic fish species was severely constrained by the limited number of suitable spawning rivers. As a consequence, the state established hatcheries on the Ton (1964) and Karakol (1969) rivers – with the brief to capture spawning fish, extract the eggs, raise the fry-fingerlings produced, and thence restock the lake. According to [7], the minimum Sevan Trout return in landings is given as 2% of releases; that means that at least 750,000 fry, each of 1 g weight, must be produced and released annually. Assuming an egg mortality of 50%, hatcheries should produce 1.5 million eggs per year. Ton hatchery produced 9 million fry annually in 1989-91. After the breakdown of the former Soviet Union the state hatchery production went down sharply. Over the period 2004-8 Ton Hatchery continued to restock the lake with Sevan Trout at much reduced rate, 446,000 fingerlings annually. Nowadays (2010) Ton Hatchery is able to release some 900,000 fry with 40% egg survival. No endemic fish fry have been produced despite of the capacity, but Rainbow Trout fingerlings have been produced on a contractual basis for the cage farmers.

Estimated IMPACT	Strong negative impact	Some negative impacts	Not visible impact	Some positive impacts
Introduction of alien fish species	Yes			
Introduction of alien food species				Yes
Over fishing	Yes			
Illegal fishing			Yes	
Disintegration of the Soviet Union 1991		Yes		
Cage culture	Yes			
Moratorium	Yes			
Hatchery failure	Yes			
Tourism		Yes		
Water level		Yes		
Irrigation	Yes			
Water pollution		Yes		
Climate change		Yes		
Radioactive leakage			Yes	
Military activities			Yes	
Mining activities	Yes			

Table 4. Impact evaluation of different natural and anthropogenic factors on fish stocks and biodiversity

During 1966-1973 over 12 million Whitefish fry were released annually from the state-owned Karakol Hatchery, but 1977-1988 fingerling production went down to 1 million per year. After privatization Karakol Hatchery has been able to produce below 2.5 million Whitefish fry annually, explaining why the collapsed Whitefish stocks are not recovering, as obviously very little or no natural reproduction takes place in the lake.

These drastic declines in restocking have undoubtedly been one contributor to the decrease in recorded fish landings at Issyk-Kul Lake.

5.3. Cage farming of fish

The cage farming of Rainbow Trout started in 1988 by Alfa Laval Avose, but was not economically viable due to the high cost of feed. Obviously large number of fingerlings escaped into the lake, when the storm was turning the experimental cages around. In 1989-

90 the company was able to produce 20 tonnes of Rainbow Trout. After the collapse of the USSR there was no development of this activity before 2006 when Ecos International commenced cage farming activities at Issyk-Kul Lake. Since that time exponential growth in trout culture has taken place.

Nowadays the existing eight cage farms and their 26 cages (as in April, 2009) are producing well over 300 tonnes of Rainbow Trout per year [37]. This production is causing pollution in the form of medicaments used for the treatment and prevention of diseases and pathogenic bacteria and parasites. By authorizing lake-based cage culture of Rainbow Trout, the authorities are allowing inevitable eutrophication. Extra nitrogen and phosphorus from unused feed will add to the primary production of algae and lower oxygen level. The second problem is the excess feed which sinks to the bottom of the lake through the net cages. At the bottom, sinking feed and faeces and urine of fish will cause the formation of hydrogen sulfide gases harming the other users and fauna of the lake. The worst, however, is the unwanted new continuous introduction of that predatory fish to the lake, because especially large specimen can and will escape the cages. After that they move around the lake and eat a lot of endemic fish species. According to fishermen (personal interviews in 2009) Rainbow Trout is the main predator in the lake, even more predacious than Pike-perch, because it comes to prey in shallow waters near the shoreline while Pike-perch often remains in the deeper waters (Figure 5).

5.4. Moratorium and illegal fishing

In order to protect the decreased fish stocks in the lake, the President of the Kyrgyz Republic declared a Moratorium for Artisanal and Commercial Fish catching for a period of 10 years (2003-2013). The need for total ban was stated to be illegal and over-fishing which was seen as the only reason to loss of fish resources and endemic species. But the moratorium can become an effective measure of restoration of fish resources in the lake only if mechanism of implementation and realization (fish inspections etc.) is developed as well. Otherwise the moratorium will not work. Despite of the total ban at least 500 people continue their activity as illegal fishermen. On average they are catching 5 to 20 kg per night, but every fourth night is stormy making artisanal fishing with small boats impossible. If fishing 100 nights per year, they are catching between 250 and 1000 tonnes per year. Should this fairly conservative estimation be true, the lake is fished at level of 0.4 - 1.6 kg/ha. So this hardly can be seen as over fishing in a lake where theoretical production capacity is estimated at 4.5 kg/ha [34]. This of course by assuming that the fish stocks have in the last ten years recovered from 2003 level after the total ban of commercial fishing.

5.5. Management and conservation possibilities

It is far too easy to blame over-fishing that some species became nearly extinct and that fewer fish are caught. More important problem is the absence of any fisheries management and lack of controlled protection of fish resources. Removal of the fishing ban is necessary, since it cannot be controlled and monitored. Exploring co-management arrangements is a

Figure 5. This kind of 7 kg Rainbow Trout eats a lot of small indigenous fish species. Photo: Azat Alamanov

better option than command and control as the resources are not available for such policy measure [15]. If more than 500 people are continuing fishing despite of the moratorium, the policy needs to be evaluated for better stewardship outcomes. Actually it is far more important to continue to fish large predatory fish, if having any concern of the survival of native non-predatory species.

However, commercial fishing needs to be reconsidered after the moratorium, in 2014, as recreational and food fishery may be far more sustainable. Due to the growing importance of Lake Issyk-Kul for recreation, fishery management might go in the direction of producing valuable sport and recreational fish to satisfy the demand of the tourists and visitors. Such recreational fishing will basically target large predatory species- Rainbow and Sevan Trout and Pike-perch, which are the favourite species for sport fishing. Recreational fishing of large fish will promote Issyk-Kul Lake as more attractive for tourists. It will also help fishery managers to shift proportions of predator and prey fish species and diminish the negative effect of alien predators towards vulnerable stocks of endemic fish species.

Rare indigenous species stocks will not improve without artificial propagation in local hatcheries. This production of fry has started through UNDP/GEF Project, but stocking the lake with small indigenous is not viable before considerable harvesting of large predators. The number of Sevan Trout is easy to regulate, as it mostly depends on the stocking rate of fingerlings into the lake and these are reproduced artificially in the hatcheries. Rainbow Trout and Pike-perch are more difficult to remove if they are able to multiply in the lake (Figure 6.). Improved stocks of small endemics could take care of these predators by eating their eggs and small fry, like small fish did in Lake Victoria by preying on eggs and fry of the Nile Perch [32].

5.6. Water pollution

Widespread mining operations are causing disruption of soils, terrain and water tables but more serious water pollution comes from illegal dumping or storing of toxic chemicals currently in use at Kumtor gold mine, in Tian Shan Mountains. It is the largest gold mine, as well as a major government revenue source, which routinely ignores national environmental legislation. Kumtor mine reportedly uses up to 10 tonnes of cyanide per day in its mining operations, and number of chemical constituents is released into the environment [38]. By sure this is affecting fish populations downstream and the health of local people using the contaminated water or fish (Table 5).

One of the worst regional environmental disasters in recent history occurred on 20 May 1998, when a truck hauling toxic chemical crashed just upstream from the mouth of the Barkuum River, which empties into Lake Issyk-Kul. As a result, 1762 kg of sodium cyanide, a chemical used in the processing of gold ore at Kumtor, were dumped into basin waters [6].

Lack of both adequate infrastructure and financial means to support public utilities (let alone any resort or tourism industry) has made it impossible to improve wastewater treatment plants. This in turn leads to further pollution and unwise use of lake waters. The gradual increase in settlements and industries around the lake has led to an increase in pollution. Although most enterprises have wastewater treatment facilities they are not efficient and some effluents still reach the lake.

Figure 6. Artificial nests are used to remove the eggs of Pike-perch from the lake to reduce the numbers of that alien predatory fish. Photo: Azat Alamanov

Agriculture, through the use of fertilizers and pesticides, also contributes to the lake pollution, but level of fertilizer application on crop fields is known to be moderate. However, the Issyk-Kul area produces 12% of total national cereal crops and over 40% of potato crops. Of the total area of orchards nationwide, 20% are in Issyk-Kul. Numbers of domestic animals in the catchment area is very high: Cattle 163,500; sheep 1,944,400; swine 32,700; poultry 623,400 and horses 48,500 [39]. Dairy product processing covers 50% of the national dairy product supply. Animal breeding is growing, with average annual sheep and cattle surplus at 5-6%. Grazing land is overloaded by 1.5 times its capacity (Figure 7). Grazing practices have changed so that all livestock owned by small proprietors are now grazing near the lake as the farmers have no transport nor money to drive their animals upland to outlaying pastures. That could cause social conflict (grazing on beaches and resort areas) and eutrophication of the lake but luckily people are commonly collecting the manure for fire or fertilizing. While the large volume of 1738 km^3 of water in the lake may have at present considerable diluting capability and with the good water mixing is also able to quickly oxygenate organic matter inputs to the lake, sheltered shallows are subject to eutrophication. As the shallows are also important spawning and feeding areas for a number of fish, such eutrophication may affect especially those cold-water fish species which require pristine waters, like Whitefish.

Figure 7. These camels are the only memory of the Silk Road at the Issyk-Kul Lake coast of which is heavily overgrazed by the domestic animals. Photo: Azat Alamanov

Eutrophication caused by birds is not often considered as a problem, but in the Issyk-Kul Lake the amount of migratory birds is such that it will affect the lake. Anywhere from 44,000 to 68,500 birds belonging to 30 to 35 species winter on the lake, and even more birds use it as stopover and feeding place during spring and autumn migration [40].

5.7. Climate change implications

Within the Issyk-Kul basin there are 834 glaciers of various sizes ranging from less than 0.1 to 11 km². For example, a typical Issyk-Kul glacier Karabatkak has shown in long-term study between 1957 and 1997 that ice loss exceeded snow mass gain by 18 m.

This thinning of ice is due to climate change, summers have been 0.6 degrees Celsius warmer, although the annual average temperature has risen only 0.2 degrees Celsius. Based on this it was calculated that before 2005 overall glaciation area near the lake will go down 32% on the northern slopes and even 77% on south-facing slopes [41].

The continuing retreat of glaciers in the Issyl-Kul catchment, the melt water from which is one of the major contributors to the lake, seems to be going in parallel with the declining lake water level [42]. Without glacial runoff overall drop in lake's water level would have been much greater.

The Kyrgyz Republic is within a high seismic activity area, and Issyk-Kul is a tectonic lake, and the lake bottom is believed to have numerous warm water springs. These explain partly why the lake never freezes over, except in the shallow Rybachinsky and Tyup bays. The water stays warmer than the air for seven months per year [39].

Hot springs at the lake and on the bottom change water quality and may facilitate winter spawning of some introduced species like Pike-perch. During the test fishing in early April 2009 we found after opening a 40 cm Pike-perch that it had preyed another (12 cm), and even that small prey had eaten a few juvenile Pike-perch, not more than 5 cm long. It was estimated that these 3rd level victims of cannibalism must have been born early February. This kind of winter spawning is not known before but could obviously take place due to the hot springs.

5.8. Other human activities

There are recent reports on the radioactive waste contamination in Central Asia [43] showing that the situation is critical especially in Kyrgyzstan, with 36 tailing sites and 25 uranium dumps in the country. Kadzhi-Say, the country's largest tailing site (containing 150,000 m³ of radioactive waste), is located barely 1.5 km from Lake Issyk-Kul. Yet although some information is available on the impact of radioactivity on humans, it is not well studied or understood what direct impact the current radioactivity levels have on the aquatic biodiversity in Kyrgyzstan. The monitoring of water bodies for radioactivity is not done consistently and to date, no assessment has been made of the uranium contamination of fish populations of indigenous species and its consequences for fish stocks in Kyrgyzstan. It is not clear, if and how much radioactive waste has already gone into the lake or still goes from incorrectly closed tailing sites and uranium dumps.

During the Soviet period, the USSR Navy operated an extensive facility at the lake's eastern end, where submarine technology was evaluated. Also Navy tested torpedoes built in Tashkent. Not known if torpedoes exploded in these experiments. If so this must have killed a lot of fish. In 2008 Kyrgyz newspapers reported that Russian Navy is planning to establish a new naval testing facility around the Karabulan peninsula on the lake. This may affect the fish stocks in the future depending on the tests undertaken.

During the Soviet era, the lake became a popular vacation resort, with numerous sanatoria, boarding houses and vacation homes along its northern shore. These fell on hard times after the break-up of the USSR, but from 2005 onwards hotel complexes are being refurbished and simple private bed- and –breakfast pensions are being established for a new generation of health and leisure visitors (Table 5).

Tourism has become one of the most dynamically developing sectors of economy of the Kyrgyz Republic. The number of arrivals of foreign tourists is expected to exceed 2 million persons per year. International tourists are primarily from Kazakhstan, Russia, and

Uzbekistan. If half of these tourists will visit the lake, there is need to accommodate an additional 1 million persons per year at the lake in hotels using natural beaches for recreation. Nowadays the lake has 343 tourist enterprises, including cafes and restaurants.

Years	Population in '000	Visitors in hotels	Visitors at homes	Total in '000
2006	430	198	296	924
2007	433	199	245	877
2008	435	194	349	978
2009	438	169	318	925
2010	441	181	227	849
2011	445	185	231	861

Table 5. Permanent population and annual visitors in the Lake Issyk-Kul area [44].

In addition, large hospitals have been built to use medicinal mud and hot springs along the coasts for medicinal purposes. Regulations exist for water system supply and to use fully purified sewerage systems. Recycled waste water could be used for irrigation. Unfortunately, some entrepreneurs have forgotten these regulations, and continue to pollute the lake as no corruption free control exists.

Asian Development Bank study [45] has concluded that available water and sanitation and waste disposal infrastructure in the Issyk-Kul area is decrepit, dysfunctional, poorly managed and negatively impacts the surrounding environment. The planned tourist influx equivalent to four times the resident population applies excessive pressure on the existing infrastructure, which results in the pollution of the lake. The proposed Issyk-Kul Sustainable Development Project initiated by the Asian Development bank (ADB) would address the environmental and institutional issues around the Lake Issyk-Kul. The Japanese International Cooperation Agency will also develop the sewerage system and sewage treatment plant in Cholpon-Ata through parallel financing with the ADB.

6. Concluding remarks

Issyk-Kul Lake is the second largest high-altitude lake in the world providing recreational and small-scale fishing activities as well as cage culture of Rainbow Trout in the Kyrgyz Republic. The original fish fauna comprised twelve indigenous species and two subspecies particular for this lake. At least 19 species have been introduced to the lake by humans, either on purpose or accidentally. The populations of several indigenous fish are seriously threatened, because many of the introduced fish species are potential predators. Issyk-Kul Marinka, Naked Osman and Sazan Carp are those commercial and indigenous species which are most seriously endangered. In 1986 a total ban was declared for catching Naked Osman, but it did not lead to positive results, indicating that anthropogenic activities were not the only reasons for the suffering of the endemic fish species.

Fishers and most of the previous papers are convinced that the predatory fish species have been the most destructive to biodiversity. Addressing the introductions, the basic rule should be that new human introduced alien species are not allowed to enter into the lake. At least any further fish introductions into the lake should be carefully evaluated to prevent unwanted changes in fish stocks. Issyk-Kul, as an oligotrophic lake of low productivity, has a low carrying capacity for fish; hence it will never become a water body which would sustain high levels of fish stocks.

Dissolution of the Soviet Union explains to some extent the collapse of the fisheries sector (including the hatchery operations) in Kyrgyzstan, but maybe not the loss of biodiversity. Rapid growth in human activities with the development of tourism industry; irrigation; water eutrophication and pollution, and climate change impacts seem to be important root causes for loss of fish stocks and biodiversity degradation.

Uptake of water for irrigation is one of the factors seen to be responsible for the present changes in the water level as water from most of the rivers has been completely diverted for irrigation before reaching the lake. Irrigation hinders also the river reproduction of many species as it prevents spawning fish from entering the rivers or fry to return to the lake. During 1960-1979 when fish catches started to decrease the total decline of the water level was 140 cm, at an average rate of seven cm annually. That loss of water level has been one important factor affecting the fish stocks and fisheries.

There are recent reports on the radioactive waste contamination in Kyrgyzstan, where the country's largest uranium tailing site is located barely 1.5 km from Lake Issyk-Kul. It is not clear, if and how much radioactive waste has already gone into the lake or still goes in from incorrectly closed tailing sites and uranium dumps. Maybe even more serious water pollution comes from illegal dumping or storing of toxic chemicals currently in use at a gold mine in Tian Shan Mountains. It is the largest gold mine, as well as a major government revenue source, which routinely ignores national environmental legislation. This mine uses daily up to 10 tonnes of cyanide in its operations, and many of toxic chemicals are released into the environment. This is surely affecting fish populations downstream and the health of local people using the contaminated water or fish for drink and food.

Existing water and sanitation and waste disposal infrastructure in the Issyk-Kul area is decrepit, dysfunctional, poorly managed and has negative impacts on the environment. The planned tourist influx equivalent to four times the resident population will apply excessive pressure on the existing infrastructure, which will result in further pollution of the lake.

Important problem is the total absence of any fisheries management and lack of controlled protection of fish stocks and diversity. Fishing ban is not helpful as it cannot be controlled and monitored. Exploring co-management arrangements is a better option than command and control as the resources are not available for such policy measure. Rare and endangered indigenous fish species will not increase without artificial propagation in local hatcheries. Stocking the lake with small indigenous species is not viable, if not first harvesting the large predators. Pike-perch would just eat small indigenous species and grow bigger and spawn more.

Over-fishing of introduced species, like Rainbow Trout and Pike-perch, could be a good thing. As popular food fish and recreational catch, they could be severely over fished. This could lead to population reduction, and several populations of endemic fish species should soon show signs of increasing numbers. So the authors should allow the local fishing communities capture large introduced fish species as much as they can rather than restricting them through moratorium.

Any new development initiatives must be consultative and participatory in order to be more consistent with local habits and cultural values. Inherited customs provide an important element for the development of locally based resource management system. Consequently, allocation and sustainable management of natural resources is one of the key issues for the local population, whose daily cash economy is directly dependent on availability of fish resources.

Before being able to define the best management ways, one needs further research in taxonomy and fish biology. The knowledge of fish stock parameters is essential for the determination of appropriate fisheries management and definition of sustainable fish yield. Impact of mining toxins and radioactive waste is important to study and to know for control measures. Water pollution is a continuous risk for this important lake, and has to be halted in the future. The Kyrgyz public must be engaged in the future through environmental education in conservation and preservation of natural and cultural riches of the Issyk-Kul area.

Author details

Heimo Mikkola

University of Eastern Finland, Kuopio Campus, Kuopio, Finland

Acknowledgement

My deepest gratitude goes to Mrs. Burul Nazarmatova who translated all grey Russian literature into English and collected a lot of relevant Soviet time data from the internet. Mrs. Mairam Sarieva from the Fisheries Department of the Kyrgyz Republic assisted me in collecting the fish catch statistics of the Issyk-Kul Lake. Some unpublished data from fish stocks was also given by Dr Muchtar Alpiev from the Kyrgyz Academy of Science. Last but not least I want to give my special thanks to Dr Ahmed Khan, Memorial University of Newfoundland, Canada, who's useful and critical suggestions improved this chapter a lot.

7. References

[1] Kodyaev G.W (1973) Morphometric characteristics of Lake Issyk-Kul. Izv. Vsesoyuz Geogr. Obschestva 105(4): 362-365 (In Russian).

[2] Berg L.S (1930) The present state of fisheries in Issyk-Kul. Issykkulskaya Ekspeditsia 1928 g. 1: 1-48 (In Russian).

[3] Molchanov I.V (1946) Lake Issyk-Kul. Gidrometeoizdat, Leningrad. 53pp (In Russian).

[4] Kadyrov G.K (1986) Hydrochemistry of Lake Issyk-Kul and its basin. Ilim, Frunze. 212pp (In Russian).

[5] Shabunin G.D (1981) Thermodynamic processes in Lake Issyk-Kul. Biol. Osnovy Ryb. Khoz. Vodoemov Srednei Azii i Kazakhstana: 394-396. Ilim, Frunze (In Russian).

[6] Baetov R (2003) Management of the Issyk-Kul basin. Mimeo 28/08/2003, 22 p.

[7] Savvaitova K.A, Petr T (1999) Fish and Fisheries in Lake Issyk-Kul (Tien Shan), River Chu and Pamir Lakes. FAO Fisheries Technical Paper 385, Rome.

[8] Foliyan L.A (1981) Zooplankton of Lake Issyk-Kul. Biol. Osnovy Ryb. Khoz.Vodoemov Srednei Azii i Kazakhstana: 389-390. Ilim, Frunze (In Russian).

[9] Pavlova M.V (1964) Zoobenthos of Issyk-Kul Lake Bays and their use by fishes. Ilim, Frunze. 95pp (In Russian).

[10] Ivanova L.M (1986) Distribution of mysids in Lake Issyk-Kul. Biol. Osnovy Ryb. Khoz.. Vodoemov Srednei Azii i Kazakhstana: 65-66. Ilim, Ashkhabat (In Russian).

[11] Konurbaev A.O, Kustareva L.A, Alpiev M.N, Kabataev D.T, Konurbaev E.S (2005) Conditions of Issyk-Kul Lake Ichtiofauna, fishery and its management. Mimeo, 24 pp. Bishkek.

[12] Berg L.S (1949) Freshwater fishes. AN SSSR. Moscow and Leningrad. 2: 478-925 (In Russian).

[13] Turdakov F.A (1963) Fishes of Kirgizia. Ilim, Frunze. 283pp (In Russian).

[14] Mikkola H (2010) Introductions causing serious problems to endemic fish. Suomen Luonto 69(4): 15-16 (In Finnish).

[15] Alamanov A, Mikkola H (2011) Is Biodiversity Friendly Fisheries Management Possible on Issyk-Kul Lake in the Kyrgyz Republic? Ambio 40: 479-495.

[16] Berg L.S (1929) Fisheries in Issyk-Kul. Izv. Inst. Opytnoi Agronomii (Moskva) 7:179-181 (In Russian).

[17] Borisov S.I (1981) Induced reproduction of trout in Issyk-Kul. Biol. Osnovy Ryb. Khoz. Vodoemov Srednei Azii i Kazakhstana : 40-42. Ilim, Frunze (In Russian).

[18] Turdakov F.A (1954) On the development of fisheries in Lake Issyk-Kul. Trudy Inst. Zool. Parazitol. Kirgiz. Fil. AN SSSR 2:31-38. Frunze (In Russian).

[19] Turdakov F.A (1961) Reconstruction of fisheries in Lake Issyk-Kul. Vestnik AN SSSR 2: 54-56 (In Russian).

[20] Nikitin A.A (1976) Introduction and induced reproduction of whitefish in Kirgizia.. Ilim, Frunze. 122pp (In Russian).

[21] Svärdson G, Nilsson N-A, Dahlström H, Tuunainen P (1968) Fish, management of fishing waters and fish farming. 302 pp. Kirjayhtymä, Helsinki (in Finnish).

[22] UNDP (2007) Access of men and women to the natural resources of Kyrgyzstan, 83 p., Bishkek

[23] Lukashov N (2007) Remains of ancient civilization discovered on the bottom of a lake. Novosti 27/12/2007 Available: http://en.rian/analysis/2007/1227/94372640-print.html. Accessed 2008 Jun 11.

[24] Zabirov R.D, Korotaev V.N (1978) Characteristics and morphometry of the lake: In: Ozero Issyk-Kul: 3-48. Ilim, Frunze (In Russian).

[25] Shnitnikov A.V (1979) Lake Issyk-Kul: nature, conservation and the prospects for lake utilization. Ilim, Frunze. 84pp (In Russian).

[26] Leontiev S.A, Sevastyanov D.V (1988) Problem of managing Issyk-Kul water level. Vodnye Res. 1: 121-128 (In Russian).

[27] Sevastyanov D.V, Smirnova N.P (1986) Lake Issyk-Kul and trends in its natural development. Izd. Nauka, Leningrad. 246pp (In Russian).

[28] Thorpe A, Van Anrooy R, Niyazov B.N, Sarieva M.K, Valbo-Jørgensen J, Mena Millar A (2009) The collapse of the fisheries sector in Kyrgyzstan: an analysis of its roots and it prospects for revival. Communist and Post- Communist Studies 42:141-163.

[29] Ball D (2004) Lake Victoria: The problems and options. 1-2. Available: http://aquarticles.com/articles/management/Ball_2Victoria.html. Accessed 2010 Oct 21.

[30] Chege N (1995) Lake Victoria: A sick giant. People & the Planet, 1-4. Available: http://cichlid-forum.com/articles/lake _victoria_sick.php. Accessed 2010 Nov 21.

[31] Kolding J, van Zwieten P, Mkumbo O, Silsbe G, Hecky R (2008) Are the Lake Victoria Fisheries Threatened by Exploitation or Eutrophication? Towards an Ecosystem-based Approach to Management. Pp. 309-355 in Bianchi, G. & Skjoldal, H.R. eds.: The Ecosystem Approach to Fisheries. CAB International.

[32] Kolding J, van Zwieten P, Mkumbo O, Hecky R, Silsbe G (2012) When prey becomes predators: eutrophication and the transition of Lake Victoria's fish community from a cichlid to a Nile Perch dominated state. Paper for the World Fisheries Congress in Edinburg 2012: G4: Impact of marine or aquatic acidification and eutrophication on fish and fisheries.

[33] FAO (1996) Precautionary approach to capture fisheries and species introductions. FAO Technical Guidelines for Responsible Fisheries. No. 2:1-54, FAO, Rome.

[34] Alpiev M.N (2006) Internal water ecosystems. In Third National Report on Saving of Biodiversity in the Kyrgyz Republic. Bishkek (In Russian).

[35] Mikkola H (2009) GCP/KYR/003/FIN " Support to fishery and aquaculture management in the Kyrgyz Republic". FAN FAO Aquaculture Newsletter 43: 6-7.

[36] Konurbaev A.O (1981) Decrease in biological productivity of Lake Issyk-Kul due to the declining water level. Biol. Osnovy Ryb. Khoz. Vodoemov Srednei Azii i Kazakhstana: 100-103. Ilim, Frunze (In Russian).

[37] Alamanov A, Mikkola H (eds.) (2009) Workshop report on Lake Issyk-Kul biodiversity friendly fisheries management regime proposal and fisheries co-management. Ecocentre, Cholpon- Ata, Kyrgyzstan, 10-12 September 2009. GCP/KYR/003/FIN FAO Field Document 1/2009, 92 p.

[38] Trilling D (2012) Kyrgyzstan: Gold Mine Could Exacerbate Central Asian Water Woes- Report. Eurasia Net's Weekly 31/01/2012 Available: http://www.eurasianet.org/node/64928. Accessed 2012 Feb 7.

[39] International Lake Environment Committee (1999) Lake Issyk-Kool. 9p. Available: http://www.ilec.or.jp/database/asi/asi-55.html. Accessed 03 Feb 2012.

[40] Van der Ven J (2002) Looking at Birds in Kyrgyz Republic, Central Asia. 180 p. Rarity, Bishkek.

[41] Dikikh A.O (2000) Glaciations in the Issyk-Kul Basin; its role as a flow source. Nature and People of Kyrgyzstan. Special Edition, Bishkek.

[42] Artemev V.V, Klinge R.K, Selivanov A.O (1992) History of formation of water resources and fluctuations in water level of Lake Issyk-Kul in late Pleistocene and Holocene. Vodnye Res. 4: 71-78 (In Russian).

[43] UNDP (2009) Uranium Tailing in Central Asia. Final Framework Document, 114 p. Bishkek.

[44] National Statistics Committee (2011) Permanent population and Tourism of the Issyk-Kul Oblast. Brief Statistical Handbook, Bishkek.

[45] Asian Development Bank (2009) Environmental Impact Assessment of Issyk-Kul Sustainable Development Project, Kyrgyz Republic. ADB TA No. 7228 KGZ, 2:1-120.

Chapter 7

Efforts to Combat Wild Animals Trafficking in Brazil

Guilherme Fernando Gomes Destro, Tatiana Lucena Pimentel,
Raquel Monti Sabaini, Roberto Cabral Borges and Raquel Barreto

Additional information is available at the end of the chapter

1. Introduction

Wildlife trafficking, including the flora, fauna and their products and byproducts, is considered the third largest illegal activity in the world, after weapons and drugs trafficking. Considering only the wild animals trafficking in Brazil, it is estimated that about 38 million specimens are captured from nature annually and approximately four million of those are sold. Based on the data of animals seized and their prices, it is suggested that this Country deals with about two billion and five hundred million dollars a year [1].

The wildlife trafficking networks, like any other criminal network, have great flexibility and adaptability and join with other categories or activities (legal or illegal), such as drugs, weapons, alcohol, and precious stones. Their products are often sent from the same regions and have similar practices such as forgery, bribery of officials, tax evasion, fraudulent customs declarations, among many others [1].

In some cases, the criminals are infiltrated in public agencies to entice public officials and, in case of problems in the target Country, they can move with ease to other destination. Moreover, people involved can be easily replaced by others more efficient, reliable and qualified for the activity. This great power of mobility and changeability is one of the major problems to map the criminal networks and their local of action [2].

Although modern techniques has been used, around the world, to help the enforcement in the combat of illegal wildlife trade [3,4,5], the trafficking structure still presents features in common with the set of network information, because it requires equipment that enables the continuous exchange of information on routes, on the most quoted animals at the black market, on new forms of fraud and on corruption pathways. The new technologies are more and more used to increase the possibility of success on criminal operations, either through the use of cell phones, computers to defraud documentation, or internet sales, among others [2].

According to the report of the National Network to Combating Wild Animal Trafficking [1], there are four methods that encourage the illegal trade in wild animal: (a) animals for zoos and private collectors, (b) for scientific use/ biopiracy, (c) for pet shops and (d) for products and byproducts.

However, it is known that identifying the site of capture isn't an easy task, because the locals where the animals are confiscated usually differs from where they were captured. Furthermore, the capture and the sale of wild animals and their byproducts are not concentrated in one only place and do not always follow the same destiny: the movement is intense, with many destinations. After being captured, the animals commonly pass through small and medium traffickers who make the connection with Brazilian and international large dealers, however, the animals can also be sold by internet, pet shops and illegal fairs [2].

Although the trafficking consequences are numerous, it is possible to group them into three main branches: (a) Sanitary, since illegal animals are sold without any sanitary control and can transmit serious diseases, including unknown ones, onto domestic breeding and people [1,2,6]; (b) economic/social, as the trafficking moves incalculable amounts of financial resources without bringing income to the public coffers [1]; and (c) Ecological, since the capture from nature done without discretion accelerate the process of extinction of species, causes damage to ecological interactions and loss of the genetic heritage. Moreover, the trafficking can also bring ecological damage arising from the introduction of exotic specimens, that, although acquired as pets, are being abandoned by their owners in various natural areas [1].

Illegal wildlife trafficking is an extremely lucrative crime with serious consequences yet relatively low penalties and few prosecutions [3]. Besides all the complicating factors inherent to the trafficking, the researchers of this subject are facing yet the lack of organized and systematized data and information [2]. In addition, the studies on trafficking and its impacts on biota are also scarce [7], what makes the task of systematization even more complex.

Thus, through this work, we presented a national view of the control and combat actions towards the wild animals trafficking in Brazil through the existing information at the corporative systems managed by the Brazilian Institute of Environment and Renewable Natural Resources - IBAMA. As specific goals, we aimed to:

-   Historically evaluate the gradual development of the Brazilian environmental enforcement related to fauna;
-   Map the Brazilian States where there are greater efforts against wild animals trafficking, as well as the most confiscated species;
-   Evaluate the major forms of admission and destination of the wild animals present at the Rehabilitation Centers;
-   List the main perspectives and recommendations of actions to combat wild animals trafficking in Brazil.

2. The enforcement for conservation of wild animals in Brazil

For the preparation of this paper we used, primarily, historical information present in four information systems (Table 1), all managed by IBAMA. This information was compiled, systematized and analyzed together with literature data.

System	Name	Objective
SICAFI	Recording, Levying and Enforcement System	Responsible for recording data and information relating to environmental enforcement activities performed by IBAMA and partners institutions
SISPASS	Recording Passeriform Amateur Breeders System	Responsible for the control of the activity of Amateur and Commercial Passeriform[1] Breeders
SISFAUNA	Fauna Management System	Responsible for the management of wild animals in captivity, including the emissions of permits, stock control, domestic trade, licenses issued and carried out transactions
SISCITES	System for the importation and exportation of specimens, biological stuff, native and exotic wildlife products and byproducts	Controls the importation and exportation of species listed in the CITES[2] appendices

Table 1. Information systems related to wildlife and managed by IBAMA.

In Figure 1 we summarized data from the Wild Animals Rehabilitation Centers – CETAS, on all the confiscated wild animals placed there during eight years. The CETAS are responsible for receiving, identifying, marking, selecting, evaluating, recovering, rehabilitating and placing wild animals. Furthermore, they are important allies to the actions for the repression of trafficking because they provide relevant information about confiscated wild animals or from voluntary delivery.

As recommended by the Brazilian Environmental Policy and showed in Figure 1, the State supervision related to illegal wildlife, under the responsibility of the Environmental Military Police, has steadily increased in number and efficiency, thanks to ongoing efforts to decentralize responsibilities in the Country. Thus, IBAMA has been able to focus on major crimes, with significant results through the dissuasion of his actions. It is important to inform that in Brazil, the fines are applied per animal. So, due to that, Minas Gerais (which had the highest participation of environment military police) has the largest number of fines (Figure 2), but it doesn't reflect the absolute value. This happens because of the type of inspection that fights against the final receptors of wild animal traffic.

---

[1] IBAMA's Normative Instruction No 15/2010

[2] CITES: Convention on International Trade in Endangered Species of Wild Fauna and Flora (www.cites.org). The Brazilian CITES Management and Enforcement Authorities are represented by IBAMA.

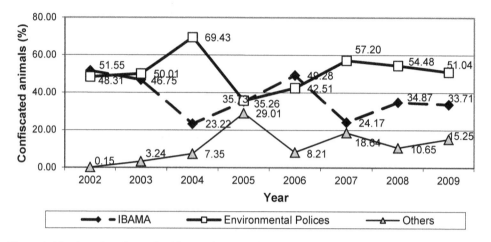

Figure 1. Number of confiscated wild animals received by CETAS

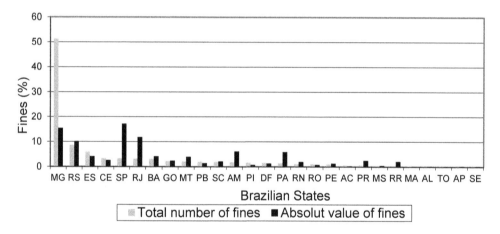

Figure 2. Percentage of Fauna fines per Brazilian State and their absolut value from 2005 to 2010.

In Figure 2, the distinction among each Brazillian State on the combat against illegal actions related to wild animals is clear. The States of Rio Grande do Sul (RS), Minas Gerais (MG), Espírito Santo (ES), São Paulo (SP) and Rio de Janeiro (RJ) were the ones with the highest numbers of fines applied between 2005 and 2010. The last four are located in the southeastern region, where is the demand from the majority of animals from traffic. The States of Sergipe (SE) and Tocantins (TO) emerged with the lowest numbers. The States with the highest absolute values applied in fines were São Paulo (SP), Minas Gerais (MG), Rio de Janeiro (RJ), Rio Grande do Sul (RS), Amazonas (AM) and Pará (PA) (both last ones are located in the rainforrest region and represent one of the main sites where some taxon are captured), unlike the States of Maranhão (MA) and Tocantins (TO), which had the lowest absolute number.

We emphasize that the animals confiscated by Brazilian environmental agencies represent only a portion of the damage [8]. The task of estimating the amount of animals withdrawn from nature per year becomes even more difficult if we consider that the possession of a wild animal captured from nature in Brazil is a common practice, despite being prohibited by law. However, IBAMA's Department of Fish and Wildlife points out that the CETAS alone received in 2008 more than 60,000 animals and were destined more than 40,000 (Figure 4). We noticed, yet, that this number is still small. It happens because most of the animals confiscated in actions of inspection are released into the wild, due to the fact that they are still in savage condition.

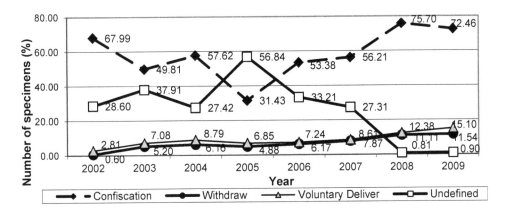

Figure 3. Number of wild animals received by CETAS and their different forms of admission.

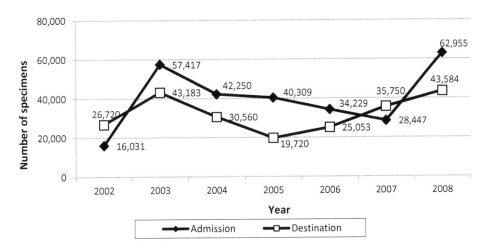

Figure 4. Relationship between admission and destination of specimens in CETAS between 2002 and 2008, in absolute numbers.

The Figure 5 and the Table 2 gather the destinations given to animals from CETAS, between 2002 and 2009. We observed that the releases, after the declining trend observed between 2004 and 2007, re-emerged as the main destination given to the confiscated animals in Brazil, reaching almost 23,000 specimens released into the wild in 2008. The placement in captivity, widely used in 2006 and 2007, has a lower incidence from 2008 on, with the publication of new normative instruments, which regulated the policy to native and exotic wild animals in captivity.

The number of deaths recorded in CETAS suffered variations over the sample period, but their values remained between 16 and 26 percent. The values of escapes/evasions remained constantly low if compared with the total number of destinations.

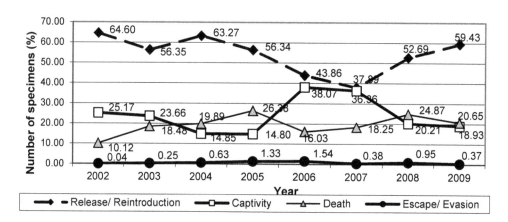

Figure 5. Destination of the animals from CETAS between 2002 and 2009.

Year	Admissions	Releases/ Reintroductions	Captivities	Deaths	Escapes/ Evasions
2002	16,031	17,260	6,725	2,705	12
2003	57,417	24,333	10,219	7,980	110
2004	42,250	19,336	4,538	6,078	191
2005	40,309	11,110	2,919	5,202	263
2006	34,229	10,988	9,537	4,015	386
2007	28,447	13,544	12,998	6,523	137
2008	62,955	22,965	8,809	10,839	413

Table 2. Number of specimens destined by CETAS between 2002 and 2008.

Figure 6. Released birds in Bahia (BA)

3. The species of confiscated animals

We found that the destination given to confiscated animals in Brazil is directly linked to the taxonomic class of animal (Figure 7). For Birds, the main form of destination was release into the wild (greater than 55%), followed by placement in captivity and death. Release into the wild was also the main destination given to reptiles (~ 60%) and mammals (~ 45%). We also noticed that reptiles obtained a lower death rate, while the exotics animals remained in captivity (~ 60%).

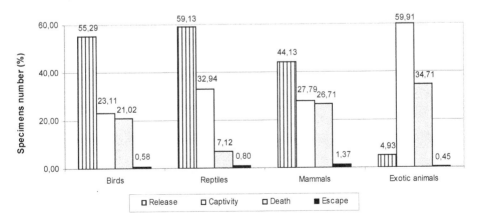

Figure 7. Destination of the animals from CETAS between 2002 and 2009, by taxonomic group.

The Table 3 presents the amount of admission and destination of animals from CETAS. Birds represented 81% of admitted specimens and 82% of released ones, between 2002 and 2009. The Birds was also the group that obtained the largest number of deaths registered (86%). In Australia and Asia, Reptilia was most targeted group of taxa for illegal trade, being also the most seized [4, 9].

Group	Entrance	Release	Captivity	Death	Escape
Birds	250,206	108,622	45,395	41,294	1,135
Reptiles	34,835	17,198	9,581	2,072	233
Mammals	17,936	7,233	4,554	4,377	225
Exotics	4,577	44	535	310	4

Table 3. Amount of specimens that entered and left the CETAS between 2002 and 2009.

We listed, in the Table 4, the 30 species most confiscated by IBAMA and accredited institutions between 2005 and 2009, according to SICAFI. The class Aves was the most representative (80%), followed by Reptilia (16.67%). The most significant families were Emberizidae (30%), Thraupidae (13.33%) and Podocnemididae (10%). The most commonly confiscated species was *Sicalis flaveola* (Saffron Finch), followed by *Saltator similis* (Green-winged Saltator) and *Sporophila caerulescens* (Double-collared Seedeater), (Figure 8).

Classif	Type	Class	Family	Specie[3]	Common name
1º	Wild animal	Aves	Emberizidae	*Sicalis flaveola* (Linnaeus, 1766)	Saffron Finch
2º	Wild animal	Aves	Thraupidae	*Saltator similis* d'Orbigny & Lafresnaye, 1837	Green-winged Saltator
3º	Wild animal	Aves	Emberizidae	*Sporophila caerulescens* (Vieillot, 1823)	Double-collared Seedeater
4º	Wild animal	Aves	Cardinalidae	*Cyanoloxia brissonii* (Lichtenstein, 1823)	Ultramarine Grosbeak
5º	Wild animal	Aves	Emberizidae	*Sporophila angolensis* (Linnaeus, 1766)	Chestnut-bellied Seed-Finch
6º	Wild animal	Reptilia	Podocnemididae	*Podocnemis expansa* Schweigger, 1812	Giant South American River Turtle
7º	Wild animal	Aves	Icteridae	*Gnorimopsar chopi* (Vieillot, 1819)	Chopi Blackbird
8º	Domestic	Aves	Phasianidae	*Gallus gallus*	Domestic

[3] Nomenclature according to the updated lists CBRO, 2011 (www.cbro.org.br) and SBH, 2011 (http://www.sbherpetologia.org.br)

Classif	Type	Class	Family	Specie[3] (Linnaeus, 1758)	Common name
	animal			(Linnaeus, 1758)	Chicken
9º	Wild animal	Aves	Thraupidae	*Paroaria dominicana* (Linnaeus, 1758)	Red-cowled Cardinal
10º	Wild animal	Aves	Emberizidae	*Sporophila lineola* (Linnaeus, 1758)	Lined Seedeater
11º	Wild animal	Reptilia	Podocnemididae	*Podocnemis sextuberculata* Cornalia, 1849	Six-tubercled Amazon River Turtle
12º	Wild animal	Aves	Emberizidae	*Zonotrichia capensis* (Statius Muller, 1776)	Rufous-collared Sparrow
13º	Wild animal	Aves	Emberizidae	*Sporophila nigricollis* (Vieillot, 1823)	Yellow-bellied Seedeater
14º	Wild animal	Aves	Emberizidae	*Sporophila collaris* (Boddaert, 1783)	Rusty-collared Seedeater
15º	Wild animal	Aves	Psittacidae	*Amazona aestiva* (Linnaeus, 1758)	Blue-fronted Parrot
16º	Wild animal	Reptilia	Alligatoridae	*Caiman crocodilus* (Linnaeus, 1758 [originally Lacerta])	Common Caiman
17º	Wild animal	Aves	Turdidae	*Turdus rufiventris* Vieillot, 1818	Rufous-bellied Thrush
18º	Wild animal	Aves	Thraupidae	*Paroaria* sp. Bonaparte, 1832	Cardinal
19º	Wild animal	Aves	---	Not specified	Bird
20º	Wild animal	Aves	Columbidae	*Zenaida auriculata* (Des Murs, 1847)	Eared Dove
21º	Wild animal	Aves	Emberizidae	*Sporophila albogularis* (Spix, 1825)	White-throated Seedeater
22º	Domestic animal	Mammalia	Bovidae	*Bos* taurus Linnaeus, 1758	Domestic Cattle
23º	Wild animal	Aves	Psittacidae	Many species	Parrot
24º	Wild animal	Aves	Fringillidae	*Sporagra magellanica* (Vieillot, 1805)	Hooded Siskin
25º	Wild animal	Reptilia	Podocnemididae	*Podocnemis unifilis* (Troschel, 1848)	Yellow-spotted Amazon River Turtle
26º	Wild	Aves	Icteridae	*Icterus jamacaii*	Campo Troupial

Classif	Type animal	Class	Family	Specie[3] (Gmelin, 1788)	Common name
27º	Wild animal	Aves	Emberizidae	*Sporophila maximiliani* (Cabanis, 1851)	Great-billed Seed-Finch
28º	Wild animal	Reptilia	Testudinidae	*Chelonoidis* sp. Fitzgerald, 1835	Tortoise
29º	Wild animal	Aves	Turdidae	*Turdus* sp. Linnaeus, 1758	Thrush
30º	Wild animal	Aves	Thraupidae	*Lanio cucullatus* (Statius Muller, 1776)	Red-crested Finch

Table 4. Most confiscated species by IBAMA and partner institutions between 2005 and 2009.

A

B   C

Figure 8. The three most confiscated species by environmental enforcement in Brasil: A. *Sicalis flaveola* (Saffron Finch), B. *Saltator similis* (Green-winged Saltator) and C. *Sporophila caerulescens* (Double-collared Seedeater).

According to studies conducted by [8] in south Brazil, the most commonly confiscated species by enforcement, between 1998 and 2000, was the Cardinal (*Paroaria coronata*), followed by the Saffron Finch (*Sicalis flaveola*). And the Emberizidae family presented the largest number of seized specimens, compelling evidence that the great interests of the illegal trade are the songbirds.

The Emberizidae family also excelled in seizures conducted in southeastern and northeastern Brazil [7,10,11]. According to the authors, that fact can be explained, preliminarily, because that family has many species and specimens, for being abundant in the Neotropics, for having easy occurrence in the sampled region, for the high quality of its singing, due to its low market value and for being easy to maintain. Generally, the birds most wanted for trafficking are the songbirds or those able to become pets, confering them high values of trade [8].

Some species listed in Table 4 are exclusively Amazonian, as the Giant South American river turtle, Six-tubercled Amazon River turtle and Yellow-spotted Amazon River Turtle, all very popular in regional cuisine and found in nature in large populations. The domestic chicken (*Gallus gallus*) and domestic cattle (*Bos taurus*) obtained national prominence in seizures, because the first is often used in arenas, being the target of actions against animal abuse, and the second is the subject of crime in embargoed areas due to deforestation, mainly in the Amazon region.

We also observed an intrinsic relationship between the passerines authorized breeding and the wild animals trafficking: the five species more seized are also the taxa of greatest interest for commercial and amateur breeders of passerines (Table 5). All other passerines listed in Table 4 are species authorized for commercial and amateur activity.

Classif.	Species[4]	Common name	Total of breeders	Total of specimens
1º	*Saltator similis* d'Orbigny & Lafresnaye, 1837	Green-winged Saltator	133.699	528.621
2º	*Sporophila angolensis* (Linnaeus, 1766)	Chestnut-bellied Seed-Finch	89.083	535.195
3º	*Sporophila caerulescens* (Vieillot, 1823)	Double-collared Seedeater	86.666	279.888
4º	*Sicalis flaveola* (Linnaeus, 1766)	Saffron Finch	83.281	444.160
5º	*Cyanoloxia brissonii* (Lichtenstein, 1823)	Ultramarine Grosbeak	46.364	108.703
6º	*Sporagra magellanica* (Vieillot, 1805)	Hooded Siskin	28.709	83.885
7º	*Turdus rufiventris* Vieillot, 1818	Rufous-bellied Thrush	27.250	57.960
8º	*Saltator maximus* (Statius Muller, 1776)	Buff-throated Saltator	19.129	53.203

[4] Nomenclature according to the updated lists CBRO, 2011 (www.cbro.org.br)

Classif.	Species[4]	Common name	Total of breeders	Total of specimens
9º	*Sporophila maximiliani* (Cabanis, 1851)	Great-billed Seed-Finch	18.142	123.832
10º	*Zonotrichia capensis* (Statius Muller, 1776)	Rufous-collared Sparrow	16.466	32.677
11º	*Sporophila lineola* (Linnaeus, 1758)	Lined Seedeater	13.868	25.317
12º	*Gnorimopsar chopi* (Vieillot, 1819)	Chopi Blackbird	12.540	21.716
13º	*Cyanoloxia cyanoides* (Lafresnaye, 1847)	Blue-black Grosbeak	11.435	23.435
14º	*Paroaria coronata* (Miller, 1776)	Red-crested Cardinal	11.310	33.110
15º	Sporophila frontalis *(Verreaux, 1869)*	Buffy-fronted Seedeater	9.301	22.073
16º	*Sporophila nigricollis* (Vieillot, 1823)	Yellow-bellied Seedeater	9.264	22.135
17º	*Molothrus oryzivorus* (Gmelin, 1788)	Giant Cowbird	8.878	18.858
18º	*Lanio cucullatus* (Statius Muller, 1776)	Red-crested Finch	6.922	13.635
19º	*Saltator fuliginosus* (Daudin, 1800)	Black-throated Grosbeak	6.756	14.533
20º	*Paroaria dominicana* (Linnaeus, 1758)	Red-cowled Cardinal	6.123	11.675

Table 5. Species of greatest interest to the passerine breeders in Brazil.

For [1], one of the ways to reduce the pressure on the populations for trafficking would be the encouragement of captive breeding programs to meet commercial demand. However, this strategy can be of great concern, since those animals cannot achieve the low prices offered by the trafficking [7].

4. The trafficking routes

In Figure 9 we grouped the main trafficking routes of wild animals in Brazil, including major airports, trade and source areas. We observed that, in general, the Brazilian fauna has been removed from the North, Northeast and Midwest of the Country and it is being sent to the Southeast, South and other regions of Northeast, by land or river, fuelling the national trade. In relation to the international illegal trade, we emphasize cities located in border regions in the North, Midwest and South of Brazil, as well as in ports and airports located in the Northern, Northeastern, Southern and Southeastern Brazilian regions.

For [1], beyond the States of Para (PA) and Amazonas (AM), which had national prominence in the amount of fines, other Amazonian frontiers must be of particular concern, such as the borders with the Guianas, Venezuela and Colombia, and the route of the Madeira River.

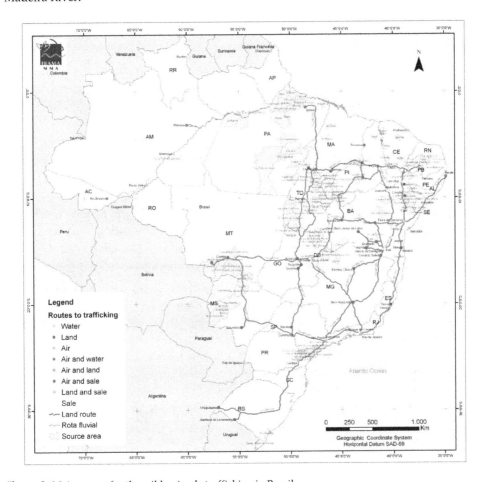

Figure 9. Main routes for the wild animals trafficking in Brazil.

The situation at the tri-border area (Brazil, Paraguay and Argentina) is also a matter of worry. According to [2], many animals are taken from the Iguaçu National Park and illegally sold during daylight or taken by peddlers to other Brazilian regions. Also in southern Brazil, the authors highlight as important areas for capturing and trading wild animals the towns of Laranjeiras do Sul (PR) and Santana do Livramento (RS), close to the border with Uruguay.

In [10], also emphasized the trafficking in the Southwest Bahia (BA) region, and they say that it is a socio-environmental problem with serious consequences to the local avifauna. According to them, the main trade in this region occurs along the BR-116 road, as well as in fairs and small shops roadside.

Specialists point the absence of alternative income for people who use the trafficking as a means of livelihood. The report elaborated by the Brazilian National Congress in 2001 [12] recommends that the Union, States and Municipalities, in an articulated manner, must develop and implement programs to generate alternative income for poor communities involved in the illegal trade of wild animals.

However, the impact of trafficking in society needs to be further studied and its actors mapped. The capture of animals in nature is part of the culture and popular tradition, being one of the main livelihoods of the poor in some regions of Brazil [10]. However, in [11] found that in many regions people are using the illegal trade of animals only as an additional source of income. Thus, mechanism for control of wildlife use and trade should be formulated that take into consideration the special ethnic conditions of each region [13].

In global scale, it is recommended a multi-pronged approach including community-scale education and empowering local people to value wildlife, coordinated international regulation, and a greater allocation of national resources to on-the-ground enforcement for effective control of trafficking and illegal trade [9]. In Brazil, we noticed that the actions against illicit related to wildlife, although increasingly more organized and efficient, still require specific structural measures, among which we may highlight:

- Improving the number and the practice of IBAMA's agents and of Environmental Military Policeman through public competition and specific and continuous training;
- Increase the volume of public resources towards the activities of control and environmental monitoring;
- Increasing the incentive for the creation, implementation and maintenance of CETAS (Wild Animals Rehabilitation Centers);
- Reviewing the penal types of Law number 9.605/1998 due to provide harsher penalties for those who engage in wild animals trafficking such as large-scale commercial activity or international and interstate trafficking;
- Increasing responsibilities and sharing information among different agencies responsible for controlling and monitoring, through formal terms and shared systems;
- Maintening permanent negotiation between the federal government and neighboring countries through bilateral agreements, so that policies or environmental standards more flexible than the Brazilian ones are not used to support the illegal activities;
- Increasing the control over the sale of wild animals by internet and their exit to abroad through joint action among different government agencies such as IBAMA, the Federal Revenue Secretariat, Ministry of Health, Federal Police, etc..;
- Promoting Specific Environmental Education Campaigns aimed at minimizing the wild animals trafficking, as well as joint efforts among the various ministries involved, including the ones of Transport, Environment, Health and Tourism.

Lastly, we hope that this paper provides important and necessary subsidies for the decision-making to combat the animal trafficking in Brazil and abroad, helping the effective protection and conservation of the nature.

5. Conclusion

We conclude that the Minas Gerais State was the largest contributor to the large volume of specimens seized in Brazil in the analized term, being *Sicalis flaveola* (Saffron Finch), *Saltator similis* (Green-winged Saltator) and *Sporophila caerulescens* (Double-collared Seedeater) the species most confiscated by environmental enforcement.

Furthermore, we noticed that releasing into the wild was the most common destination for mammals, birds and reptiles seized. The Wild Animals Rehabilitation Centers are essential support structures for the environmental enforcement actions related to fauna in Brazil.

Author details

Guilherme Fernando Gomes Destro, Tatiana Lucena Pimentel, Raquel Monti Sabaini,
Roberto Cabral Borges and Raquel Barreto
*Coordination of Enforcement Operations, Brazilian Institute of Environment and Renewable Natural
Resources – IBAMA. SCEN, Trecho II, Ed. Sede, Brasília/DF, Brazil*

6. References

[1] RENCTAS (2011) Rede Nacional de Combate ao Tráfico de Animais Silvestres. 1º Relatório Nacional sobre o Tráfico de Fauna Silvestre. Available: http://www.renctas.org.br/. Acessed 2010 jul 15.

[2] Hernandez, E.F.T.; Carvalho, M.S. de. (2006) O tráfico de animais silvestres no Estado do Paraná. Acta Scientiarum: Human and Social Sciences, j. 28, n. 2: 257-266.

[3] Wasser, S.K.; Clark, W.J.; Drori, O.; Kisamo, E.S.; Mailand, C.; Mutayoba, B.; Stephens, M. (2008) Combating the Illegal Trade in African Elephant Ivory with DNA Forensics. Conservation Biology, j. 22, n. 4: 1065–1071.

[4] Alacs, E.; Georges, A. (2008) Wildlife across our borders: a review of the illegal trade in Australia. Australian Journal of Forensic Sciences. j. 40, n. 2: 147–160.

[5] Johnson, R.N. (2010). The use of DNA identification in prosecuting wildlife-traffickers in Australia: do the penalties fit the crimes? Forensic Scienci, Medicine and Pathology, j. 6: 211–216.

[6] Pavlin, B.I.; Schloegel, L.M.; Daszak, P. (2009) Risk of Importing Zoonotic Diseases through Wildlife Trade, United States. Emerging Infectious Diseases, j. 15, n. 11: 1721-1726.

[7] Borges, R.C.; Oliveira, A. de; Bernardo, N.; Martoni, R.; Costa, M.C. da. (2006) Diagnóstico da fauna silvestre apreendida e recolhida pela Polícia Militar de Meio Ambiente de Juiz de Fora, MG (1998 e 1999). Revista Brasileira de Zoociências, j. 8, n. 1: 23-33.

[8] Ferreira, C.M.; Glock, L. (2004) Diagnóstico preliminar sobre a avifauna traficada no Rio Grande do Sul, Brasil. Biociências, j. 12, n. 1: 21-30.

[9] Rosen, G.E.; Smith, K.F. (2010) Summarizing the evidence on the international trade in illegal wildlife. EcoHealth, j. 7: 24–32.

[10] Souza, G.M. de; Soares Filho, A. de O. (2005) O comércio ilegal de aves silvestres na região do Paraguaçu e Sudoeste da Bahia. Enciclopédia Biosfera, j. 1: 1-10.

[11] Rocha, M. da S.P.; Cavalcanti, P.C. de M.; Souza, R. de L.; Alves, R.R. da N. (2006) Aspectos de comercialização ilegal de aves nas feiras livres de Campina Grande, Paraíba, Brasil. Revista de Biologia e Ciências da Terra, j. 6, n. 2: 204-221.

[12] BRASIL (2001) Comissão Parlamentar de Inquérito destinada a investigar o tráfico ilegal de animais e plantas silvestres da fauna e da flora brasileiras – cpitrafi. Relatório Final. Available: <http://www2.camara.gov.br/atividade-legislativa/comissoes/comissoes - temporarias/parlamentar-de-inquerito/51- legislatura/cpitrafi/relatorio/relatoriofinal.pdf>. Acessed 2010 sep 20.

[13] Yi-Ming L., Zenxiang, G.; Xinhai, L.; Sung, W.; Niemela, J. (2000) Illegal wildlife trade in the Himalayan region of China. Biodiversity and Conservation, j. 9: 901–918.

Genetics and Hereditary

Tree Species Diversity and Forest Stand Structure of Pahang National Park, Malaysia

Mohd Nazip Suratman

Additional information is available at the end of the chapter

1. Introduction

Information on composition, diversity of tree species and species-rich communities is of primary importance in the planning and implementation of biodiversity conservation efforts. In addition, the diversity of trees is fundamental to the total tropical rainforest diversity as trees provide resources and habitat structure for almost other forest species (Cannon et al., 1998). According to Singh (2002), biodiversity is essential for human survival and economic well being and ecosystem function and stability. UNEP (2001) reported that habitat destruction, over exploitation, pollution and species introduction are identified as major causes of biodiversity loss. Hubbel et al. (1999) mentioned that disturbances created by these factors determine forest dynamics and tree diversity at the local and regional scales. These disturbances have been considered as an important factor structuring communities (Sumina, 1994).

In forest management operations, inventories on biodiversity are used to determine the nature and distribution of biodiversity region at the region being managed. Quantification of tree species diversity is an important aspect as it provides resources for many species (Cannon et al., 1998). Being a dominant life form, trees are easy to locate precisely and to count (Condit et al., 1996) and are also relatively better known, taxonomically (Gentry, 1992).

While Pahang National Park provides both fully-protected habitats and long-term maintenance of biological diversity, the structure and composition of its flora still remain rather insufficiently known. To protect forests from declining, it is essential to examine the current status of species diversity as it will provide guidance for the management of protected areas. Therefore, using Kuala Keniam forest as an example, a study was conducted to describe quantitatively stand structure of the forests of Kuala Keniam within Pahang National Park, and to determine the level of species composition, diversity and distribution in this area. Information from this quantitative inventory will provide a valuable reference for forest assessment and improve our knowledge in identification of ecologically useful species as well as species of special concern, thus identify conservation efforts for sustainability of forest biodiversity.

2. Materials and methods

2.1. Description of study area

The data for this study were collected from Kuala Keniam forest, Pahang National Park, Malaysia (latitude 4° 31' 07.17" N, longitude 102° 28' 31.26" E) which ranges about 120 – 200 m above sea level. Kuala Keniam is located at the protected lowland dipterocarp forests within the national park in the state of Pahang. The area is administered by the Department of Wildlife and National Park (DWNP) Malaysia in collaboration with the Universiti Teknologi MARA (UiTM) which operates a research station in the area.

The weather in Pahang National Park is characterized by permanent high temperatures ranging from 20°C at night and 35°C in the day with a high relative humidity (above 80%). Periods of sunshine in the morning are usually followed by heavy thunderstorms in the afternoon, sometimes accompanied by severe gusts of wind. The highest rainfall occurs in October to November with about 312 mm of rainfall. The lowest rainfall occurs in March with only about 50 mm of rain. Sedimentary rocks account for about 83% of National Park. The last formation of sedimentary rocks belongs to the Cretaceous-Jurassic era which exists in Kuala Keniam and its vicinity. The rocks are thick cross-bedded sandstone deposits with subordinate conglomerates and mudstones. The topography consists mainly of lowland, undulating and riverine areas and gently rolling hills with slopes of between 5° to 45°.

The overall vegetation type in Pahang National Pahang is lowland dipterocarp forests in which is characterized by high proportion of species in the family of Dipterocarpaceae with Meranti (*Shorea* spp.) and Keruing (*Dipterocarpus* spp.) as the dominant species. Lowland dipterocarp forest is one of the most rich-species communities in the world, with more than 200 species ha^{-1} (Okuda *et al.*, 2003). Other vegetation communities in Pahang National Park range from the humid rainforests of the lowland, to the montane oak and ericaceous forests in the higher elevation. The highest peak is Mount Tahan 2,187 m, which also the highest point in Peninsular Malaysia. Tahan River and Tembeling River are the headstream tributaries of Pahang National Pahang with the presence riparian tree species, i.e., Gapis (*Saraca multiflora*), Keruing neram (*Dipterocarpus oblongifolius*), Merbau (*Intsia palembanica*), Kasai daun bersar (*Pometia pinnata*) and Melembu (*Pterocambium javanicum)*, along river banks. The rainforest consists of tall evergreen trees which attain heights between 30 – 50 m (i.e., Tualang - *Koompassia excelsa*).

2.2. Sampling design and data collection

A topographic map was used to locate the existing forest trails and baselines in the forest area. A total of five transect lines of 100 m in length and 20 m in width (abbreviated as T1, T2, T3, T4 and T5 thereafter) were established in east-west direction using a compass (Table 1, Figure 1).

Each transect line was gridded into five plots, each 20 m × 20 m in size, as workable units. These transect lines were perpendicular to the existing baseline in the forest area and constructed 5 m after the line. The topographic position, including the gradient was measured at each plot. The slope was measured using a clinometer. A tape measure was used to mark the transect lines at the intervals of 20 m. All trees with a diameter at breast height (DBH, 1.3 m above the ground) above 10 cm were measured, tagged and identified by species. The DBH was measured using a DBH tape. If field identification was not possible, the botanical specimens were taken to the herbarium section of the Forest Research Institute Malaysia (FRIM) for identification.

Transect	No. of plots	Area (m²)	Slope (°)	Topography
T1	5	20 × 20	5 – 35	Steep lower slope with riverine areas
T2	5	20 × 20	3 – 20	Gentle to mid-slope
T3	5	20 × 20	0 – 10	Mainly flat and gentle slope
T4	5	20 × 20	3 – 30	Mid-slope with riverine areas
T5	5	20 × 20	0 – 10	Mainly flat and gentle slope

Table 1. General features of sample plot within the five transect lines of the study area.

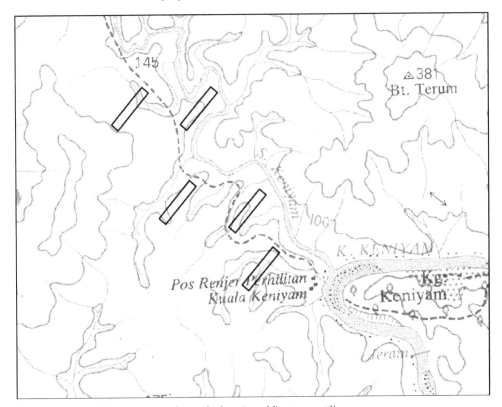

Figure 1. A map of the study area shows the location of five transect lines.

2.3. Data analysis

The means of basal area, genera, species and stem per hectare were calculated for each transect line. One-way analysis of variance (ANOVA) was used to test the differences between the means of these parameters using SAS system (SAS Institute, 2000). The relative dominance of species in each transect line was identified on the basis of relative basal area. The relative basal area of a species on transect lines was calculated as the basal area of a species divided by total basal area of the site and multiplied with 100. The dominant and co-dominant species of each site were identified based on this value. The species with the highest relative basal area was defined as dominant and that with the second highest relative basal area was defined as co-dominant.

In this study, the stand structure was described based on the distribution of species in the study sites and distribution of trees by diameter classes. Therefore, the tree data were grouped into 5 cm diameter classes e.g., the class boundaries were 10 – 14.9, 15 – 19.5 cm, etc. These gave a frequency of trees in each diameter class and were then used to draw bar char graphs.

2.4. Basal area

Basal area is a measure of tree density that defines the area of a given section of land that is occupied by the cross-section of tree. Basal area (BA) is calculated using the following equation that converts the DBH in cm to the basal area in m².

$$BA = \pi r^2$$

$$= 3.142 \times \left(\frac{dbh}{200}\right)^2$$

Where

BA = tree basal area (m2)
r = radius (cm)

2.5. Species diversity, richness and evenness indices

A variety of different diversity indices can be used as measures of some attributes of community structure because they are often seen as ecological indicators (Magurran, 1988). Diversity indices provide important information about rarity and commonness of species in a community. The indices can be used to compare diversity between habitat types (Kent and Coker, 1992). The comparison can be between different habitats or a comparison of one habitat over time. Different diversity, species richness, species evenness indices were calculated for each transect as well as pooled data for all transects.

a.   Shannon-Weiner diversity index (H') (Shannon and Weiner, 1949) is calculated using the following equation:

$$H' = \sum_{i=1}^{S} p_i \ln p_i$$

Where

H'= the Shannon-Wiener index
p_i= the proportion of individuals belonging to species i
ln=the natural log (i.e., 2.718)

b.   The species richness (number of species per unit area) was calculated using Margalef index of species richness (Margalef, 1958) as follows:

$$SR = \frac{S - 1}{\ln(N)}$$

Where

SR=the Margalef index of species richness
S =the number of species
N =the total number of individuals

c.   The Whittaker's index of species evenness (Whittaker, 1972) was calculated using the following equation:

$$E_w = \frac{S}{\ln N_i - \ln N_s}$$

Where

E_w=the Whittaker's index of evenness
N_i=the abundance of most important species
N_s=the abundance of the least important species

d. α-diversity was measured based on unified indices (exponential Shannon-Weiner index and Simpson's diversity) as follows:

$$N_1 = \exp^{H'}$$

Where

N_1=the number of equally common species
H'=the Shannon-Weiner index

e.   Simpson's diversity (D) (Simpson, 1949) was calculated using the following equation:

$$D = 1 - \lambda$$

Where

D=the Simpson diversity

λ= the Simpson's concentration of dominance calculated as $\sum p_i^2$.

f. The Whittaker's index of β-diversity (Whittaker, 1972) was calculated as:

$$\beta w = \frac{S_c}{\bar{S}}$$

Where

β_w= the Whittaker's index of β-diversity
S_c=the total number of species
$\bar{S}$= the average number of species per sample

g. Bray-Curtis index (C_N) (Bray and Curtis, 1947), a similarity coefficient, is used to measure similarity between transect lines.

$$C_N = \frac{2jN}{(aN + bN)}$$

Where

C_N=the Bray-Curtis index
aN=individual numbers of plot A
bN=individual numbers of plot B
jN= the sum of less individual numbers of each species common in plots A and B

3. Results and discussion

3.1. Stand structure analysis of different sites

Information on the basal area, stem, species and genera densities are efficient expression for revealing forest stand structure and spatial distribution of trees present in the landscape. These four parameters are presented in Table 2. In this study, the means of basal area ha^{-1}, stem ha^{-1}, species ha^{-1} and genera ha^{-1} were measured in every plot (20 m × 20 m) and were averaged to provide an estimate for each transect line. From the analysis of variance, it was found that the difference in the means of these parameters among transects were not statistically significant at P≤0.05.

The mean of basal area obtained in the present study ranged from 17.2 m^2 ha^{-1} (T4) to 34.3 m^2 ha^{-1} (T3) (Table 2), which is lower compared to those recorded in other tropical rainforests. Examining the structure and composition of lowland tropical rainforests in north Borneo, Burgess (1961) recorded a basal area of 73.6 m^2 ha^{-1} (≥ 10 cm DBH) over a small area (0.08 ha) at Gum Gum Sabah. In another study in an evergreen forest of Andaman Islands, basal area of 44.6 m^2 ha^{-1} has been recorded in 4.5 ha sampled area (Padalia et al., 2004). A much lower basal area of 29 m^2 ha^{-1} and 5.6 m^2 ha^{-1} have been recorded in logged over forest of Sungkai, Perak (Suratman et al., 2007) and secondary forests of Sungai Sator, Kelantan (Suratman et al., 2009), respectively. Both are secondary forests and were put under a selection system of timber extraction in the past, and are considered to be of poor species.

Variables	T1	T2	T3	T4	T5
Mean basal area (m² ha⁻¹)	25.2	24.8	34.3	17.2	33.3
	(25.2)	(13.3)	(15.1)	(16.0)	(12.2)
Mean no. of stems ha⁻¹	510	430	505	315	480
	(123.3)	(105.2)	(51.2)	(219.8)	(186.6)
Mean no. of species ha⁻¹	370	370	450	280	405
	(105.2)	(105.2)	(30.6)	(190.7)	(125.5)
Mean no. of genera ha⁻¹	340	340	435	250	365
	(72.0)	(72.0)	(51.8)	(165.8)	(109.8)
Total no. of species per individual	0.50	0.69	0.63	0.79	0.64

Note: The values in parentheses are standard deviation. All means for the first four parameters above are not significantly different at $P \leq 0.05$.

Table 2. The stand structure of Kuala Keniam forest.

The density and size distribution of trees contribute to the structural pattern characteristic of rainforests. In primary tropical rainforests, the density of trees varies within the limits and depends on many factors. The means number of species and stems per hectare on different transects varied from 280 (T4) – 450 (T3) and 315 (T4) – 510 (T1), respectively (Table 2), indicating a mixed nature of distribution of species and individuals in the forest at each transect, a characteristic of the tropical rainforests. The factors controlling tree density include the effects of natural and anthropogenic disturbance and soil condition (Richards, 1952). From the field observation, the reserve area of the primary forest in the study sites is generally homogenous, with no evidence of major disturbance, and appeared to be a representative example of the lowland forest of Kuala Keniam.

Information on the density-dependent status of species in the study site is important for conservation and management. Studies have classified the density of trees ha⁻¹ in tropical forests ranges from low values of 245 stems ha⁻¹ (Ashton, 1964; Campbell et al., 1992; Richards, 1996) intermediate values of 420 – 617 stems ha⁻¹ (Campbell et al., 1992) in Brazilian Amazon and high values of 639 – 713 stems ha⁻¹ in Central Amazon upland forests (Ferreira et al., 1998). In the present study, the density of stems per hectare ranged from 315 – 510 stems ha⁻¹, reflecting spatial variability in the sampled sites. The range fell within intermediate category in the above studies. In the Neotropics, the maximum richness is found up to 300 stems ha⁻¹ (Gentry, 1988). A much lower result was reported for forests in Africa where the species richness is about 60 stems ha⁻¹ (Bernhard-reversat et al., 1978).

Tree species composition in tropical areas varies greatly from one place to another mainly due to variation in biogeography, habitat and disturbance (Whitmore, 1998). In the tropical rainforests, the tree species per hectare ranges from about 20 to a maximum of 223 (Whitmore, 1984). Philips and Gentry (1994) reported a range of 56 – 282 species ha⁻¹ (>10 cm DBH) in mature tropical forests. In the present study, a range of 280 to 450 species ha⁻¹ has been recorded in the lowland rainforest of Kuala Keniam (Table 2). In the very rich rainforests, the number of species in rainforests could be as high as 400 species ha⁻¹ (Nwoboshi, 1982). When compared to some rainforests around the world, the lowland rainforest of Kuala Keniam could be considered to be species rich. Tropical rainforests in

South America harbour 200 – 300 species ha^{-1} (Richards, 1996). In the tropical evergreen forest of Andaman Islands, India, Padalia *et al.* (2004) found that 58 tree species ha^{-1} were recorded belong to 176 genera and 81 families.

The mean numbers of genera per hectare varied from 340 to 435 genera ha^{-1}. These values are much higher than that obtained by Sagar *et al.* (2003) at a dry tropical forest region of India (4 – 22 genera ha^{-1}). T4 had the highest total number of species per individual when compared to the other four sites of study. The difference could be due to genetic and site difference. A study on vegetation types in Yunnan, Chiangcheng *et al.* (2007) found that slope direction had influence on the tree diversity at different altitudes. The tree diversity on the sunny slope was lower than that on shady slope. The difference in terrain, gradient and slope direction causes the difference soil, water and microclimate which may cause of differences in species adaptability.

3.2. Dominant tree species

On the basis of relative basal area, the five sites differed in the combination of dominant and co-dominant species (Appendix). *Elateriospermum tapos* was dominant in T1 and co-dominant in T4. *Koompassia malaccensis* dominated at the T2 and co-dominated at the T3. *Xanthophyllum lelacarum* was dominant in T3 while *Shorea leprosula* was dominant in T4. *Dyera costulata* and *Dipterocarpus costulatus* were dominated and co-dominated at T5, respectively. Thus, the species exhibit local dominance. These data revealed that T1 represented *Elateriospermum-Intsia* community; T2, *Koompassia-Pentaspadon* community; T3, *Xanthophyllum-Koompassia* community; T4, *Shorea-Koompassia* community; and T5, *Dyera-Dipterocarpus* community. Two tree species, i.e., *Alphonsea elliptica* and *Syzygium* sp., are common on all transects.

3.3. Species diversity

The five transect lines yielded a total of 448 stems and 198 species of trees ≥ 10 cm DBH. These species represent 116 genera and 44 families (Appendix). The number of species and individual varied from 50 to 64 species and 63 to 102 individuals per transect of 100 m × 20 m size, respectively. Table 3 shows the summary statistics for various indices of diversity, richness and evenness. It is generally recognized that the area and environmental heterogeneity have strong effects on species diversity (Rosenzweig, 1995; Whitmore, 1998; Waide *et al.*, 1999). The Shannon-Weiner index (H') was used to compare species diversity between transects. The H' for T1–T5 were 3.42, 3.91, 3.97, 3.84 and 3.91, respectively, indicating that among transects, T3 was the most complex in species diversity whereas T1 is the simplest community in terms of species composition. The Shannon-Weiner diversity index (range between 3.42 – 3.91) obtained for trees more than 10 cm DBH in this study was lower than those recorded in the tropical rainforests of Barroo Colorado Island, Panama [4.8](Knight, 1975) and Silent Valley, India [4.89](Singh *et al.*, 1981). In a more recent study in Shenzen, China, Wang *et al.* (2006) recorded a lower range of Shannon-Weiner index (i.e., 1.92 – 3.10) for trees ≥ 2 cm DBH in a subtropical forest. However, a comparison of diversity indices obtained in the present study with the ones above is difficult due to vast differences in the area sampled, plot size, and the standard diameter class taken.

Variables	T1	T2	T3	T4	T5
Shannon-Weiner index (H')	3.42	3.91	3.97	3.84	3.91
Margalef index of species richness (SR)	10.81	13.02	13.65	11.83	13.15
Whittaker index of evenness (E_w)	16.04	36.66	35.72	44.60	31.35
The number of equally common species (N_1)	30.72	49.72	53.11	46.55	49.75
Simpson's diversity (D)	0.93	0.98	0.98	0.98	0.97
Whittaker index of β-diversity (β_w)	3.51	3.88	3.56	4.46	3.77

Table 3. Pattern of tree species diversity in Kuala Keniam forest.

Similar patterns were found for species richness, which was computed using Margalef index of species richness (SR) and the number of equally common species (N_1). The SR ranged from 10.81 to 3.97 and the N_1 ranged from 30.72 to 53.11. Whittaker index of evenness (E_w) ranged from 16.04 to 44.60, the highest value was recorded at T4 and the lowest at T1. In the present study, Simpson's diversity (D) was not a very sensitive indicator of diversity as four of five sites (T2 – T5) had somewhat similar values. Whittaker index of β-diversity (β_w) was used to compare habitat heterogeneity within a transect. The β_w value was the highest for T4 (4.46) and the lowest for T1 (3.51). Further analysis indicated that the number of species per individual had a direct positive influence on β-diversity (Figure 2). According to Condit *et al.* (1998), species richness is positively associated with species abundance. This relationship suggests that large population is less prone to extinction than small ones (Preston, 1962). Based on the relationship between abundance and diversity, habitats supporting larger numbers of individuals can support more populations and more species than habitat supporting small number of individuals.

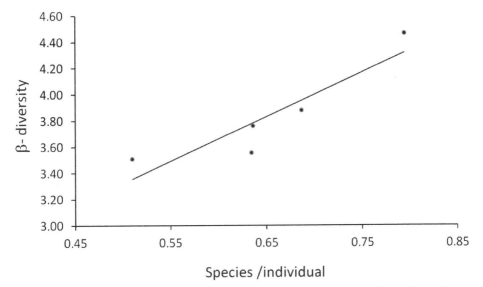

Figure 2. Relationship between β-diversity (β_w) and species/individual (S_n) according to $\beta_w = 1.624 + 3.393 S_n$, $r^2 = 83$, $p = 0.03$.

3.4. Similarity between transects

The similarity based on Bray-Curtis index (C_N) was calculated between the pair of transects, and abundance similarity matrix was constructed (Table 4). The Bray-Curtis similarity index was used because it is often a satisfactory coefficient for biological data on community structure (Clarke and Warwick, 1994). Comparison of C_N values among the five transects data indicates that the species composition of T1 was fairly different from those of the other four sites. T3 had a high species similarity to T4 and T5, and T4 had a high species similarity to T5. T2 was similar to some degree to T4 and T5.

Transect	T1	T2	T3	T4
T2	0.15			
T3	0.14	0.19		
T4	0.13	0.13	0.27	
T5	0.14	0.11	0.37	0.29

Table 4. Similarity coefficient among the five transects of Kuala Keniam forest.

3.5. Family-wise distribution

A total of 44 tree families were encountered in the forest of Kuala Keniam (Figure 3). The maximum number of tree species belongs to the family of Euphorbiaceae which accounts for 23.9% of the total individuals encountered in the study site. *Elateriospermum tapos* is the most widely occurring species from this family. Other trees from this family such as *Macaranga lowii, Mallotus leucodermis* and *Pimelodendron griffithianum* are among the important part of floristic composition in the study area. The other dominant families are Myristicaceae, Burseraceae, and Leguminosae which account for 8.3%, 5.4% and 4.5% of the total individual encountered in the study site, respectively. The fifth most dominant family is Myrtaceae with 4.2%. Earlier study also indicated that Euphorbiaceae was the dominant family in Sungkai forest with 27% of tree species belong to this family (Suratman *et al.*, 2007). Two other studies conducted in India for tree species also support the fact that Euphorbiaceae is the dominant family in Bay Islands (Dagar and Singh, 1999) and Andaman Islands (Padalia *et al.*, 2004). The dominant plant family in Neotropical lowland forests and Africa is Leguminosae (Gentry, 1988) and in Southeast Asia the dominants are Dipterocarpaceae (Richards, 1952; Whitmore, 1998).

3.6. Diameter class distribution

The stand structure of lowland rainforests of Kuala Keniam forest was studied based on the distribution of tree diameter class. The diameter distribution of trees is very variable and some forests have large numbers of trees of 40 – 60 cm DBH (Richards, 1952). In this study, the distribution of trees clearly displays the characteristic of De iocourt's factor procedure (inverse J distribution) where stems frequencies decrease with the increase in DBH (Figure 4). This generally indicates that stands are developing and regeneration in the forest is

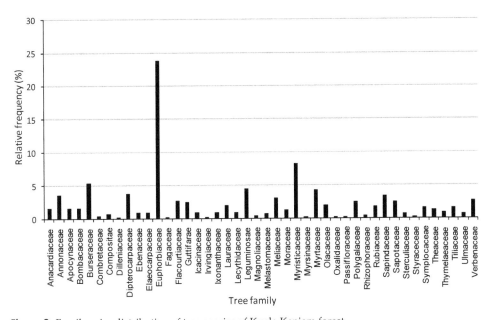

Figure 3. Family-wise distribution of tree species of Kuala Keniam forest.

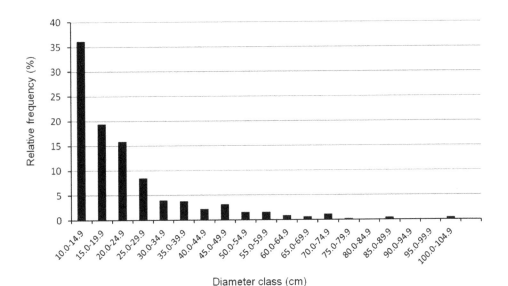

Figure 4. Diameter distribution of trees at Kuala Keniam forest.

present. Natural regeneration is dependent on the availability of mother trees, fruiting pattern and favourable conditions. As shown in the figure, the presence of growth of the forest is indicated by the movement of trees in various diameter classes. Higher number of stems for smaller diameter classes, with 36% of trees fell within the 10 – 14.9 cm, 19% fell within 15 – 19.9 cm, 16% fell within 20 – 24.9 cm, 9% fell within 25 – 29.9 cm and 4% fell within 30 – 34.9 cm. The histogram shows a less or an absent number of stems in diameter classes from 79.9 cm onwards. Under natural conditions, an old, big emergent tree may fall down and create gap. Forest regeneration via natural succession will take place if the area is not too far away from mature primary forest trees serving as source for the recalcitrant seeds.

4. Conclusion

The forests of Kuala Keniam are protected primary forests which comprises of natural vegetation and are dictated by a combination of biotic and abiotic factors like topography, altitude, geology, climatic etc. as well as historical conditions of geology and climate. The density and size distribution of trees contribute to the structural pattern characteristics of the forest. The study indicated that the forests of Kuala Keniam are characterized by a uniform distribution of individuals with mixed species composition, and the sites are represented by different combinations of the dominants and co-dominant species. The distribution of trees displays the characteristic of De iocourt's factor procedure (inverse J distribution) where stems frequencies decrease with the increase in DBH, indicating stable populations in which regeneration of forest in this area is present.

Appendix

List of species, family and the relative basal area of Kuala Keniam forest

Species	Family	T1	T2	T3	T4	T5
Aglaia sp.	Meliaceae	-	2.65	0.23	0.37	0.17
Agrostistachys longifiolia	Euphorbiaceae	-	0.26	-	-	-
Aidia densiflora	Rubiaceae	-	-	0.21	-	0.24
Alphonsea elliptica	Annonaceae	2.37	0.55	1.48	5.91	1.41
Alphonsea jengkasii	Annonaceae	-	-	-	-	0.16
Alseodaphne intermedia	Lauraceae	-	-	0.84	-	-
Anisoptera laevis	Dipterocarpaceae	-	-	4.35	-	-
Antidesma coriaceum	Euphorbiaceae	-	-	0.47	-	-
Antisdesma sp.	Euphorbiaceae	-	0.28	-	-	-
Aporosa arborea	Euphorbiaceae	0.51	-	-	-	-
Aporosa aurea	Euphorbiaceae	-	0.28	-	-	-

Species	Family	T1	T2	T3	T4	T5
Aporosa falcifera	Euphorbiaceae	-	-	-	1.63	0.21
Aporosa globifera	Euphorbiaceae	-	0.25	-	-	-
Aporosa microstachya	Euphorbiaceae	-	-	-	-	0.22
Aporosa nigricans	Euphorbiaceae	-	0.29	-	-	-
Aporosa prainiana	Euphorbiaceae	-	-	-	0.24	0.12
Aporosa symplocoides	Euphorbiaceae	-	-	-	-	0.17
Archidendron ellipticum	Leguminosae	-	-	0.32	-	-
Ardisia sp.	Myrsinaceae	-	0.21	-	-	-
Aromadendron elegans	Myristicaceae	-	1.42	-	-	-
Artocarpus griffithii	Moraceae	-	-	0.48	-	-
Artocarpus lowii	Moraceae	-	-	-	-	0.73
Austrobuxus nitidus	Euphorbiaceae	-	-	-	-	5.10
Baccaurea brevipes	Euphorbiaceae	-	-	-	0.24	-
Baccaurea kunstleri	Euphorbiaceae	-	-	-	0.42	-
Baccaurea minor	Euphorbiaceae	-	0.72	2.77	-	-
Baccaurea reticulata	Euphorbiaceae	-	-	-	0.38	3.46
Barringtonia macrostachya	Lecythidaceae	-	-	0.28	0.37	0.20
Beilschmiedia lucidula	Lauraceae	-	-	0.99	-	-
Blumeodendron kurzii	Euphorbiaceae	-	-	0.82	-	-
Buchanania sessifolia	Anacardiaceae	-	-	0.79	-	0.28
Callicarpa maingayi	Verbenaceae	-	-	-	-	0.39
Callophyllum sp.	Guttiferae	-	0.40	-	-	-
Canarium littorale	Burseraceae	1.20	-	2.00	-	-
Carallia brachiata	Rhizophoraceae	-	-	-	-	0.75
Casearia clarkei	Flacourtiaceae	-	-	0.61	0.94	0.13
Casearia sp.1	Flacourtiaceae	-	0.91	-	-	-
Cheilosa malayana	Euphorbiaceae	-	0.21	-	0.35	-
Chisocheton sp.	Meliaceae	0.25	-	-	-	-
Cinnamomum iners	Lauraceae	-	1.55	-	-	-
Croton levifolium	Euphorbiaceae	-	-	-	-	0.29
Cryptocarya densiflora	Lauraceae	-	-	0.61	-	-
Cryptocarya infectoria	Lauraceae	-	-	-	-	0.45
Cryptocarya kurzii	Lauraceae	-	0.63	0.50	-	-
Dacryodes costata	Burseraceae	-	-	0.24	-	-
Dacryodes rostrata	Burseraceae	0.59	-	0.13	0.31	4.79

Species	Family	T1	T2	T3	T4	T5
Dacryodes rugosa	Burseraceae	0.48	-	0.19	1.08	0.18
Dialium indum L. var. *indum*	Leguminosae	0.74	-	-	-	-
Dialium platysepalum	Leguminosae	-	0.82	-	-	-
Dillenia reticulata	Dilleniaceae	-	6.69	-	-	-
Diospyros buxifolia	Ebenaceae	-	-	0.18	-	-
Diospyros maingayi	Ebenaceae	-	-	-	-	1.87
Diospyros sumatrana	Ebenaceae	-	-	-	-	0.28
Diplospora malaccensis	Rubiaceae	-	0.28	-	-	0.57
Dipterocarpus costulatus	Dipterocarpaceae	-	-	-	-	12.63
Durio griffithii	Bombacaceae	0.28	-	1.67	0.33	0.68
Durio lowianus	Bombacaceae	-	-	-	5.03	-
Dyera costulata	Apocynaceae	-	-	0.84	2.97	13.38
Dysoxylum flavescens	Meliaceae	-	-	1.71	-	-
Dysoxylum sp.	Meliaceae	1.19	-	-	-	-
Dysoxylum sp1.	Meliaceae	-	-	2.91	-	-
Elaeocarpus nitidus	Elaeocarpaceae	-	-	0.30	1.13	-
Elaeocarpus palembanicus	Elaeocarpaceae	-	-	-	-	0.13
Elaeocarpus petiolatus	Elaeocarpaceae	-	-	-	0.68	-
Elateriospermum tapos	Euphorbiaceae	16.11	-	4.48	7.73	3.40
Erythrospermum candidum	Flacourtiaceae	-	0.34	-	-	-
Flacourtia rukam	Flacourtiaceae	0.33	-	-	-	-
Garcinia bancana	Guttiferae	-	-	-	-	3.96
Garcinia griffithii	Guttiferae	-	-	-	0.75	3.18
Garcinia nervosa	Guttiferae	-	-	-	-	1.79
Garcinia parvifolia	Guttiferae	-	-	1.03	-	-
Garcinia pyrifera	Guttiferae	-	-	0.40	-	-
Gironniera nervosa	Ulmaceae	0.26	-	-	-	-
Gironniera subaequalis	Ulmaceae	-	-	0.77	-	-
Gonocaryum gracile	Icacinaceae	-	-	-	0.27	-
Gonystylus maingayi	Thymelaeaceae	-	4.24	0.24	2.89	-
Gordonia penangensis	Theaceae	-	0.69	0.98	-	1.77
Guioa sp.	Sapindaceae	-	-	-	-	-
Homalium longifolium	Flacourtiaceae	-	-	1.95	-	-
Hopea sulcata	Dipterocarpaceae	-	-	0.84	-	-
Horsfieldia fulva	Myristicaceae	-	-	1.12	-	-

Species	Family	T1	T2	T3	T4	T5
Horsfieldia sucosa	Myristicaceae	0.74	-	-	-	-
Horsfieldia tomentosa	Myristicaceae	0.19	-	-	-	-
Hosfieldia polyspherula var. sumatrana	Myristicaceae	-	-	-	-	0.57
Hunteria zeylanica	Apocynaceae	-	-	3.41	-	-
Hydnocarpus woodii	Flacourtiaceae	1.72	-	-	-	-
Intsia palembanica	Leguminosae	12.80	-	-	-	-
Irvingia malayana	Irvingiaceae	-	-	-	0.58	-
Ixonanthes icosandra	Ixonanthaceae	-	3.28	0.61	-	0.81
Kibatalia maingayi	Apocynaceae	0.31	-	-	-	-
Knema furfuracea	Myristicaceae	-	-	-	1.46	0.96
Knema hookeriana	Myristicaceae	1.70	0.27	-	-	-
Knema intermedia	Myristicaceae	1.62	0.60	-	-	-
Knema laurina	Myristicaceae	0.38	-	-	-	-
Knema patentinervia	Myristicaceae	-	1.53	1.35	2.51	0.30
Knema scortechinii	Myristicaceae	1.78	0.43	-	-	-
Knema stenophylla	Myristicaceae	-	-	-	-	0.57
Koompassia excelsa	Leguminosae	-	2.08	-	2.79	-
Koompassia malaccensis	Leguminosae	0.97	14.87	7.90	-	-
Lasianthus sp.	Rubiaceae	-	-	-	0.34	-
Lithocarpus curtisii	Fagaceae	-	-	-	3.48	-
Litsea machilifolia	Lauraceae	-	-	-	0.26	-
Macaranga hypoleuca	Euphorbiaceae	-	-	-	0.94	0.33
Macaranga lowii	Euphorbiaceae	-	-	1.06	0.78	2.57
Magnolia liliifera	Magnoliaceae	-	-	1.07	-	-
Mallotus leucodermis	Euphorbiaceae	6.22	1.05	-	-	-
Mallotus oblongifolius	Euphorbiaceae	0.24	-	0.24	-	-
Mallotus sp.	Euphorbiaceae	1.70	-	-	-	-
Mangifera griffithii	Anacardiaceae	-	-	-	1.33	-
Medusanthera gracilis	Icacinaceae	-	-	0.21	-	0.45
Meiogyne monosperma	Annonaceae	-	0.27	-	-	-
Memecylon minutiflorum	Melastomaceae	-	0.26	-	-	-
Memecylon pubescens	Melastomaceae	-	0.87	-	-	-
Mesua ferrea	Guttiferae	-	-	0.21	-	-
Mesua lepidota	Guttiferae	1.00	-	-	-	-

Species	Family	T1	T2	T3	T4	T5
Mesua racemosa	Guttiferae	-	-	-	-	0.47
Mezzettia elliptica	Annonaceae	-	-	-	-	2.54
Microcos antidesmifolia	Tiliaceae	-	0.72	-	-	-
Microcos fibrocarpa	Tiliaceae	0.17	-	-	-	-
Microcos laurifolia	Tiliaceae	-	1.26	-	-	-
Microcos tomentosa	Tiliaceae	-	0.16	-	-	-
Monocarpia marginalis	Annonaceae	-	-	0.35	-	-
Myristica gigantea	Myristicaceae	-	-	-	-	1.81
Nauclea officinalis	Rubiaceae	-	-	-	1.89	-
Neoscortechinia kingii	Euphorbiaceae	-	-	0.23	-	-
Nephelium costatum sub-species oppoides	Sapindaceae	-	0.41	-	-	-
Nephelium cuspidatum	Sapindaceae	4.77	-	-	1.05	-
Nephelium maingayi	Sapindaceae	-	0.16	-	1.62	-
Nothaphoebe umbelliflora	Lauraceae	0.90	-	-	-	-
Ochanostachys amentacea	Olacaceae	1.10	-	0.12	-	0.47
Palaquium clarkeanum	Sapotaceae	-	-	-	0.75	-
Palaquium gutta	Sapotaceae	-	-	2.01	4.02	1.22
Palaquium hexandrum	Sapotaceae	-	-	0.64	-	-
Palaquium hispidum	Sapotaceae	-	2.71	-	-	-
Palaquium maingayi	Sapotaceae	-	-	2.83	-	-
Palaquium microcarpum	Sapotaceae	-	0.57	-	-	-
Parartocarpus venenosus	Moraceae	-	-	0.39	-	-
Paratocarpus bracteatus	Moraceae	-	-	-	1.72	-
Paropsia vareciformis	Passifloraceae	-	0.23	-	-	-
Payena dasyphylla	Sapotaceae	-	-	0.94	-	-
Payena lanceolata var. lanceolata	Sapotaceae	-	-	-	5.07	-
Payena maingayi	Sapotaceae	-	1.88	-	-	-
Pellacalyx saccardianus	Rhizophoraceae	-	-	-	0.67	-
Pentace curtisii	Tiliaceae	0.17	-	-	-	-
Pentace strychnoidea	Tiliaceae	-	-	-	-	0.98
Pentaspadon velutinus	Anacardiaceae	7.40	7.76	-	-	-
Pimelodendron griffithianum	Euphorbiaceae	0.62	-	3.51	0.72	0.14
Planchonia grandis	Lecythidaceae	-	-	-	2.71	-
Polyalthia jenkensii	Annonaceae	-	0.56	-	-	0.13

Species	Family	T1	T2	T3	T4	T5
Polyalthia rumphii	Annonaceae	-	1.08	-	-	-
Pometia ridleyi	Sapindaceae	2.65	-	-	-	-
Pseudoclausena chrysogyne	Meliaceae	-	0.29	-	-	-
Pternandra echinata	Melastomaceae	-	-	-	0.38	-
Pterocymbium javanicum	Sterculiaceae	1.44	-	-	-	-
Ptychopyxis caput-medusae	Euphorbiaceae	3.29	-	-	-	-
Pyrenaria acuminata	Theaceae	-	-	-	0.75	-
Santiria griffithii	Burseraceae	-	-	0.18	-	-
Santiria laevigata	Burseraceae	1.03	0.96	10.47	-	-
Santiria sp.	Burseraceae	0.17	-	-	-	-
Santiria tomentosa	Burseraceae	-	-	-	-	1.64
Sapium baccatum	Euphorbiaceae	-	6.52	-	-	-
Saraca declinata	Leguminosae	-	0.46	0.25	-	-
Sarcotheca griffithii	Oxalidaceae	-	-	-	-	0.24
Scaphium linearicarpum	Sterculiaceae	1.13	-	-	-	-
Scaphium macropodum	Sterculiaceae	-	-	-	-	0.26
Schoutenia accrescens	Tiliaceae	-	0.20	-	-	-
Shorea leprosula	Dipterocarpaceae	5.18	-	-	12.67	6.73
Shorea multiflora	Dipterocarpaceae	-	-	-	3.26	-
Shorea ovalis	Dipterocarpaceae	-	-	0.28	-	6.19
Shorea parvifolia	Dipterocarpaceae	3.15	-	2.86	-	0.28
Species A	Meliaceae	0.73	-	-	-	-
Streblus elongatus	Moraceae	0.88	-	-	-	-
Strombosia ceylanica	Olacaceae	-	-	-	-	0.62
Strombosia javanica	Olacaceae	-	1.33	-	-	-
Strombosia sp.	Olacaceae	-	-	-	8.59	-
Styrax benzoin	Styraceceae	-	0.50	-	-	-
Symplocos fasciculata	Symplocaceae	2.37	1.75	-	-	-
Symplocos sp.	Symplocaceae	-	-	-	1.00	-
Syzygium chloranthus	Myrtaceae	0.63	-	-	-	-
Syzygium densiflora	Myrtaceae	-	-	-	0.25	-
Syzygium duthieanum	Myrtaceae	0.32	-	-	-	-
Syzygium griffithii	Myrtaceae	-	-	0.13	3.46	0.33
Syzygium lineatum	Myrtaceae	0.90	-	-	-	-
Syzygium protulata	Myrtaceae	0.69	-	-	-	-

Species	Family	T1	T2	T3	T4	T5
Syzygium pustulatum	Myrtaceae	-	-	0.24	-	-
Syzygium sp.	Myrtaceae	0.79	1.38	9.23	0.92	1.32
Tarenna mollis	Rubiaceae	1.13	-	-	-	-
Teijsmanniodendron coriaceum	Verbenaceae	-	-	1.11	-	2.35
Terminalia citrina	Combretaceae	-	7.03	-	-	-
Timonius wallichianus	Rubiaceae	1.35	-	-	-	-
Vernonia arborea	Compositae	-	0.59	-	-	0.64
Vitex pinnata	Verbenaceae	1.35	-	-	-	-
Xanthophyllum griffithii	Polygalaceae	-	-	0.35	-	1.53
Xanthophyllum lelacarum	Polygalaceae	-	-	10.08	-	-
Xerospermum laevigatum	Sapindaceae	-	0.50	-	-	-
Xerospermum noronhianum	Sapindaceae	-	3.94	-	-	-
Xylopia magna	Annonaceae	-	7.76	-	-	-
Xylopia malayana	Annonaceae	-	0.16	-	-	0.43

Table 5. List of species, family and the relative basal area of Kuala Keniam forest.

Author details

Mohd Nazip Suratman
Faculty of Applied Sciences, University of Technology MARA, Shah Alam, Malaysia

5. References

Ashton, P. S. 1964. A quantitative phytosociological technique applied to tropical mixed rainforest vegetation. Malays. For., 27, 304–307.

Bernhard-reversat, F, Huttel, C., and G. Lemee. 1978. Structure and functioning of evergreen rain forest ecosystem of the Ivory Coast. In: Tropical Forest Ecosystems: A State-of-Knowledge Report, UNESCO, Paris, pp. 557 – 574.

Bray, J. R. and J. T. Curtis. 1957. An ordination of upland forest communities of southern Wisconsin. Ecological Monographs 27:325-349.

Burgess, P. F. 1961. The structure and composition of lowland tropical rain forest in north Borneo. Malays. For., 24: 66–80.

Changcheng, T., Xuelong, J., Hua, P., Pengfei, F. and Z. Shoubiao. 2007. Tree species diversity of black-crested gibbons (*Nomascus concolor*). Acta Ecologica Sinica. 27(10): 4002 – 4010.

Campbell, D.G., J.L. Stone and A. Rosas Jr. 1992. A comparison of the phytosociology and dynamics of three floodplain (Varzea) forest of known ages, Rio Jurua, western Brazilian Amazon. Botanical Journal of the Linnean Society 108: 231-237.

Cannon, C.H., Peart, D.R., and M. Leighton. 1998. Tree species diversity in commercially logged Bornean rain forest. Science 28, 1366–1368.

Condit, R., Hubbell, S.P., La Frankie, J.V., Sukumar, R., Manokaran, N., Foster, R.B. and P.S. Ashton. 1996. Species–area and species–individual relationships for tropical trees a comparison of three 50 ha plots. J. Ecol. 84, 549–562.

Dagar, R. J. C. and N.T. Singh. 1999. Plant Resources of the Andaman and Nicobar Islands, Vol. 81, 211–0165.

Ferreira, L. V. and G.T. Prance. 1998. Species richness and floristic composition in four hectares in the Jaú National Park in upland forests in Central Amazonia. Biodivers. Conserv , 7, 1349 – 1364.

Gentry, A.H. 1988. Changes in plant community diversity and floristic composition on environmental and geographic gradients. Annals of the Missouri Botanical Garden. 75 1-34

Gentry, A.H. 1990. Floristic similarities and differences between southern Central America and upper and central Amazonia. In: Gentry, A.H. (Ed,) Four Neotropical Rainforests. Yale University Press, New Haven, CT, pp. 141 – 157.

Gentry, A.H., 1992. Tropical forest biodiversity: distributional patterns and their conservational significance. Oikos 63, 19–28.

Hubbell, S.P., Foster, R.B., O'Brien, S.T., Harms, K.E., Condit, R., Wechsler, B., Wright, S.J. and S. Loode Lao. 1999. Light-gap disturbance, recruitment limitation, and tree diversity in a Neotropical forest. Science 283, 554–557.

Kent, M., and P. Coker. 1992. Vegetation description and analysis: a practical approach. Belhaven Press, London.

Knight, D. H. 1975. A phytosociological analysis of species rich tropical forest on Barro Colorada Islands, Panama. Ecol. Monogra., 45, 259–284.

Magurran, A.E. 1988. Ecological Diversity and Its Measurement. Princeton University Press, Princeton, N J.

Margalef, F. R. 1958. Information theory in ecology. Gen. Syst., 3:36–71.

Okuda, T., Suzuki, M., Adachi, N., Quah, E. S., Hussein, N. A. And N. Manokaran. 2003. Effect of selective logging on canopy and stand structure and tree species composition in a lowland dipterocarp forest in Peninsular Malaysia. Forest Ecology and Management 175: 297–320.

Philips, O.L. and A.H. Gentry. 1994. Increasing turnover through time in tropical forests. Science. 263, 954 – 958.

Preston, F.W. 1962. The canonical distribution of commonness and rarity, Part I. Ecology, 43:185 – 215.

Richards, P. W. 1952. The Tropical Rain Forest, Cambridge University Press.

Richards, P. W. 1996. The Tropical Rain Forest: An Ecological Study, Cambridge University Press, Cambridge.

SAS Institute Inc. 2000. SAS/STAT®User's guide, version 8, vol. 1, Cary, NC, USA.

Shannon, C. E. and W. Weiner, 1949. The Mathematical Theory of Communication, University of Illinois Press, Urbana, USA.

Simpson, E. H. 1949. Measurement of diversity. Nature, 163: 688.

Singh, J. S., Singh, S. P., Saxena, A. K. and Y.S. Rawat. 1981. Report on the Silent Valley Study, Ecology Research Circle, Kumaun University, Nainital, 86.

Singh, J.S., 2002. The biodiversity crisis: a multifaceted review. Curr. Sci. 82, 638–647.

Sumina, O.I., 1994. Plant communities on anthropogenically disturbed sites on Chukotka Peninsula, Russia. J. Veg. Sci. 5, 885–896.

Suratman, M.N., M. Kusin, and S.A.K. Yamani. 2007. Study of tree species composition in Sungkai Forest. Paper presented at the National Biodiversity Seminar, Department of Wildlife and National Parks. 20 – 21 Nov. 2007, Seremban.

Suratman, M.N., M.S. Daim, S.A.K. Yamani, M. Kusin and R. Embong. 2009. Report of Flora and Fauna Surveys for Sungai Sator Forest, Jeli, Kelantan. Environmental Impact Assessment (EIA) Report for the Development of UMK campus. 26pp.

UNEP, 2001. India: State of the Environment. United Nations Environment Programme.

Waide, R.B., Willig, M.R., Steiner, C.F., Mittelbach, G., Gough, L., Dodson, S.I., Juday, G.P. and R. Parmenter. 1999. The relationship between productivity and species richness. Annu. Rev. Ecol. Systematics, 30, 257 – 300.

Wang, D.P., Ji, S.Y., Chen, F.P., Xing, F.W. and S. L. Peng. 2006. Diversity and relationship with succession of naturally regenerated southern subtropical forests in Shenzhen, China and its comparison with the zonal climax of Hong Kong. Forest Ecology and Management 222: 384-390.

Whitmore, T. C. 1998. An Introduction to Tropical Forests, Clarendon Press, Oxford and University of Illinois Press, Urbana, 2nd, Ed. pp. 117.

Whitmore, T.C. 1984. Tropical Rain Forests of the Far East. Clarendon Press, Oxford.

Whittaker, R. H. 1972. Evolution and measurements of species diversity. Taxon, 21: 213–251.

Chapter 9

The Influence of Geochemistry on Biological Diversity in Fennoscandia and Estonia

Ylo Joann Systra

Additional information is available at the end of the chapter

1. Introduction

The Earth's crust is predominantly composed of a relatively small number of chemical elements. Only eight of them: oxygen, silicon, aluminum, iron, calcium, magnesium, sodium and potassium are present in amounts exceeding one weight percent and together they comprise almost 99% of the entire crust [1, 2]. Some elements are exceedingly rare in the Earth, or have short-lived radioactive isotopes. For example, promethium (Pm) occurs in the crust in only very small concentrations in certain uranium ores, being produced as result of nuclear fission, with the longest-lived isotope having a half-life of only 17.7 years. Similarly, technetium is a relatively light radioactive metal with atomic number 43, having no stable isotopes, while the longest-lived radioactive isotope (Te-98) has a half-life of 4.2 million years. Technetium only occurs naturally, in trace amounts, in uranium ores produced by nuclear fission [3]. When such elements are excluded, there remain 90 naturally occurring chemical elements [4] to form the geochemical basis of the life of the Earth.

2. Distribution of chemical elements in the biosphere of the Earth

The Earth's biosphere is consists of crust, hydrosphere and atmosphere. The Earth's crust is predominantly composed of 9 chemical elements, each having abundances, expressed in atomic weight, of more than one percent: oxygen (53.39%), hydrogen (17.25%), silicon (16.11%), aluminum (4.80%), sodium (1.82%), magnesium (1.72%), calcium (1.41%), iron (1.31%) and potassium (1.05%). The next most abundant elements are: carbon (0.51%), titanium (0.22%), chlorine (0.10%), fluorine (0.07%) and sulfur (0.05%). Together these 14 elements comprise 99.81% of the crust in terms of atomic weight, the remaining 76 elements representing only 0.19% or on average 0.0025 atomic weight percent each [5]. Some of these elements are therefore referred to as rare earth elements (REE). Zinc, copper, nickel, chromium and manganese, which are common in everyday use, also represent trace elements. The number of minerals recognized is also in accordance with crustal element abundances. A total of 2909 containing oxygen minerals are known, 1921 minerals contain hydrogen, 906 contain silicon, 714 contain aluminum, 560 minerals contain Na, 555 contain, Mg, 272 contain K, 272 also contain C and 172 minerals contain Ti. The chemically active elements Cl and F occur in 220 and 221 elements respectively. However, due to their greater chemical activity, relatively more minerals have been described containing Ca (867), Fe (883) and S (761) [5].

There are some differences between mean cosmic abundances and those of the Earth's crust, but the higher abundances of light elements are shared by both environments, implying that the process of nuclide synthesis obeys the same rules. The Oddo-Harkins rule states that for any two neighboring elements, the abundance of the element with an even atomic number is higher than that of the odd one. Of the 28 first elements of Earth's crust the even elements in the periodic table constitute 86.36 weight percents, while odd elements comprise only 13.64%. There are some exceptions from the rule. The noble gases (atomic numbers in brackets): helium (2), neon (10), argon (18), krypton (36), xenon (54) abundances are much lower than precited by the Oddo-Harkins rule, because they have stable nuclei and are chemically inert, and hence do not participate in chemical reactions or form compounds. The noble gases are predominantly present in the atmosphere, with argon abundance being nearly 1%, while the others are present in small amounts, because they can easily escape into space [2].

It must also be noted that especially high abundances are associated with elements for which numbers differ by 6 or multiple 6 and plus 2, such as O (8), Si (14), Ca (20), Fe (26), Sr (38), Sn (50), Ba (60). Of the most abundant elements in the Earth's crust elements, 13 have atomic numbers between 1 and 20, with only iron having a higher number, of 26 [6]. On the diagram of cosmic nuclear abundances plotted against atomic mass numbers, a significant peak indeed appears in the region of iron. This implies that that Earth's core may also contain an abundance of elements of lower atomic number, including iron. The dominance of the lighter elements is easily understood, because less energy is required for the formation. Heavy nuclei have higher electric charges than light nuclei, while for reactions between them also tend to require higher temperatures [7, 8].

Chemical elements as a rule are represented by several different isotopes, for which atoms have the same atomic number, but differ in terms of atomic mass. Tin (Sn) has the largest number (10) of naturally occurring stable isotopes, while cadmium and tellurium both have 8 isotopes. Many other elements have from 2-6 isotopes and only 22 elements, including F, Na, P, V, Mn, Au, have one single stable isotope. The highest isotopic abundances also tend to correlate with those isotopes for, which the atomic mass can be divided by 4. Oxygen isotopes have the following order of abundance: ^{16}O – 99.76%, odd ^{17}O – 0.04%, even ^{18}O – 0.2%; for silicon, isotope abundances are: ^{28}Si – 92%, ^{29}Si – 5% and ^{30}Si – 3%. In the case of carbon isotopes, ^{12}C represents 99% and ^{13}C about 1%; both of these are non-radioactive, the third isotope ^{14}C is radioactive and occurs in trace amounts in the atmosphere [3].

The 90 naturally occurring elements are found throughout the Earth's crust, in bedrock, groundwater, unconsolidated sediments and soils. More than 70% of Earth's surface is covered by oceans and seas, which provide habitats and diverse environments for numerous species of very different size. One ton of seawater contains the following elements by mass: Cl (19.4 kg), Na (10.8 kg), Mg (1,29 kg), S (0.9 kg), Ca (411 g), K (392 g), Br (67 g) and 10 trace elements (in order of abundance from 8.1-0.003 g): Sr, B, F, Li, Ru, Ba, I, As and Cs [2]. The composition of atmosphere by volume is: nitrogen (78.09%), oxygen (20.95%), argon (0.95%), CO_2 (0.03%), and neon (0.0018%); trace gases are in order of abundance (from 0.0052 to 0.000 008%): He, methane, Kr, nitrous oxide, O, O_3 and Xe. The abundance of ozone is notable in that it increases with altitude [2].

The biosphere contains innumerable species of plants, animals, fish, birds, insects and other forms of biological life. None of them can exist without water and numerous chemical elements as nutrients.

3. Chemical elements as nutrients for all forms of biological life

All plants and animals need a range of different chemical elements as nutrients for normal and healthy growth and development [9]. However, not all of the most widely distributed elements are needed in large amounts. Of the 90 naturally occurring elements on Earth, there are 11 elements found throughout the atmosphere, hydrosphere and within bedrock and soils, that are essential for all plants and animals, namely oxygen, hydrogen, carbon, nitrogen, calcium, magnesium, sodium, potassium, sulfur, phosphorus and chlorine. These elements are also known as macronutrients, of which the first four can be easily obtained from air (C, N, O) or water (H, O), while the remaining 7 elements are more common in bedrock or are present at required concentrations in groundwater. All these elements are essential to life and for example, an adult human requires a daily intake of 100 mg or more of each of these elements [10].

The sources of mineral elements are soils, drinking water and food – each kind of food has its own distinct composition of macro- and micronutrients. For example, phosphorus, calcium, sodium and magnesium are present in every kind of food, but sulfur is usually absent from dried produce [11]. The abundances of the main macronutrients in marine and terrestrial plants, marine and terrestrial animals, and in bacteria (in dry matter) are remarkably similar to one another, which indicate that all forms of living life require the same macro-elements (Table 1).

All macro-biogenic elements are essential for normal life functions and are vital for manufacturing cells and tissues in plants and animals. There is some variation in the actual element abundances according to specific environmental conditions and plant and animal requirements. Terrestrial animals need stronger skeletal support than marine organisms and their proportional abundances of phosphorus and fluorine are consequently 3 and 7 times higher. Other elements, such as potassium, sodium, sulfur and magnesium are more readily available in marine environments than on land and their concentrations in marine plants and animals are accordingly higher. Bacteria do not differ very markedly from other groups, indicating that all life forms on the Earth are essentially composed from the same chemical elements. The human being is not exception, a typical 70 kg human body consisting of the following major elements: oxygen (61%), carbon (23%), hydrogen (10%), nitrogen (2.6%), calcium (1.4%), phosphorus (1.1%), potassium (0.2%), sulfur (0,2%), sodium (0.14%), chlorine (0.12%) and magnesium (0.027%). However, not all of these widely distributed chemical elements are needed in large amounts by plants and animals. Four elements in particular: silicon, aluminum, iron and titanium, are only required in very small amounts as micro-nutrients for the regulation of specific processes [3].

Macronutrients, grams in 100 g	Marine plants	Terrestrial plants	Marine animals	Terrestrial animals	Bacteria	Average of all groups
Carbon – C	34.5	45.4	40.0	46.6	54.0	44.1
Oxygen – O	47.0	41.0	40.0	18.6	23.0	33.7
Hydrogen – H	1.5	3.0	7.5	10.0	9.6	6.32
Nitrogen – N	4.1	5.5	5.2	7.0	7.4	5.06
Potassium – K	5.2	1.4	0.5-3.0	0.74	11.5	4.12
Calcium – Ca	1.0	1.8	0.15-2.0	0.02-8.5	0.51	1.44
Sodium – Na	3.3	0.12	0.4-4.8	0.4	0.46	1.38
Phosphorus – P	0.35	0.23	0.4-1.8	1.7-4.4	3.0	1.29
Sulphur – S	1.2	0.34	0.5-1.9	0.5	0.53	0.75
Magnesium – Mg	0.52	0.32	0.5	0.1	0.7	0.43
Chlorine - Cl	0.47	0.2	0.5-9.0	0.28	0.23	0.29

Table 1. Macro-elements content in marine and terrestrial plants and animals, and bacteria (in grams to 100 g dry matter), modified after Barabanov [12]

The distribution of macro- and micronutrients in different types of bedrock, soils, plants and animals have been studies intensively for more than 50 years, but due to the large ranges in composition, no consensus has been reached on the precise abundances of essential trace element concentrations in living organisms [9-10, 13-22]. The reason for this is that micronutrients are present and needed in very small amounts. For example, selenium is an important microelement for humans but daily intake must not exceed 28-55 μg [22]. The most complete list of micronutrients is given by Emsley [15], which was published on the Internet by Uthman [20]. For an average 70 kg human body mass the abundances of 59 elements have been determined, including 11 major elements. Of 48 trace elements 43 are considered essential, while the role of thorium, uranium, samarium, beryllium and tungsten is not known. Vanadium is the least abundant biologically necessary element in the human body (0.11 mg) whereas rubidium is the most abundant element in the body (0.68 g) that lacks any biological role. Silicon (1.0 g) may or may not have a metabolic function. Only four elements exceed the 1.0 g level: Fe (4.2 g), F (2.6 g), Zn (2.3 g) and Si (1.0 g), and another four have levels greater than 0.1 g: Rb (0.68 g), Sr (0.32 g), Br (0.26 g) and Pb (0.12 g). The abundances of the remaining elements fall within the interval between 72–0.11 mg [15, 20].

Table 2 shows 29 trace element concentrations in marine and terrestrial plants and animals, and also in bacteria for some elements. The situation with respect to microelements is nearly the same as for major elements. The contents of silicon, iron, zinc, aluminum, rubidium, nickel, cobalt, molybdenum in each group are quite comparable, although some other elements show greater differences in concentration than in the case of macro-nutrients. The abundances of bromine, strontium, iodine, boron, titanium lithium and chromium are higher in marine plants and animals, which is to be expected given that sea water contains more of these elements [2] and that uptake from solution is relatively easy. Excess concentrations of many of the trace elements present in soils, including: As, Cd, Co, Cu, Cr, Hg, Mo, Ni, Pb, Se, Zn, U, Th and Zn, may be toxic to plants and animals or may affect the quality of foodstuff for human consumption [14]. They are potentially toxic to plants and animals, but can have adverse effects at relatively low, insufficient level [21]. In many countries, including Estonia, there are governmental regulations concerning maximum permitted concentrations for toxic and dangerous elements in agricultural lands, industrial and urban environments, and in drinking water.

Microelements, mg in 100 g dry matter	Marine plants	Terrestrial plants	Marine animals	Terrestrial animals	Bacteria	Average of all groups
Silicon – Si	150-2000	20-500	7-100	12-600	18	342.5
Iron – Fe	70	14	40	16	25	33.0
Bromine – Br	74	1.5	6-100	0.6	–	32.3
Zinc – Zn	15	10	0.6-150	16	–	29.1
Strontium – Sr	26-140	2.6	2-50	1.4	–	28.3
Iodine – I	3-150	0.042	0.1-15	0.043	–	21.0
Manganese – Mn	5.3	63	0.1-6	0.02	–	17.8
Aluminum – Al	6	50 (0.05-400)	1.5	0.4-10	–	15.7
Fluorine – F	0.45	0.005-4	0.2	15-50	–	8.79
Boron – B	12	5	2-5	0.05	–	5.13
Copper – Cu	1	1.4	0.4-5	0.24	4.2	1.91
Rubidium – Rb	0.74	2	2	1.7	–	1.61
Titanium – Ti	1.2-8	0.1	0.02-2	0.02	–	1.43
Barium – Ba	3	1.4	0.02-0.3	0.075	–	1.16
Arsenic – As	3	0.02	0.0005-0.03	≤ 0.02	–	0.76
Nickel – Ni	0.3	0.3	0.04-2.5	0.08	–	0.49
Lead – Pb	0.84	0.27	0.05	0.2	–	0.34
Tin – Sn	0.1	< 0.03	0.02-2	< 0.015	–	0.28
Vanadium – V	0.2	0.16	0.014-0.2	0.015	–	0.21
Silver – Ag	0.025	0.006	0.3-1.1	0.0006 (?)	–	0.18
Lithium – Li	0.5	0.01	0.1	<0.002	–	0.15
Cobalt – Co	0.07	0.05	0.05-0.5	0.003	–	0.099
Selenium – Se	0.08	0.02	–	0.17	–	0.090
Molybdenum – Mo	0,045	0.09	0.06-0.25	0.02	–	0.078
Cadmium – Cd	0.04	0.06	0.015-0.3	≤ 0.05	–	0.070
Chromium – Cr	0.1	0.023	0.02-0.1	0.0075	–	0.048
Antimony – Sb	–	0.006	0.02	0.0006	––	0.009
Mercury – Hg	0.003	0.0015	–	0.0046	–	0.003
Tungsten – W	0.0035	0.007	0.00005-	?	–	0.0016

Table 2. The content of chemical microelements in marine and terrestrial plants and animals, and bacteria, modified after [12]

The role of chemical elements in the natural environment first became the subject of study during the early decades of the twentieth century, with the first significant findings emerging in the middle of the century. In some countries, as in Great Britain, the importance of geochemical investigations has been recognized and has been supported by government-funded research. The discovery of irregularities in the distribution of diseases in the early 1960's, particularly with respect to the incidence of cancer in southwest England, initiated a detailed geochemical survey of Britain by a multi-disciplinary research team [23]. Some 20 years later, following numerous international conferences and workshops in environmental geochemistry and health, the main results of investigations were published in a special book [14], which presented for the first time, information concerning all aspects of environmental geochemistry and the distribution of harmful elements in the environment in which people live [23].

Such geochemical studies represent an important component of the environmental and geological research programs in USA, Canada, Australia, Russia, Japan, China, Germany, Poland, Sweden, Norway, Finland and many other countries.

4. Methods of studying the influence of geochemistry on biodiversity in Fennoscandia and Estonia

Specialized studies within areas covered by old-growth forests were initiated during the late 1980's in the Republic of Karelia. Much of the forest in this part of the former Soviet Union was intensively logged during the fifty years or so following the 2nd World War. As demand by the forestry and paper industry increased, there was increasing recognition of the need to preserve areas of taiga forest within a network of National Parks. As a result, the bilateral Russian-Finnish project "Inventory and studies of biological diversity in Republic of Karelia" was initiated in 1997-2000 by the Karelian Research Centre of the Russian Academy of Sciences. This project included investigations into bedrock, Quaternary sediments and soil geochemistry, vegetation, forests, mires, fish populations, algal flora and zooplankton, bird fauna, mammals and insect populations in different environments of the Republic of Karelia, with particular emphases on areas of high nature conservation value [24]. The author of the current paper was involved in the project as a geologist and as a result of 30 years experience, is familiar with bedrock composition and geochemistry in all kinds of terrain. Fieldwork during the years 1962-1984 in North Karelia and another parts of Karelia principally comprised many thousand kilometers of geological traverses on foot, from the White Sea to Russian-Finnish border, where significant areas of old growth forest remained [Figure 1].

The research group at the Karelian Research Center of Russian Academy of Sciences was characterized by close collaboration between specialists in different fields, including forestry, vascular plants, mosses, lichens, mammal species, bird fauna, insects, algal flora, periphyton-, zooplankton-, macrozoobenthos-, and fish communities. The interactions between these specialists during the field studies resulted in much new information being obtained and published at first in Russian, as progress reports of field studies, and finally in

Russian and English. The final report [24] provides extensive material for researchers in widest wide range of ecological and biological fields. Geological reviews of the Karelian Region and more detailed local studies, with numerous chemical analysis of bedrock have been published in Systra [27-36] and in [37]. The general geochemical characteristics for soils in the Republic of Karelia were published by Fedorets et al. [38], while a geochemical study of the bedrock of Valaam Island, Lake Ladoga, was done by Sviridenko and Svetov [39], forest litter and underlying mineral material were studied by Shiltsova et al. [40], and the distribution of vascular plants in Republic of Karelia was reported by Kravchenko et al. [41].

Detailed fieldwork was done in the Onega Synclinorium and the central part of the Zaonezhje Peninsula, the Paanajärvi National Park, and along the western and northwestern shore of Lake Ladoga. The results of previous research also facilitated characterization of the Western White Sea area and Central Karelia. Later, in 2002-2003 and 2008-2009, LAPBIAT and LAPBIAT-2 financial support made it possible to visit key areas for studying the influence of bedrock geochemistry on vegetation in subarctic Lapland in Finland, in the surroundings of Kilpisjärvi and Kevo, and in the Oulanka National Park, Kuusamo. The influence of carbonate rocks on vegetation in the Kilpisjärvi Region in Lapland, at latitude 69° N, was already noted by Pesola [42] at the beginning of the last century, leading to the creation in 1916 of the first protected area in Finland (now Malla Strict Reserve). Subsequently, in 1956, the Oulanka River valley near the Arctic Circle, the Oulanka National Park was declared. On the Russian side of the border the Paanajärvi National Park is larger still (1040 sq. km) and was created in 1992. The geochemical control of vegetation type is clearly marked, especially in the old forests, which are in near pristine condition and the chemical composition of bedrock is well studied. A comprehensive description of the bedrock of Finland is given in Lehtinen et al. [43, 44] and bedrock chemical characteristics in Rasilainen et al. [45].

The composition of bedrock and soils in Estonia are also well studied [46-49]. Environmental protection and the study of vegetation, birds and mammals has has a history extending back more than 100 years. Based on data from about 1550 analyzed samples, the geochemical atlas of the humus horizon of Estonian soils was prepared by Petersell et al. [49]. The 33 geochemical maps in this study, and the "Atlas of Estonian Flora" with 1353 species maps of vascular plant distribution [50] was used for accessing the influence of bedrock on vegetation.

The area considered here embraces the eastern part of Fennoscandian Shield from Kilpisjärvi and Halti Mountain and Kevo in Finnish Lapland (69°45′ N) to northern and northwestern Ladoga Lake (61°15′ N) and Estonia (59°40′- 57°30′ N). The ancient shield is represented geologically by crystalline metamorphic and igneous bedrock of Archean and Early Proterozoic ages. In Estonia these crystalline rocks are completely covered by Late Proterozoic and Paleozoic sedimentary rocks. Crystalline and sedimentary rocks differ in their respective physical and mechanical properties, and chemical composition [2]. The geology of Finland is well known and covered by comprehensive geological and geochemical maps [25-26, 45].

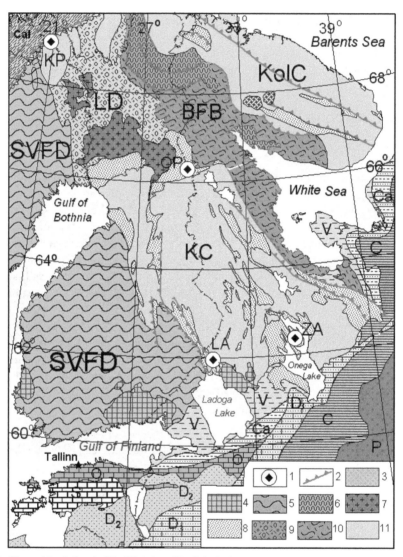

Legend: 1 – Places of detailed study: KP – Kilpisjärvi, OP – Oulanka-Paanajärvi area, ZA – Zaonezhje peninsula, Onega synclinorium, LA – northwest Ladoga Lake shore. 2 – thrusts; 3 – Devonian nepheline syenite; 4 – rabakivi granites; 5 – Svecofennian folded domain (SVFD); 6 – Lapland-Belomorian Granulite Belt; 7 – Kemijärvi granites.
Early Proterozoic volcanic-sedimentary rocks: 8 – Lapland Domain (LD); 9 – cover on the Archean basement. Archean: 10 – Belomorian Fold Belt (BFB); 11 - Archean cratons: Kola Craton (KolC) and Karelian Craton (KC).
Neoproterozoic-Paleozoic sedimentary cover (from youngest): P – Permian and younger rocks:
C – Carbon (carbonate rock, coal, clay, sandstone); D_3 – Upper Devonian (carbonate rocks, sandstone, clay); D_2 – Middle Devonian (sandstones); Cal – thrusted Scandinavian Caledonides; S and O – Silurian and Ordovician (carbonate rock, oil shale); Ca – Cambrian (sandstone, claystone); Ediacaran, Vendian complex (arkose, sandstone, claystone).

Figure 1. Geological map of the eastern part of Fennoscandian Shield and northwest marginal zone of the Russian Platform. Compiled by Y.J.Systra with using geological map of Finland [25] Fennoscandia [26] and unpublished author's map „Tectonics", Karelia (1996).

For all of the areas studied, sufficient data were available concerning bedrock chemical composition. In the Onega region some 20 complete and partial soil profiles were analyzed for major and trace element chemistry, including sampling from both bedrock and soil horizons, in order to quantitatively assess the influence of bedrock on soil composition. Geochemical data were carefully compared with the distribution of each group of plant and animal distribution; for rare and endangered species the Red Data books of Eastern Fennoscandia [51] and the Republic of Karelia [52] were useful.

Comparisons between geochemistry in different environments and the diversity of vegetation and, especially with the distribution of rare and protected plants, showed that biodiversity is influenced by different geological and geographic features, such as latitude and its solar aspect, relief and orientation of landforms, the presence of fault zones as conduits for mineralized groundwater and springs, the color of exposed bedrock and the presence of migration corridors. Nevertheless, the main control on biodiversity is usually geochemical, where the bedrock and overlying regolith contains an abundance of essential macro- and micronutrients. The Fennoscandian Shield and the surrounding Russian Platform marginal zone with its 150-700 m thick sedimentary cover sequence is an excellent area for studying the interdependence of geochemistry and biological diversity. Extensive areas have never been used for agriculture and many are currently protected from commercial exploitation. The biodiversity of the region is also relatively young, in that the continental ice sheet last withdrew from Southern Estonia between 13 500–11 700 years ago, and from northern Fennoscandia between 9000–6500 years ago. Indeed, natural exchange and colonization plants and animals, especially birds, between Oulanka National Park, (Finland) on the western side of the Maanselkä topographic divide and Paanajärvi National Park in Russian Karelia to the east, is continuing to the present day.

5. Results

The results of the study are presented separately for each region, because their geological and geochemical conditions are very different. The Republic of Karelia, in the Russian Federation, has a total area of 172 400 km^2 [53], and occupies the southeastern part of the Fennoscandian Shield. The Karelian part of the Shield comprises 3 major northwesterly trending geological domains, the Karelian Craton in the centre, the Archean Belomorian fold belt to the northeast and the Svecofennian fold terrane to southwest [35]. The exposed part of the Karelian Craton is about 600 km long and 300 km wide, with the southeastern edge being covered by Late Proterozoic Ediacaran and Paleozoic sedimentary rocks, which are contiguous with those in Estonia. The Karelian Craton formed during two distinct orogenic events, the first of which is represented by the Archean (3.8-2.5 Ga) basement, composed of Paleoarchean gneiss, gneissose diorite and migmatites, and narrow Meso- and Neoarchean greenstone belts, consisting of volcanic-sedimentary rocks. The younger phase of evolution of the craton is recorded by Paleoproterozoic (2.5-1.8 Ga) volcanic-sedimentary sequences which are preserved in synclinorial structures. The Archean and Proterozoic rocks differ greatly in geochemical characteristics: basement granites and migmatites are rich in SiO$_2$, but contain only very small amounts of micronutrients, while Paleoproterozoic rocks are more diverse in composition, including many types of volcanic rocks, sandstone, quartzite, carbonate rocks, and shungite-bearing black schists (Figure 2). The areas selected for detailed study in the Republic of Karelia were in the Paanajärvi national park near the Arctic Circle, supplemented by brief studies in the adjoining Oulanka National Park in Finland the northwestern shoreline of Lake Ladoga, focusing on the influence of the Mesoproterozoic Valaam gabbro-dolerite sill on soil composition and vegetation, and the Onega Synclinorium, each of which represent Paleoproterozoic bedrock having favorable compositions for supporting diverse vegetation.

Estonia is geologically situated on the northwest margin of the East-European Platform, where the Precambrian crystalline basement is covered by Ediacaran and Paleozoic sedimentary cover, varying in thickness from 120-150 m in northern Estonia to 500-780 m in southern parts of the country. The basement surface and all sedimentary cover layers are tilted gently southwards at a gradient of nearly 2.5-3 m per kilometre. Cambrian, Ordovician and Silurian sediments are exposed in Northern Estonia as east-west trending belts, whereas to the south these older sequences are covered by younger Middle and Late Devonian deposits. As everywhere in Fennoscandia the bedrock is usually covered by a thin veneer of Quaternary soft sediments, which contain material derived from both local bedrock as well as clasts transported from the crystalline bedrock of Finland and Sweden. Accordingly, since the southern marginal zone of the Fennoscandian shield is represented mostly by migmatites and gneisses, more than 90% of the transported glacial deposits consist of granitic material. This in turn has an influence on soil composition and vegetation in Estonia.

6. Paanajärvi-Oulanka protected reserves

Conservation areas near the Arctic Circle (66°10′- 66°30′ N) include two significant and contiguous national parks straddling the national borderline between Finland and Russia. The older and smaller Oulanka National Park (277.2 km²) was established in Finland in 1956, while the larger Paanajärvi National Park (1045 km²) in Russian Karelia was declared in 1992. Both parks share similar geological features and vegetation types. The Paanajärvi–Oulanka area has a significantly higher level of biodiversity compared to surrounding areas, due to a combination of mountainous terrain, the east-west orientation of deep valleys, microclimate and distinctive bedrock composition (Table 3).

Paleoarchean basement gneisses and granites (Table 3, No 2906, No 2969) are rich by SiO_2, but contain limited amounts of Mg and Ca and biologically needed micronutrients and (Figure 2). Small metamorphosed Meso- and Neoarchean ultramafic bodies (Table 3, No 2957-2) and thin komatiite layers between basalts contain appropriate microelements, but have very restricted distribution and do not significantly influence soil composition. The nutrient richness of soils is instead due to the Paleoproterozoic intermediate, acid, mafic, ultramafic volcanic rocks and different types of sedimentary rocks, including dolomite marbles and carbon-rich black schist. Effusive rock types such as felsic quartz porphyry (Table 3, No 3299) also contains higher concentrations of the trace metals Ti, Fe, K, Ba, Mn and V than granites. Intermediate basaltic andesite (Table 3, No 3319) has still higher abundances of Fe, Mg, Ca, Co, Cr, Cu and Ni. The mafic volcanic diabase (Table 3, No 2932-3) and leucogabbro (Table 3, No 3337) have similar compositions, although there are some differences in macro-element abundances; both rock types nevertheless contain adequate amounts of necessary microelements. Because mafic volcanic rocks are widely distributed in both national parks, there is no deficiency in microelements in the region. Although conglomerates (Table 3, No 2942) and quartzites (Table 3, No 3167-10) are SiO_2-rich, they nevertheless contain some Co, Cr, Mn, Ni and V. Quartzite sampled from a fault zone is brecciated and mineralized with fuchsite (Cr), Ca, Ba and carbonates, as indicated by the higher weight losses on ignition (LOI) and Ca abundances. The dominant source of Mg and Ca in soil is dolomitic marble (Table 3 No 2944), which is common throughout the bedrock of the area and consists of about 95% dolomite $(Ca,Mg)CO_3$: CaO 32.22%, MgO 19.51% and CO_2 43.12% (LOI).

All of the above-mentioned rock types are well represented in Quaternary glacial and glaciofluvial deposits, as glacial erosion for some 50-60 km along the deep Oulankajoki River-Paanajärvi Lake-Olanga River valley was entirely within bedrock of the Paleoproterozoic volcanic and sedimentary cover sequence. Fragments of marble are common in till and fluvioglacial sediments everywhere in the studied area, their influence on soil composition can be observed some km to east, outside of the Paanajärvi syncline, in areas underlying by Archean basement (Figure 2).

This favorable soil composition is the main reason for the unusual biological diversity of the Paanajärvi-Oulanka national parks. About 600 species of vascular plants have been identified, 67 of which have never have been reported from other parts of Karelia [24, 30]. There are 298 species of, 42 of which are listed in the Red Book of Karelia [52], while 443 species of lichen species have been recorded [56], of which 10% are rare or endangered. This number of taxa represents twice that of the corresponding figure for surrounding terrain. The Oulanka and the Paanajärvi national parks are underlain by Paleoproterozoic (2.5-1.9 Ga) volcanic and sedimentary rocks – including carbonates, while the bedrock in the surrounding region consists of diorites, granites and gneisses of the Karelian Craton [27-30, 35].

The mountainous nature of the terrain, including the higher peaks, which are over 450 m high and lie above the tree line, means that the area lies within the mountain tundra belt. Alpine-Arctic vegetation, including *Loiseleuria procumbens* L., *Phyllodoce caerulea* (L.) Bab., *Diphasiastrum alpinum* L. and other species is present on the highest mountains Nuorunen (576 m), Mäntytunturi (550 m) and Kivakka (499.5 m). Along the Kiutaköngäs rapids in the Oulanka River valley, a relict community of *Dryas octapetala* L. has survived since the Ice Age, together with *Calipso bulbosa* L. In contrast, there are no more than 30-35 species of more northerly affinity. On the Paleoproterozoic volcanic and sedimentary rocks, soli acidity is near-neutral, and soils are enriched in all essential trace elements. Bedrock geochemistry is the main reason for the high biological diversity here, principally due to the presence of carbonate rocks; the dolomitic marbles are responsible for buffering the neutral soil composition. Most occurrences of *Cypripedium calceolus* L. (Lady's Slipper) are confined to carbonate rock exposures, where up to 2000 individuals may be found an area of some hundred square meters. Slope aspect and exposure also has a marked effect on microclimate. Thus, southerly species inhabit south-facing slopes and deep valleys which show a greenhouse effect, whereas northern species persist under favorable conditions on cold north-facing slopes and in deep shady valleys.

Component	2906	2969	2957-2	3299	3319	2932-3	3337	2942	3167-10	2944
SiO_2	69.66	74.64	44.74	69.40	55.00	50.46	51.10	84.64	85.36	3.54
TiO_2	0.27	0.10	0.29	0.77	0.65	0.73	0.43	0.10	0.12	0.04
Al_2O_3	15.19	14.67	6.90	12.86	13.00	14.84	16.26	8.40	2.80	0.26
Fe_2O_3	0.60	0.18	4.22	2.29	2.40	3.51	1.47	0.77	0.23	0.17
FeO	2.15	1.01	7.90	3.23	7.54	9.05	5.60	0.21	0.73	0.93
MgO	1.31	0.45	25.60	0.50	7.10	8.80	7.60	1.35	0.30	19.51
CaO	1.82	1.96	4.28	1.96	7.43	6.09	11.43	0.14	4.76	32.22
K_2O	2.66	0.93	0.03	4.19	2.13	0.85	1.52	2.86	0.79	0.05
Na_2O	5.00	5.45	0.13	3.50	2.50	2.53	2.15	0.03	0.02	0,03
P_2O_5	–	–	0.06	–	–	0.124	–	0.072	–	–
S	0.01	0.04	–	–	–	0.02	–	0.01	–	0.04
LOI	0.61	0.26	5.08	0.67	1.61	2.54	2.00	1.10	3.83	43.12
Ba	536	626	–	1584	806	–	985	–	4475	–
Co	20	20	–	–	63	63	55	16	–	–
Cr	109	146	796	78	260	135	114	104	598	–
Cu	16	8	–	–	64	32	24	8	–	–
Mn	349	202	1550	891	1550	837	1085	39	202	1465
Ni	20	20	632	–	95	150	142	39	–	102
V	31	10	78	62	47	125	292	31	61	–
Li	–	–	–	35	39	–	10	–	–	–
Rb	–	–	–	71	66	–	22	–	–	–
Zn	–	–	–	146	–	–	–	–	–	—

Table 3. Chemical composition of bedrock in the North-Karelian synclinal zone in Russia (represents the Oulanka–Paanajärvi parks area). Analyses from Systra [27, 28]. Macro-element oxides in weight % (sulphur as element), microelements in ppm, analyses made in Laboratory of Institute Geology, Karelian Research Centre, Russian Academy of Sciences. (–) component not analyzed, Archean bedrock: No 2906 – biotite gneiss, No 2969 – gneissose plagiogranite, No 2957-2 – metapyroxenite. Paleoproterozoic bed-rock: No 3299 – quartz porphyry, No 3319 – andesitic basalt, No 2932-3 – diabase, No 3337 – leucogabbro, No 3167-10 – quartzite with fuchsite from fault zone, No 2942 – Proterozoic basal quartz rich conglomerate, No 2944 – dolomite marble

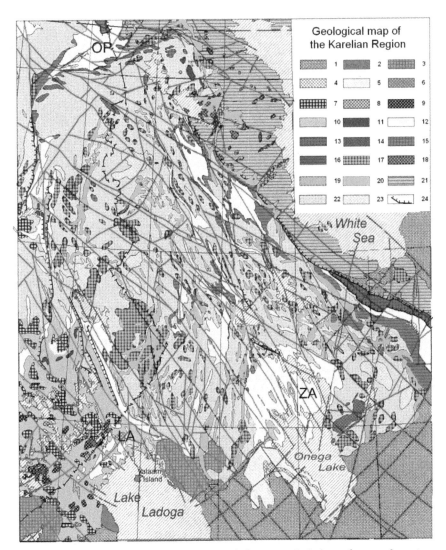

Legend: Post-Svecofennian bedrocks: 1 – Ediacara-Paleozoic platform cover; 2 – Ladoga aulacogene; 3 – post-orogenic rabakivi granites; 4 – Vepsian dolerite sills; 5 – Vepsian sedimentary rocks. Rocks of the Svecofennian (1.92-1.77 Ga): alkaline diorites and gabbro; 7 – granites; 8 – gabbro; 9 – diorites; 10 – Kalevian volcanic-sedimentary rocks; 11 – 1.98-1.95 Ga ultramafic rocks; 12 – Ludicovian and Jatulian sedimentary and volcanic rocks, including carbonates. Sumian and Sariolian (2.5-2.3 Ga) bedrocks: 13 – komatiitic basalts; 14 – layered mafic-ultramafic rocks; 15 – paligenetic granites; 16 – sedimentary-volcanic bedrocks. Neo- and Mesoarchean (2.6-3.0 Ga) bedrocks: 17 – granites; 18 – diorites; 19 – gabbro; 20 – sedimentary-volcanic rocks of greenstone belts; 21 – Belomorian Foded Belt, undivided. The oldest rocks in the region: 22 – basement granitic gneisses in the Karelian Craton; 23 – relics of the oldest sedimentary and volcanic sequences; 24 – thrust zones. Red lineaments: the main fault zones of Karelian Region. The places of study the bedrock geochemistry influence to vegetation: OP – Oulanka-Paanajärvi national parks are with surrounding territory; ZA – Zaonezhje Peninsula in the Lake Onega; LA – northeastern shore area of the Lake Ladoga and Valaam Island.

Figure 2. Geological map of the Karelian Region. Compiled by Y.J.Systra with using geological maps of Finland [25], Fennoscandia [26] and author's unpublished map "Tectonics, Karelia"(1996).

Figure 3. On left Kivakka Hill (499.5 m) in the Paanajärvi NP presents one the best layered peridotite-gabbronorite massifs in Europe. Strong bedrock from magma is not eroded so quickly as surrounding gneisses and diorites. On the top is developed mountain tundra zone with typical vegetation. On the photo below is flowering one of protected northern species - *Loiseleuria procumbens* L. On the southern slope of hill found places for nesting some southern species of birds

Numerous springs occur on the deep valley slopes, with discharge rates usually about 0.2-3 L/s. The chemical composition of groundwater in the upper vadose zone above the water-table is determined principally by topographic relief. In the eastern part of Paanajärvi, the summits of the hills are at 305-342 m above sea level, which is from 100-185 m above the level of lakes and rivers, which lie at elevations between 136-208 m [55]. Groundwater infiltrates bedrock to some extent, yielding ultra-fresh hydro-carbonate-Ca waters, with very low mineral contents, less than 100 mg/L. The temperature of spring water is 3.5-4.2° C, and pH is within the range 7.39-6.76. Although the chemical components occur at very low concentrations in spring waters, the cumulative annual flux can be considerable, even at relatively low discharge rates. This is illustrated by calculations for spring no. 6 (discharge rate = 0,5 l/s and (spring no. 9 (discharge rate = 0.3 l/s) respectively: macro-elements Ca (216 and192 kg), SO_4^{2-} (86.7 and 54.6 kg), MgO (34.7 and 14.2 kg), Na (22 and 18 kg), K (17.3 and 10.4 kg), PO_4^{3-} (0.47 and 0.19 kg), NO_3^- (0.63 and 0,095 kg), Cl (11 and 6.6 kg) and microelements: F (2 and 1.8 kg), Fe (0.93 and 1.2 kg), Zn (1.2 and 0.32 kg), Al (140 and 643 g), Ba (227 and 457 g), Sr (219 and 304 g), Mn (14.7 and 45.5 g), I (22 and 12.3 g), Cr (8.4 and 8.5 g), V (9.6 and 2.9 g), Ni (3.9 and 2.4 g), Cu (8.5 and (6.0 g), Co (0.6 and 0.6 g), Se (14.9 and 0.95 g), Sb (10.1 and 4.7 g), Cd (0.16 and 7.8 g), Pb (2 and 15.3 g), Hg (1.9 and 0.28 g), As (0.47 and 0.47 g) [57]. Small mires often form near springs, typically covering an area of several hundred square meters. Some of the water discharge is through capillary flow and in vapour phase, further enriching the soil in nutrients required by vegetation. It is only in such mires, with cold water, that cold-resistant calciphilous species are found, such as: *Saxifraga hirculus* L., *Epilobium davuricum* Hornem., *E. alsinifolium* Vill., *Angelica archangelica ssp. norvegica* (Rupr.) Nordh., *Juncus triglumis* L.

One prominent cliff section – Ruskeakallio (= Brown Cliffs) on the northern shore of lake Paanajärvi, composed by albitite dyke with numerous carbonate veins, is about 60m high and more than 300 m long. This vertical sunny wall with hanging gardens is unique, with its botanical rarities including *Gypsophila fastigiata* L.,*Aspenium ruta-muraria* L., *Draba cinerea* Adams, *D.hirta* DC, *Potentilla nivea* L., *Androsace serpentronalis* L., *Hackelia deflexa* (Wahlenb.) Opiz and more than 20 additional rare plants. During the last 150 years many generations of famous Scandinavian, and after the last war, Russian botanists have visited the Ruskeakallio cliffs. Now it is one of the most picturesque destinations in the Paanajärvi National Park, although it is not permitted to set foot on the shore, in order to protect the unique vegetation. The Paanajärvi National Park is also the only locality in Eastern Fennoscandia from which the lichen *Usnea longissima* Ach. has been recorded during the last 50 years [51,56]. A total of 97 vascular plants present at Paanajärvi have been listed in Red Book of Karelia [52]. One of the indications that the soils at Paanajärvi are compositionally favorable is the abundance of old spruce forest, which covers more than 60% of the terrain. Pine forests prevail on the coarse-grained glaciofluvial Quaternary gravels in the Oulankajoki River and Olanga River valleys, which formed through the action of very powerful melt-water-streams [32] during melting of the last ice sheet, about 10 000 years ago.

Figure 4. Ruskeakallio (Brown Cliff) is 60 m high vertical wall with unique vegetation (above left). One of very rare plant *Gypsophila fastigiata* L. (above right). Lady's slipper (*Cypripedium calceolus* L.) is common in both Oulanka and Paanajärvi National Parks (below left). *Bartsia alpina* L. - common plant for Paanajärvi and Oulanka (below right).

The deep valley of the containing Lake Paanajärvi and through which the Oulankajoki River flows has been incised into the Maanselkä Uplands, the highest points of which are between 400-600 m above sea level. This rugged topography has given rise to a special migration corridor at 109 m above sea level on Lake Pääjärvi to elevations of 200 m above sea level along the western border of the Oulanka National Park. Over the last 20 years a number of new species of birds and plants have been found on both sides of the Maanselkä topographic divide. Lake Paanajärvi has a maximum depth of 128 m, and its deepest part is thus only 8 m above the mean level of the White Sea. Nevertheless, at no stage during its history has Lake Paanajärvi been in contact with the White Sea, and the lake is inhabited by relict populations of smelt, sea trout, arctic char, whitefish, grayling and other typical Karelian lacustrine fishes. The small crustaceans *Mysis*, *Monoporeia* and *Pallacea* provide an important food source for valuable fish species.

Favorable geochemistry, hilly relief, and soils and floral diversity also contribute to the unusual diversity of birds, animals, fishes and insect species for this the latitude near the Arctic Circle.

7. Onega Synclinorium and Zaonezhje peninsula

Onega Synclinorium in one of the largest Paleoproterozoic structural features preserved in the Karelian Craton and covers about 10 thousand square km surrounding Lake Onega and Zaonezhje peninsula (Figure 1, 2). The Paleoproterozoic succession in the Onega Synclinorium begins usually with Jatulian conglomerate, sandstone and dolomite marble, and includes a remarkable salt and gypsum horizon, discovered during drilling in 2008, at depth of about 2 km, and finishes with Ludicovian black shungite-bearing schist, mafic lava and numerous sills, of age 2.2-1.8 Ga. The metamorphic grade these rocks are relatively low.

The synclinorium represents two separate folding episodes intersected by numerous NW-trending fault zones, parallel to the axial plane of the open later generation folds. The Zaonezhje peninsula sequence contains sandstones, carbonate and mafic volcanic rocks, shungite-bearing black schists, shungite-bearing aleurolites etc. The main types of bedrock are given in Table 4 and some more in [37]. Archean granite-gneisses are rich by SiO_2 as well lydit, carbon-bearing silicate bedrock, but gneisses content verysmall amounts of microelements, while shungite-bearing black schists are more rich by Cr, Cu, V, Zn. As usual carbonate rocks have high concentrations of MgO and CaO, higher Ba, Sr and sometimes Mn content. More rich by metallic microelements are mafic and ultramafic volcanites, picrite-basalts, pyroxene diabase and also tuff-aleurite and shungite bearing rocks.

The synclinorium represents two separate folding episodes intersected by numerous NW-trending fault zones, parallel to the axial plane of the open later generation folds. The Zaonezhje peninsula sequence contains sandstones, carbonate and mafic volcanic rocks, shungite-bearing black schists, shungite-bearing aleurolites etc. The main types of bedrock are given in Table 4 and some more in [37]. Archean granite-gneiss (Table 4, No 1) is rich by SiO_2, but contains small amounts of CaO and MgO, and all trace elements. Typical andesite basalt (Table 4, No 2) contains some more FeO, F_2O_3, CaO and MgO, Mn, V and Zn, but other trace elements stay in the low level. Gabbro-dolerite (Table 4, No 3), picrite-basalt (Table 4, No 4) and pyroxene diabase (Table 4, No 5) contain more FeO, CaO, MgO and the most of needed for vegetation trace elements. Dolomite marble (Table 4, No 6 and No 7) are the main sources of the Ca and Mg for soils. In Onega Synclinorium are widely developed carbon (shungite)-bearing tuff-aleurolites (Table 4, No 8, shungite-bearing black schist (Table 4, No 9), shungite-bearing aleurolite (Table 4, No 10) and lydit (Table 4, No 11), which usually contain notable amounts of microelements: Ba, Co, Cr, Cu, Mn, Ni, V, Sr and Zn. In the places, where shungite–bearing rocks are near the surface, the soils are black color and sunlight keeps soils warm earlier than on the other soils. Especially it works on agricultural territories for getting crops of vegetables and potatoes.

Component	1	2	3	4	5	6	7	8	9	10	11
SiO_2	71,38	57.20	48.08	43.54	47.49	11.78	17.90	40.28	29,11	53.92	93.89
TiO_2	0.32	0.98	2.03	1.42	1.51	0.03	0.07	6.60	0.60	1.95	0.08
Al_2O_3	14.50	14.07	13.77	8.97	13.21	0.26	2.08	12.31	7.56	14.20	1.18
Fe_2O_3	1.33	1.21	3,38	1.74	2.46	0.12	1.52	2.58	6.23	1.93	1.08
FeO	0.86	8.62	10.94	10.58	12.73	0	1.09	20.18	0	11.26	1.05
MgO	1.47	4.09	5.97	18.22	7.74	20.33	15.80	4.11	1.36	3.58	0.03
CaO	0.87	5.89	8.59	8.56	5,65	25.95	23.10	3.00	0.56	1.58	0.17
Na_2O	3.98	4.92	2.05	0.49	2.23	0.02	0.98	1.11	4.78	3.15	0.10
K_2O	2.85	0.63	1.19	0.15	0.40	0.04	0.16	0.10	0.25	1.53	0.52
P_2O_5	0.12	0.19	–	0.18	0.17	0.12	–	0.20	0.02	0.30	–
H_2O	0.17	0.15	0.78	0.47	0.69	0.04	–	0.29	0.20	0.23	0.10
LOI	1.78	1.54	3.83	5.43	5.89	41.06	–	8.54	51.44	5.83	1.36
Ba	–	-	52	82	140	280	127	384	306	169	108
Co	–	47	43	76	51	10	43	55	60	68	18
Cr	–	34	230	710	130	8	62	314	142	274	299
Cu	–	32	77	86	120	12	54	102	42	88	80
Mn	225	1178	1472	1627	1395	124	1318	2325	620	1163	155
Ni	–	63	70	290	75	12	96	201	184	182	49
V	–	117	240	300	240	32	75	107	260	252	221
Sr	–	–	170	98	97	–	319	103	120	187	84
S	–	–	–	500	500	200	–	–	1700	–	10100
Zn	24	80	65	100	–	48	56	122	184	103	140

Table 4. Chemical analyses of main bedrock types of Onega synclinorium (macro-elements in weight %, microelements – in ppm). Sample bedrock and place of sampling: 1 – graniit, 5 km to north from Kumsa River; 2 – andesite basalt, Kumsa River, Central Karelia; 3 – gabbro-dolerite, Unitsa Bay, Lake Onega; 4 – picrite-basalt, Rovkozero; 5 – pyroxene diabase, Radkola neck; 6 –dolomite marble, Pyalozero village; 7 – carbonate bedrock, Zaonezhje; 8 – shungite-bearing tuff-aleurolite, Nigrozero; 9 – shungite-bearing schist, Mednyje Jamy village, Zaonezhje peninsula; 10 – shungite-bearing aleurolite, Nigrozero; 11 – lydit, Pustoshi

In western limb of the Onega Synclinorium were studied soils in reduced and full profiles directly overlying on different bedrock, which shows that soils chemical composition is influenced by bedrock and Quaternary sediments composition (Table 5). In soil overlying shungite schist (carbon rich black schist)) the concentration of most elements is higher (Table 5, No 15, 16) than in bedrock (Table 5, No 17). This indicates that the soil has been from the surrounding environment. About 70% of the Republic of Karelia is covered by granites, granite gneisses and migmatites of the Karelian Craton, which are rich by Si, Al, Ca, Na, K, but contain very low concentrations of vital micronutrients. For example chromium concentrations in soil is some less than in bedrock, Ni, Cu and S concentrations are reduced in soil lower horizons. Shungite, consisting of native carbon, is burned on ignition resulting in high weight losses (LOI). The direct influence of bedrock geochemistry on soils is less pronounced, when soil thickness over bedrock exceeds 0.5 m. Most of soils are enriched in SiO_2, Fe_2O_3, Na_2O, K_2O and Cr from Quaternary soft sediments (Table 5).

Soil and bedrock	SiO_2	Al_2O_3	Fe_2O_3+FeO	MgO	CaO	Na_2O	K_2O	LOI	Ni	Co	Cr	Cu	V	Zn
1.5-20cm	77.23	10.74	2.33	0.94	1.44	2.77	1.95	2.17	24	8	27	8	39	16
2.20–45cm	77.20	10.05	2.44	1.06	1.56	2.97	1.86	1.38	32	8	41	8	61	56
3. Granite	**71.38**	**14.50**	**2.19**	**1.47**	**8.87**	**3.98**	**2.85**	**1.78**	–	–	–	–	–	24
4. 0–5cm	16.97	3.19	1.50	0.99	1.22	0.70	0.57	74.05	24	0	0	6	28	112
5. 5–16 cm	44.91	8.68	3.91	2.25	3.12	2.23	1.24	31.82	71	16	34	72	34	104
6. 6–32 cm	50.46	10.09	6.59	3.28	3.09	2.42	1.48	21.08	55	16	137	144	224	72
7.Peridotite	**41.26**	**5.28**	**11.64**	**26.8**	**3.79**	**0.07**	**0.02**	**8.46**	**1493**	**95**	**1680**	**32**	**168**	**803**
8. 0–4 cm	22.92	5.51	2.23	1.01	2.79	0.82	0.61	62.26	32	24	41	7	73	473
9. 4–16 cm	59.29	11.99	5.62	1.80	2.50	2.79	1.23	12.71	79	24	89	24	162	264
10.16-45cm	59.31	11.58	5.66	1,97	2.10	2.74	1.32	13.56	24	16	75	24	246	225
11.Diabase	**49.20**	**12.94**	**10.45**	**9.75**	**5.39**	**3.42**	**0.80**	**4.20**	**134**	**47**	**226**	**32**	**221**	**104**
12. 0-5cm	14.22	2.10	4.16	11.7	14.5	0.43	0.42	51.35	24	16	27	16	34	136
13. 5-15cm	17.50	2.39	2.55	16.3	16.4	0.51	0.45	42.40	16	16	7	24	67	72
14. Marble	**1.04**	**0.04**	**1.59**	**21.3**	**28.7**	**0.05**	**0.01**	**46.16**	**10**	**8**	**5**	**10**	**6**	**62**
15. 0-15cm	69.29	9.05	6.35	1.56	1.15	0.90	1.99	8.58	71	8	96	88	302	145
16.15-71cm	78.38	7.59	2.91	1.06	0.59	0.39	1.94	6.21	39	8	89	32	263	104
17.Shungite	**59.80**	**3.8**	**1.39**	**0.60**	**0.21**	**0.05**	**1.55**	**31.86**	**40**	**8**	**102**	**40**	**225**	**16**

Table 5. Composition of reduced soil profiles on bedrocks in Onega Synclinorium (macro-elements and LOI in wt%, micro-elements in ppm). Analyses made in laboratories Institute of Geology and Forest Research Institute, the Karelian Research Centre, Russian Academy of Sciences, Petrozavodsk, Russia. Soil sampling intervals in cm above bedrock (in bold): 1-3 – soils and Archean basement granite, 5 km to north from Kumsa River; 4-7 – soils and peridotite, Lake Konchozero: 8-11 – soil and gabbro-diabase, Hirvas village; 12-14 – soil and dolomite marble, Pyalozero village; 15-17 – soil and shungite schist, Zazhogino quarry, Zaonezhje peninsula, Lake Onega '

The Archean basement granites with soils were studied at northwestern limb of the Onega structure (Figure 2). Soils on Archean plagio-microcline granite (Table 5, No 3) are enriched by SiO_2, Fe_2O_3, Cr, V and Zn (Table 5, No 1-2). On the diabase, peridotite and marble the organic matter content in soil humus horizon is higher, up to 58-65%, if counting differences in LOI, and SiO_2 content is reduced to 59-14%. Soils on peridotite (Table 5, No 7) have tendency of diminishing of elements in soils characterizes Fe, MgO, CaO, SiO_2, Al_2O_3, Na_2O, K_2O, Ni, Cr, Cu, V, Co, but S and Zn show opposite tendency (Table 5, No 4-6). Soils on gabbro-diabase in Hirvas Village are more acid, contain SiO_2 59% in horizons 4-16 and 16-45 cm, and less macro-element (Table 5, No 8-10) than bedrock (Table 5, No 11). Usually the element concentration diminishes gradually from bedrock to humus horizon: A_2O_3, FeO, MgO, K_2O, V, Cu, S, but for CaO, Co and Zn trend is opposite. Zn content grows from 104 ppm in bedrock to 473 ppm in humus horizon (Table 5, No 8-11). Dolomite marble theoretical composition is CaO 30,4%, MgO 21,7%, CO_2 47.9%, quite close to that is content of these elements in dolomite – $CaMg(CO_3)_2$ from Pyalozero Village (Table 5, No 14), in soil intervals only MgO and CaO are originated from bedrock, all other nutrients are added from Quaternary sediments (Table 5, No 12-13). It is likely that the abundance of CaO, MgO and necessary micronutrients has resulted in the formation of a rich humus horizon, which is the basis for biodiversity. Phosphorus and sulphur are enriched in the humus horizon, due to the activity of animal and microbial life in the soil [58, 59], even if they are absent from the underlying bedrock, as for example, sulphur in the case of ultramafic magmatic rock peridotite (Table 5, No 4-7).

Quaternary sediments are not usually transported by ice over distance greater than 50 km, and typically less than 5 km. For the Onega Synclinorium this means, that favorable volcanic and carbonate rock influences are present everywhere, including fluvioglacial eskers and deltas. It must be noted that the most of bedrock contains notable amounts of Mg, Ca, Na, Fe, Mn; mafic and ultramafic volcanic and intrusive rocks, shungite (carbon)-bearing aleurolites, in addition to essential metallic micronutrients such as Co, Cu, Cr, Ni, V, Zr etc. About 800 vascular plants species have been recorded here as well as many rare plants from the Red Data Book for Fennoscandia [51] and Republic of Karelia [52], which is twice the number of vascular plant species (500) in the very large (5206 square km) Vodlozero National Park, which is located at the same latitude and shares the climatic conditions, but which is situated 50 km to east, where thick Quaternary till and glaciofluvial deposits overlie Archean granites. To the east of the Vodlozero NP, in the Archangelsk Region of Russia is the Kenozero NP, which has similar physiographic conditions, but differs geologically due to the presence of Paleozoic carbonate rock and till derived from mixed sources, including granite from the Karelian Craton and local limestone. The number of vascular plant species present here is 534, of which 61 is listed into Red Book.

8. Western and Northwestern Lake Ladoga area and the Valaam Archipelago

The northeastern marginal zone of the Svecofennian Domain near Lake Ladoga is characterized by mantled granite gneiss domes. Narrow synclinal zones between basement domes comprise equivalents of the same Paleoproterozoic volcanic and sedimentary rocks, including dolomite marbles that occur in the Paanajärvi – Oulanka area and in the Onega

Synclinorium. In the atlas of soils of the Republic of Karelia pH-values in the upper parts of soil horizons in the northwestern part of Lake Ladoga and in the central part of Onega Synclinorium are much higher (pH = 3.5-5.0) than in Central Karelia, where bedrock consists mostly of granites and diorites, associated with pH values of 3.5 or less [38]. Due the small number of samples from the Paanajärvi National Park the influence of carbonate bedrock is not apparent in the Soil Atlas. Early Proterozoic Jatulian and Ludicovian rock units in the Svecofennian Domain in the Ladoga Zone are overlain by younger Kalevian sediments. The Precambrian basement is divided into separate blocks by numerous fault zones, which are reflected topographically as deep narrow valleys in which some northern species of mosses and lichens are found. The total number of vascular plant species in the area reaches 750, of which 550 are also present in the Valaam archipelago. The number of lichens and lichenicolous fungi exceeds 800 and 269 mosses have also been identified here [24, 52]. The main reason for such biodiversity is the extensive occurrence of carbonate and mafic volcanic rocks with their high concentrations of essential micro-nutrients.

The central part of Lake Ladoga coincides with the Mesoproterozoic sedimentary-volcanic bedrock of Ladoga Graben (aulacogene), which includes the thick dolerite sill outcropping on the islands of the Valaam Archipelago. Lake Ladoga is the largest freshwater body in Europe and shows a significant effect on microclimate. The coastline is much more favourable for agriculture and natural vegetation compared to terrain some 100-150 m above lake level, where snow melts some weeks in springtime and where autumn frosts occur several weeks earlier. The bedrock of the entire Valaam Archipelago is composed of a gabbro-dolerite sill, 220 m in thickness, associated with monzonite-quartz syenite, granite porphyry and granophyre [39]. A study of heavy metals and sulfur distribution in the soils of Valaam Archipelago has also recently been completed [40].

Comparison of the distribution of heavy metals (Table 6) in the main types of bedrock, namely gabbro-dolerite No 3211-6 and an average composite from monzonite – quartz syenites No 1488-3-12 [39] and soil horizons, shows good correlation between bedrock and soil composition (Table 6). The whole archipelago is formed from gabbro-dolerite and monzonite-syenite sill, while the surrounding lake is more than 100 m deep [61]. Therefore only small amounts of material would have been transported from the western and northwestern shores of the lake to the Valaam Island during glacial time. The local doleritic bedrock might therefore be expected to be enriched in microelements. Zinc is often concentrated in the upper parts of soil profiles (Table 5), while Cr, Cu, Ni and V abundances are close to those in bedrock. Only Pb shows some higher concentrations in soil horizons, but this may be due the small number of studied analyses in bedrock.

V. Koval'skij [13] gave next concentration intervals for normal life and development: Co – 7-30, Cu – 15-60, Zn – 30-70, Mn – 400-3000, Mo – 1,5-4, I – 2-40, B – 3-30, Sr – 0-10 [16]. At levels in excess of 500 ppm, zinc in soil interferes with the uptake of essential metals: Fe, Mn and B [3]. Anyway, so rare bedrock had significant influence to plant diversity. The Valaam Archipelago Nature Park has at least 590 vascular plant species in addition to introduced plant species, which grow here in great abundance. The finding of 61 Red Data Book species indicates that Valaam Archipelago Nature Park is of considerable floristic value. Some species, as *Plantanthera chlorantha* (Cust.) Reichenb., *Potentilla neumanniania* Reichenb., *Corydalis intermedia* (L.) Merát, *Cotoneaster integerrimus* Medik and *Myosotis ramosissima* Rockel ex Schult do not occur in other protected areas of Karelia [60]. The content of Ca in Valaam bedrock usually does not exceed 6-7% and is connected with other elements in hard mineral lattice. At weathering Na and Ca ions are carried out the first order, so in soil forms deficit of Ca and in archipelago does not grow beautiful plant *Cypripedium calceolus* L.

Bedrock or soil	Cd	Co	Cr	Cu	Mn	Ni	Pb	V	S	Zn	FeO+ Fe₂O₃	Fe
Gabbro-dolerite	–	24	82	8	953	16	–	67	–	137	12.2%	–
Monzonite-syenite	–	20	34	12	973	9	16	54	–	166	10.9%	–
Forest litter	0.74	2.6	10	14	746	7.7	26	–	1240	170	–	11900
Soil beneath litter	0.76	6.4	23	15	612	4.9	25	–	612	129	–	38000

Table 6. Heavy metals contents in bedrock [39] and average abundances in forest litter and the uppermost mineral soil horizon beneath the litter [40] of Valaam Island. Sum Fe oxides in weight%, microelements in ppm.

9. Kilpisjärvi area, Finnish Lapland, NW Finland

This area belongs geologically to the eastern marginal zone of the Scandinavian Caledonide nappes, far to the north of Arctic Circle (69° 03′). The relief on the eastern slope of Scandinavian Mountain Ridge lies between 472 m (Lake Kilpisjärvi) and 1328 m above sea level, vegetation is typical for tundra zone, trees are represented with tundra birch, willows and some aspen trees on the southern slope of Saana Mountain (1029 m) near the Lake Kilpisjärvi, from coniferous grows only juniper, which may reach very high age, preliminary study shows, that more than 1000 years.

The oldest gneissose granodiorites, felsic, mafic and ultramafic volcanics and mica gneisses belong to the 2.7-2.8 Ga Neoarchean basement, which represents the autochthonous foreland to the Caledonian Orogen. osition, cratonised ago. This deformed and metamorphosed basement is unconformably overlain by Early Cambrian basal conglomerate, with silty and quartz-rich intercalations. Dolomitic marble is typical of the Jerta Nappe, which has been thrust somewhat to the southeast and forms a layer from 1-40 m thick , in some places strongly folded, and which may cover extensive parts of the hill slopes. Although there are 3-4 more nappes, with a total thickness of 450 m near the Kilpisjärvi Lake and up to 1250 m near the Halti Mountain [62-64], the most interesting rock unit with respect to the diversity vegetation is the dolomitic marble of the Jerta Nappe and the numerous small springs on the southern slope of Saana Mountain. The important influence of carbonate rocks on vegetation was first noted here by A. Pesola [42]. There are now three botanical protected areas that include all large dolomite outcrops in the Kilpisjärvi area: the Malla Strict Nature Reserve, 3000 ha, has been under protection since 1916, and was declared as a strict reserve in 1938, the Saana Protected Territory and the Annjalonji Protected Area were declared in 1988. Although there are many other hills with the same elevation and slope aspect, most of the rare and protected species are absent, where the bedrock is other than dolomite.

Figure 5. *Rhododendron lapponicum* L.flowering on the Paleozoic marbles of the Malla Strict Protected Territory (above). Slope of Pikku-Malla Hill with dolomite marble layer (below, light layer).

On the upper part of the Pikku-Malla fell these dolomite outcrops coincide with most of the rare plant finds: *Erigeron acer* L., *E. uniflorus* , *Rhododendron lapponicum* L., *Polystichum lonchitis* L., *Pseudorchis albida* (Fernald), *Silene uralensis* Rupr., *Veronica fruticans* Jacq. These and other protected rare and endangered vascular plant species are only found in the presence of carbonate rocks. During the flowering of *Dryas octapetala* L. the folded dolomite layers on the hill-slopes resemble natural flowerbeds. The influence of dolomite continues down slope and also in the spring waters. Numerous small springs discharge in the lower part of slopes and are associated with great floral diversity. In winter time the springs may be frozen but during the growing season, discharge flow rates of 0.2 l/s were recorded for Spring 1 and 0.1 l/s for Spring 2. Respective mineral components for Spring 1 and Spring 2 (where data are available)are: Ca(45 and 47 kg), Mg (18 and22.2 kg), K (3.7 and 1.8 kg), Na (7.4 and 2.4 kg), Si (9.5 and 2.7 kg), S (26.6 and 23.8 kg), $SO_4{}^{2-}$ (71.9 and 68.6 kg), $NO_3{}^-$ (1.3 and 0.3 kg), Cl (4.3 and 1.4 kg), Sr (262 and 180 g), Al (55.2 and 5.5 g), Ba (23.2 and 30.6 g), B (7,3 and 4.8 g), Li (2.4 and 3.4 g), Zn (6.1 and 3.1 g), Rb (3.3 and 2.1 g), Mo (0.95 and 2.1 g), V

(4.3 and 0.36 g), Cu (3.1 and 0.86 g), U (0.72 and 2.4 g), Co (190 mg), Bi (63 and 110 mg), Mn (63 and 47 mg), As (252 mg, Spring 1 only), Sb (31.5 mg, Spring 1 only) and Th (95 and 16 mg). Elements such as Ag, Be, Cd, Cr, Ni, P, Pb, Se, Tl, Fe, Br, F have concentrations below detection limits. Water samples were analyzed at the Geological Survey of Finland. Because discharge from such springs is low, much of the water evaporates, leading to enrichment of essential elements in soils. On the soils developed over Precambrian granodiorite near these springs, some rare species were found, such as *Saxifraga aizoides* L., *Pseudorchis albida* (Fernald).

Figure 6. *Saxifraga aizoides* L. is flowering near the spring on the southern slope of Saana Mountain (above). *Polystrichum lonchitis* (L.) Roth is very rare plant in the Northern Finland, in Kilpisjärvi area is possible to meet it only on carbonate bedrock (below left). Another protected species *Pseudorchis albida* L. also likes Ca-rich soils (below right).

About 470 vascular plants species have been recorded in the surroundings of Kilpisjärvi, most of them occurring on exposures of dolomitic marble. Pelitic Cambrian schists contains more microelements than quartzite and may be a more favored substrate for some species, including *Dryas octapetala* L.

10. Areas surrounding Kevo Lake and Kevo Subarctic Research Station

The bedrock of this area consists of Archean migmatites and gneisses, mostly hornblede-bearing gneisses, which are divided into blocks with fault zones. Surficial deposits are well-sorted glaciofluvial gravels with clasts of granite and gneiss, which are thus rather poor sources of nutrients. Scots pine (P. sylvestris) is common in valleys and on hillslopes, but the Kevo area (69 °45′ N) is some 150-200km beyond the northern limit for spruce. The region is characterized by typical northern tundra species, and lacks rare species, which need more abundant macro- and micronutrients. Common species are *Loiseleuria procumbens* L., *Phyllodoce caerulea* (L.) Bab., *Diphasiastrum alpinum* L., *Pinguicula vulgaris* L. etc.

11. Influence of bedrock and Quaternary sedimentary geochemistry on biodiversity in Estonia

Estonia is located along the southern shore of the Gulf Finland and the Precambrian crystalline basement is everywhere covered by Ediacaran and Paleozoic sedimentary cover. The sedimentary cover together with basement is tilted gently southwards at a gradient of nearly 3 m per kilometre and its thickness increases from 125-140 m in the north to 600 m and more in southern and southwestern Estonia. Cambrian, Ordovician and Silurian bedrock are exposed in northern Estonia as east-west trending belts, whereas to the south these older sequences are covered by younger rocks. The resistant Ordovician limestone forms cliffs up to 56 m high, known as the North Estonian Klint (Figure 7).

During the last 400 000 years, the region was repeatedly covered by ice during several glacial events, which advanced towards the south or southeast from the Scandinavian Mountains, transporting metamorphic and igneous rock material to Estonia. The Svecofennian domain is covered mostly by granites, migmatites and gneisses, with mafic rocks comprising only 3.5% of till and boulder material [65]. Northern Estonia belongs to the zone of glacial erosion, where the thickness of Quaternary cover seldom exceeds 5-10 m, being in many places on alvars less than 1 m thick; eskers and glaciofluvial deltas may however, exceed 20 m in thickness. Southern Estonia is in contrast characterized by moderate sedimentary accumulation, with till cover in the Otepää and Haanja Uplands commonly exceeding 100 m and in the ancient buried valley of Abja a local maximum thickness of 207 m is attained. Local carbonate rock cobbles and pebbles predominate in the thick till sequences in central Estonia, but near the southern border the crystalline Fennoscandian bedrock becomes prevalent. These have a strong influence on soil composition, the Ca content in the humus horizon falling to 0.2% or less, while in carbonate bedrock Ca contents are between 1-8%. The Mg content of sediments derived from carbonate rocks is usually 0.5-0.8%, whereas in southern Estonia it is commonly less than

0.09%. The minimum content of Mn in soils for healthy growth and development of plants is 400 ppm [16], but much of Estonia has Mn concentrations less than 230 ppm [49].

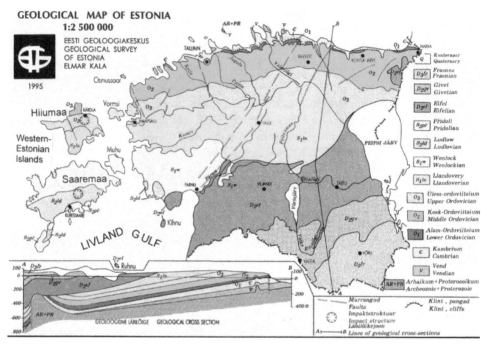

Figure 7. Geological map of Estonia. Compiled by E.Kala, 1995. Is published with permission of the Geological Survey of Estonia

Cambrian claystones and sandstones outcrop in the narrow zone between the klint and Baltic Sea shore. A number of micro- and macronutrients: K, Mg, P, B, Co, Cr, Cu, Ni, V and Zn have higher concentrations in these clays (Table 7, Es-1) than in carbonate rocks and sandstones. The influence of the clays is however restricted to this narrow zone. The Ordovician sequence begins with the Pakerordi Stage sandstone and the kerogenous dark-brown argillite horizon, which has a thickness of 7.7 m in northwestern Estonia, diminishing to about 2.0 m in northeastern Estonia. This argillite is notable for high concentrations of U (up to 400 ppm), K, As, B, Co, Cr, Cu, Hg, Mo, Ni, Pb, V and Zn (Table 7, Es-2). Soils are enriched in F, Mn. Mo, P, U and Y, but only within a narrow zone to the south, about 10 km wide, parallel to the klint [49].

The central and northern part of the Estonian mainland and the Western Islands are composed of Ordovician and Silurian carbonate rocks: marl, limestone and dolostone. Clayey limestone (Table 7, Es-17) contents some more SiO_2, Al_2O_3, MgO, K_2O, Ba, F, Rb and Sr. Limestone (Table 7, Es-3 and Es-9) contain maximum amount of CaO (50.51-52.67%), some SiO_2, Al_2O_3, MgO, Mn, Sr, sometimes also F. Dolostone of Estonia (Table 7, Es-16, Es-18 and Es-4) usually contain some per cents impurities due the short distance from Fennoscandian Shield, as source of sandy material. MgO content in dolostone often exceeds 20%, CaO content 28,87-30.87%.

Limestone and dolostone as bedrock and as pebbles and cobbles in gravel enrich all soils of Central Estonia and Western Estonia Islands with needed Ca and Mg.

Components	Es-1	Es-2	Es-3	Es-16	Es-9	Es-17	Es-18	Es-4	Es-15	Es-5
SiO_2	59.24	52.14	4.84	4.17	1,21	9.83	4.69	2.84	63.00	95.1
TiO_2	0.88	0.76	0.077	0.084	0.022	0.089	0.062	0.041	0.20	0.23
Al_2O_3	17.38	13.15	1.10	1.20	1.25	2.45	0.88	0.75	18.04	1.81
Fe_2O_3	4.29	0.85	0.06	0.49	0.5	0.38	0.17	0.1	0.59	0.11
FeO	2.60	3.02	0.44	2.58	–	0.38	0.30	0.35	0.14	1.33
MgO	2.58	1.11	0.85	16.10	1.25	5.36	20.16	20.42	0.76	0.05
CaO	0.84	0.22	50.51	30.87	52.67	40.79	28.87	29.35	0.29	0.047
K_2O	5.84	7.95	0.51	0.59	0.16	1.72	0.413	0.07	14.91	0.07
Na_2O	0.13	0.10	0.08	0.04	<0.03	0.043	0.04	0.26	0.05	1.03
P_2O_5	0.31	0.13	0.42	0.232	0.043	0.015	0.026	0.012	0.077	0.022
Cl	0.13	0.02	0.03	0.05	0.04	0.03	0.06	0.07	0.04	0.01
S	0.05	2.19	0.12	0.132	0.19	0.21	0.02	0.11	0.04	0.01
LOI	4.95	19.91	40.39	42.49	42.61	37.70	44.02	45.33	1.37	0.29
As	1.3	37	<1	<1	0.80	<1	<1	<1	2.6	<3
B	150	53	21	–	8.9	–	–	12	40	23
Ba	420	379	29	41.6	20.09	46.88	31.92	14	119	147
Cd	<1	<1	<1	0.04	<5	0.01	0.065	<1	<1	<1
Co	20	12	1.8	1.43	<3	0.98	1.28	1.5	4.0	2.6
Cr	78	80	9	9.68	8.25	7.28	10.67	9	9	31
Cu	25	105	3	3.95	4.0	3.64	1.95	<4	6	10
F	1195	570	545	350	<100	210	180	<100	1200	<100
Hg	<0.01	0.162	<0.01	<10	–	<10	<10	<0.01	0.023	<0.01
Mn	320	158	460	2060	371.4	229	399	214	16	98
Mo	<2	56	<2	0.41	<1	0.57	0.31	<2	1.0	<2
Ni	40	98	4	5.11	3.99	5.37	2.91	4	6	9
Pb	11	77	5	2.84	13.94	3.54	1.63	4	9	7
Rb	176	118	10	10.53	2.77	15.16	8.92	7	63	20
Se	<4	2.3	<6	<1	0.50	<1	<1	<3	<3	<4
Sn	3,9	3.2	<1	0.13	<1	0.25	0.14	<1	7.2	0.73
Sr	91	53	178	59.64	113.1	137.4	51.43	32	10	19
V	109	509	9	14.00	3.40	9.60	20.8	11	14	11
Zn	176	133	4.0	14.88	8.46	8.88	7.94	5	38	4

Table 7. Macro- and micronutrients content in the main bedrock groups of Estonia (compiled after reference analyses [47, 48] macro-elements and LOI in weight %, micro-components in ppm. Samples: Es-1 – claystone (clay), Ca_2, Kunda; Es-2 – kerogenic argillite, O_1, Tallinn; Es-3 – limestone,O_2, Tallinn; Es-16 – dolostone O_2, Maardu, Tallinn; Es-9 – limestone, O_3, Vasalemma, Harjumaa; Es-17 – clayey limestone, S_1, Valgu, Läänemaa; Es-18 – dolostone, S_1, Anelema quarry, Pärnumaa; Es-4 – dolostone, S_1, Mündi quarry; Järvamaa; Es-15 – metabentonite, Kinnekulle bed, O_3, Pääsküla, Tallinn; Es-5 – sandstone, Middle Devonian, Suur-Taevaskoda outcrop, Ahja River.

The entire region, except in areas of thick Quaternary cover, is enriched in Ca (1.25-6.06%), Mg (0.44-2.12%), and sporadically in Mn and F. High fluorine contents are typical for felsic volcanic ash (metabentonite) layers (Table 7, Es-15). All other micronutrients in carbonate rocks are present in small concentrations (Table 7), which is also reflected in the compositions of soils developed on such bedrock. The central and northern part of the Estonian mainland and the Western Islands are composed of Ordovician and Silurian carbonate rocks: marl (clayey limestone), limestone and dolostone. Clayey limestone (Table 7, Es-17) contents some more SiO_2, Al_2O_3, MgO, K_2O, Ba, F, Rb and Sr. Limestone (Tabel Es-3 and Es-9) contain maximum amount of CaO (50.51-52.67%), some SiO_2, Al_2O_3, MgO, Mn, Sr, in some places also F. Dolostones in Estonia (Table 7, Es-16, Es-18, Es-4) are usually secondary origin, so they contain some silicate minerals impurities due the short distance to Fennoscandian Shield. MgO content stays between 16.10 and 20.42%, CaO – 28.87-30.87%, some higher is Ba, Sr, sometimes Mn and F concentrations. The entire region, except in areas of thick Quaternary cover, is enriched in Ca (1.25-6.06%), Mg (0.44-2.12%), and sporadically in Mn and F. High fluorine content is typical for felsic volcanic ash (metabentonite) layers (Table 7, Es-15). All other micronutrients in carbonate rocks are present in small concentrations (Table 7) as they are also in soils developed on such bedrock [49].

Southern Estonia is covered mostly by sandstones, which usually contain more than 95% quartz (Table 7, Es-5). As might be expected, concentrations of macro- and micronutrients in bedrock and Quaternary cover are low. The Sakala and Otepää Uplands contain material transported by ice from northern Estonia – mostly carbonate rocks - and from the Fennoscandian Shield – crystalline rock, mostly granite and migmatite. Many elements, such as B, Cd, Co, Cr, Cu, F, Hg, K, Mo, Na, Ni, Sn, Sr, U, V and Zn, occur at very low concentrations in Fennoscandian bedrock and on the geochemical maps of the humus horizon of soil of Estonia they are distributed more evenly than elements derived from local bedrock. Because Estonian soils thus represent mixing of two components of differing composition, the influence of soils on vegetation is generally not so marked when compared with soils in Fennoscandia, where differences in vegetation are much greater. Nevertheless, *Asplenium ruta-muraria* L., *A. septentrionale* L., *Equisetum hyemale* L., *E. x moorei* Newman, *etc* grow only on the carbonate rocks of northwestern Estonia, near the Baltic Sea shore.

The most comprehensive listing of Estonian vascular plants and their distribution [50], has made it possible to assess how bedrock composition influences vegetation. The list records a total of 1353 plant species, about 50 of have uncertain occurrence, while some 700 are distributed more or less evenly, with no preference for bedrock type. A further 160 rare plant species likewise show no particular correlation with rock type, while 67 species are only found in proximity to the Baltic coast. There are 55 species that are endemic to carbonate bedrock and 35 species that grow exclusively on the sandstones of southern Estonia and 76 species that are completely absent from sandstone terrain. There are 137 species that occur predominantly on carbonate rocks, and compared to only 58 species that grow preferentially on sandstone.

The Western Estonian Islands emerged from the Baltic Sea later than the continental part of Estonia, which is also reflected in vegetation diversity: 46 vascular plant species are absent from Saaremaa, but another 50 species occur only on Saaremaa and have not been found on the mainland. The soil in Estonia has formed from two main sources, local sedimentary bedrock and crystalline material transported by glaciers from Fennoscandia. Geochemical maps of the humus horizon show that the concentrations of some elements - B, Ba, Cd, Co, Cr, K, Na, Ni, Sn, U, Th, V, Zn - show little or no influence from local bedrock. Conversely, abundances of Ca, Mg and F in soil are correlated with bedrock composition, Mo and P concentrations in particular being closely associated with the kerogenic argillite near the North Estonian Klint; bedrock controls on Fe and Hg distribution is not so evident. In summary, material derived from the Fennoscandian Shield provided most of the following microelements in Estonian soils: B, Ba, Cd, Co, Cr, K, Na, Ni, Sn, U, Th, V and Zn. Recommended maximum permissible concentrations have been defined and legislated for most elements, but natural minimum concentrations should also be considered, below which there may be adverse influences on plant and animal health and life. For Mo this lower limit is 1.5 ppm and for Mn it is 400 ppm, but normal ecological functions require a concentration of 3000 ppm [16]. In Estonia the average Mo content is 2.5 ppm, but almost a third of the country has concentrations less than 1.2 ppm, Concentrations of Mn are between 75-2400 ppm, but half of the country records concentrations less than 400 ppm.

Southern Estonia is covered mostly by sandstones, which usually contain more than 95% quartz (Table 7, Es-5). As Might be expected, concentrations of macro- and micronutrients in bedrock and Quaternary cover are low. The Sakala and Otepää Uplands contain material transported by ice from northern Estonia – mostly carbonate rocks - and from the Fennoscandian Shield – crystalline rock, mostly granite and migmatite. Many elements, such as B, Cd, Co, Cr, Cu, F, Hg, K, Mo, Na, Ni, Sn, Sr, U, V and Zn, occur at very low concentration in Fennoscandian bedrock and on the geochemical maps of the humus horizon of soil of Estonia they are distributed more evenly than elements derived from local bedrock.

Comparison of average microelement concentrations in bedrock and humus horizons in Karelia, Russia (Precambrian Fennoscandian Shield) and Estonia (mainly Paleozoic sedimentary bedrock) clearly shows that soils on Precambrian bedrocks in Karelia contain much more microelements than sedimentary bedrock and soils on sedimentary bedrocks of Estonia (Table 8). It means that magmatic processes are bringing out from crust deeper levels enriched by many trace metals as Cd, Co, Cr, Cu, Ni, V, Zn and others. At studying geochemistry of soil humus horizon in Zaonežje peninsula in 2000 only in single probes Cu exceeded 2-3 times the highest permitted concentrations.

Bedrock/ Elements	Cd	Co	Cr	Cu	Mn	Ni	V	Zn
Karelian bedrock	0.74	41	168	50	942	99	124	85
Estonian bedrock	0.04	2.8	11.7	4.6	481	5.0	12.3	15.9

Table 8. Comparison of average micronutrients content in bedrocks of Karelia Tables 3-5, Karelia, Russia and in bedrocks of Estonia (Table 7), excluded are bedrock analyses with extreemly high concentrations as peridotite in Karelia, clay and kerogenous argillite in Estonia

The comparison shows clearly that soils on metamorphic or magmatic rocks of Karelia content much more trace elements, than sedimentary bedrocks in Estonia. Only Mn content in limestone may be reach the same level than in volcanic or intrusive bedrocks, all other elements have in magma concentrations 10-20 times higher.

12. Discussion

The geochemistry of bedrock and Quaternary sediments strongly influences soil nutrient content and at the same time biodiversity. Where beneficial nutrients are abundant in soils, southern species may spread far beyond their normal range, as in the case of *Fragaria vesca* L. and *Rebes rubrum* L. which are found near the Arctic Circle in the Paanajärvi National Park. The importance of micronutrients for wild animals and birds is also well known. In Northern Karelia in autumn, when blueberries are ripe, forest birds such as the capercaillie (*Tetrao urgallus*), black grouse (*Tetrao tetrix* L.) and even ducks tend to be selective in feeding, preferentially choosing plants growing on gabbroic and ultramafic rock, where berries are larger and contain more micronutrients. Larger mammals, including moose (*Alces alces* L.), brown bear (*Ursus arctos* L.), and reindeer (*Rangifer tarandus* L.) feed on mushrooms in summer and autumn time, because of the need for microelements. Animals that have access to all necessary nutrients are healthier and stronger, and hence better equipped to defend territory and offspring. Areas with abundant flowering plants are favored by insects, and therefore also by birds which then make nests in these places.

Geochemistry is therefore a valuable tool in assessing regional biodiversity in national parks, and in protection of species or areas of natural significance, or in designating areas for construction and urban development and agricultural use. A relationship between bedrock geochemistry and endemic diseases is established in many places [14, 66, 67, 68]. We must not forget that all life is built from the same building blocks, known as elements [3].

Author details

Ylo Joann Systra
Department of Mining, Chair of Applied Geology, Tallinn University of Technology, Tallinn, Estonia

Acknowledgement

Field work in Karelia was done in part with financial support from Ministry of the Environment of Finland. We acknowledge the support of the European Community - Research Infrastructure Action under the FP6 "Structuring the European Research Area" Programme, LAPBIAT (RITA-CT-2006-025969) for studies in the Kilpisjärvi, Kevo and Oulanka areas in 2008-2009 and project No SF0140093s08 from the Estonian Ministry of Education and Research. This study is also part of the Estonian Science Foundation Grants No 7499 and No 8123. The author would like to thank Peter Sorjonen-Ward for correcting the English of the manuscript and numerous colleagues from different institutes of Karelian Research Center, who took part in the Finnish-Russian biodiversity project 1997-2000.

13. References

[1] Mason BH (1966) Principles of Geochemistry. 3rd edit., New York, John Wiley and Sons, Inc.329 p.

[2] Mason BH (1995) Geochemical Distribution of the Elements. In: New Encyclopedia Britannica, 15, Macropedia. pp. 939-950.

[3] Emsley, J (2003) Nature's Building Blocks, An A-Z. Guide to the Elements. Oxford University Press. 539 p.

[4] Albarède F (2009) Geochemistry. An Introduction. 2nd ed. Cambridge University Press. 342 p.

[5] Bulakh AG (2002) General mineralogy. Sankt-Petersburg University Publisher. 356 p.(in Russian).

[6] Saukov AA (1975) Geochemistry. Nauka, Moscow. 480 p. (in Russian).

[7] Tayler RJ (1972) Origin of the Chemical Elements. 3rd edition, London & Winchester. 170 p.

[8] Tayler RJ (1995) Origin of the elements. In: The New Encyclopedia Britannica, 15, Macropedia. pp. 934-938.

[9] Thornton I (1983a) (ed) Applied Environmental Geochemistry. Academic Press, London). 501 p.

[10] Crounse RG, Pories WJ, Bray TB, Mauger RL (1983). Geochemistry and Man: Healt and Disease. 1. Essential Elements. In: Thornton I (ed.) Applied Environmental Geochemistry. Academic Press, London. pp. 267-308.

[11] Food and Nutrition III (1997) Tables of the chemical composition of food. Tallinn Technial University. 87 p.

[12] Barabanov, VF 1985 Geochemistry. Nedra, Leningrad. 423 p. (in Russian).

[13] Koval'skij, VV 1982 Geochemical environment and life. Moscow. 282 p. (in Russian).

[14] Thornton I (1983b) Geochemistry Applied to Agriculture. In: Thornton I (ed.) Applied Environmental Geochemistry, Academic Press. London. pp. 231-266.

[15] Emsley, J (1998) The Elements, 3rd ed. Oxford, Clarendon Press. 251 p.

[16] Trofimov VT, Ziling DG, Baraboshkina TA et al. (2000) Ecologic functions of the lithosphere. Moscow University Press. 432 p (in Russian).

[17] Kabata-Pendias A, Pendias H (2001) Trace Elements in Soils and Plants. 3rd Edition. CRC Press, Boca Raton. 413 p.

[18] Merian E, Anke M, Ihnat M, Stoeppler M, (eds) (2004) Elements and their Compounds it the Environment. 2nd Edition. Wiley-VCH. V. 1–3. 1773 p.

[19] Nies D.H. 2004. Essential and Toxic Effects of Elements on Microorganisms. In: *Elements and their Compounds in the Environment*. V.1, 2nd edit. Merian E. et al. (eds), pp 257-304.Wiley-VCH Verlag CmbH & Co.KGaA.

[20] Uthman E (2000) Elemental Composition of the Human body: http://www.abti.net/ABEBackup/Elemental_Composition_of_ the Body.doc 27.02.2012.

[21] Plant JA, Raiswell R (1983) Principles of Environmental Geochemistry. In: Thornton I. (ed) Applied Environmental Geochemistry. Academic Press, London. pp.1-39.

[22] Himeno S, Imura N (2002) Selenuim in Nutrion and Toxicology. In: Sankar B. (ed.). Heavy Metals in the Environment. New York, NY, USA: Marcel Dekker Inc.pp.587-629

[23] Webb JS (1983) Foreword. In: Thornton I (ed) Applied Environmental Geochemistry. Academic Press, London. pp. vii-viii.

[24] Gromtsev, A.N., Kitaev, S.P., Krutov, V.I. et al. (eds) (2003) Biotic diversity of Karelia: conditions of formation, communities and species. Russian Academy of Sciences, Karelian Research Centre, Petrozavodsk. 244 p.

[25] Korsman K, Koistinen T, Kohonen J, Wennerström M, Ekdahl E, Honkamo M, Idman H, Pekkala Y (eds.) (1997) Bedrock map of Finland 1:1 000 000. Geol. Survey of Finland, Espoo.

[26] Koistinen T, Stephens MB, Bogatshev V, Nordgulen Ø, Wennerström M, Korhonen J. (comps.) (2001) Geological map of Fennoscandian Shield 1:2 000 000. Espoo, Trondheim, Uppsala, Moscow. Geol surveys of Finland, Norway, Sweden, Ministry of Natural Resources, Russia, 2 sheets.

[27] Systra YJ (1987) Geological formations and tectonics of the Kukasozero-Hankusjärvi-Kuzhjärvi area. In: Early Precambrian of Karelia (geology, petrology, tectonics). Petrozavodsk, Karelian Branch Academy of Sciences USSR. pp.35-54 (in Russian).

[28] Systra YJ (1990) Structural evolution of the eastern part of Paanajärvi Synecline. In: Precambrian of Northern Karelia. Petrozavodsk, Karelian Research Center, Academy of Sciences USSR. pp. 40-61 (in Russian).

[29] Systra, YJ (1991) Tectonics of the Karelian Region. St. Petersburg, Nauka. 176 p. (in Russian).

[30] Systra YJ (1998a) Geological diversity: the main reason for the biodiversity of the Paanajärvi National Park. Oulanka Reports 19, 23-26.

[31] Systra Y (1998b) Vegetation map of the Oulanka-Paanajärvi national parks. In: Hautala H, Rautiainen L (eds) Oulanka-Paanajärvi. Acticmedia, Kuusamo.1 sheet.

[32] Systra, YJ (1998c) Role of geological-geomorphological factors in the formation of biodiversity in the Paanajärvi National Park. In: Biodiversity inventories and studies in the areas of Republic of Karelia bordering on Finland (express inform. materials). RAS, Karelian Research Center, Forest Research Institute. Krutov VI, Gromtsev AN (eds). Petrozavodsk. pp.27-32 (in Russian).

[33] Systra YJ (2000) Northern shore of Lake Ladoga. Geological characteristics of the territory. In: Biodiversity inventories and studies in Zaonezhskij peninsula and on the northern shore of Lake Ladoga (express inform. materials) Gromtsev AN, Krutov VI (eds) Petrozavodsk. pp. 194-197 (in Russian).

[34] Systra YJ (2003) Geological characteristics. In: Gromtsev, A.N., Kitaev, S.P., Krutov, V.I. et al. (eds.). 2003. Biotic diversity of Karelia: conditions of formation, communities and species. Petrozavodsk, Karelian Research Center of RAS. pp. 7-12

[35] Systra YJ (2004a) Main features of the Karelian region structures. In: Sharov, V.N. (ed.). Deep structure and seismicity of the Karelian region and its margins. pp.14-29. Karelian Research Centre, Russian Academy of Sciences, Petrozavodsk.

[36] Systra YJ (2004b) Geological background for biodiversity in the eastern Fennoscandia, Estonia and Latvia. In: Parkes, M.A. Natural and Cultural Landscapes - The Geological Foundation, Royal Irish Academy, Dublin. pp. 73-76.

[37] Golubev AI, Systra YJ (2000) Zaonezhskij peninsula. Geological characteristics of the territory. In: Biodiversity inventories and studies in Zaonezhskij peninsula and on the northern shore of Lake Ladoga (express inform. materials) Gromtsev AN, Krutov VI (eds) Petrozavodsk. pp. 9-15 (in Russian).

[38] Fedorets NG, Bahmet, ON, Solodovnikov AN, Morozov AK (2008) Soils of Karelia: geochemistry atlas. Moscow, Nauka. 47 p (in Russian).

[39] Sviridenko LP, Svetov AP (2008) Valaam gabbro-dolerite sill and geodynamics of the Ladoga Lake trough. Petrozavodsk, Institute of Geology, Karelian Research Centre, Russian Academy of Sciences. 123 p. (in Russian).

[40] Shiltsova GV, Morozova RM, Litinski PJu (2008) Heavy metals and sulphur in Valaam archipelago soils. Petrozavodsk, Karelian Research Center, RAS, Forest Research Institute. 109 p. (in Russian with English conclusions and abstract).

[41] Kravchenko AV, Gnatiuk EP, Kuznetsov OL (2000) Distribution and Occurrence of Vascular Plants in Floristic Districts of Karelia. Petrozavodsk, Karelian Research Centre of RAS, Forest Research Institute, Institute of Biology. 76 pp.

[42] Pesola VA (1928) Kaltsiumkarbonaatti kasvimaantieteellisenä tekijänä Suomessa. Ann.Soc. "Vanamo" 9 (1), 246 pp. (in Finnish with summary: Calcium carbonate as a factor in the distribution of plants in Finland).

[43] Lehtinen M, Nurmi P, Rämö T (eds) (1998) Suomen kallioperä – 3000 vuosimiljoonaa. Helsinki, Geological Society of Finland. 373 p (in Finnish).

[44] Lehtinen M, Nurmi P, Rämö T (eds) (2005) Precambrian Geology of Finland: Key to the Evolution of the Fennoscandian Shield. Elsevier BV, Amsterdam. 736 p.

[45] Rasilainen K, Lahtinen R, Bornhorst TJ (2008) Chemical characteristics of Finnish Bedrock – 1:1 000 000 scale bedrock map units. Geological Survey of Finland. Report of Investigations 171. Espoo. 94 p.

[46] Raukas A, Teedumäe A (eds.) (1997) Geology and mineral resources of Estonia. Tallinn, Estonian Academy Publishers. 436 p.

[47] Kiipli T, Batchelor RA, Bernal JP, Cowing Ch, Hagel-Brunnström M. et al. (2000) Seven sedimentary rock reference samples from Estonia. Oil Shale, 17, 3, 215-223.

[48] Kiipli T (2005) Database of the Chemical Quality of Carbonate Rocks. In: Yearbook of the Geological Survey of Estonia 2004, Tallinn, pp. 34-37 (in Estonian).

[49] Petersell V, Ressar H, Carlsson M, Mõttus V, Enel M, Mardla A, Täht K (1997). The Geochemical Atlas of The Humus Horizon of Estonian Soil. Tallinn-Uppsala: Geol. Survey of Estonia & Geol.Survey of Sweden. 37 maps and Explanatory text. 75 p.

[50] Kukk T, Kull T (eds.) (2005) Atlas of the Estonian Flora. Tartu: Estonian University of Life Sciences. 527 pp.

[51] Kotiranta H, Uotila P, Sulkava S & Peltonen S-L (eds) (1998) Red Data Book of East Fennoscandia. Ministry of Environment, Finnish Environment Institute & Botanical Museum, Finnish Museum of Natural History, Helsinki. 351 p.

[52] Red Book: Republic of Karelia (2007) Ministry of agriculture, fish husbandry and ecology the Republic of Karelia. Petrozavodsk, Karelia. 368 p. (in Russian)

[53] Atlas Karel'skoi ASSR (1989) Peihvasser VN (ed) Moscow. 40 p (in Russian).

[54] Söyrinki N, Saari V (1980) Die Flora im Nationalpark Oulanka, Nord-Finnland (The flora of Oulanka National Park, northern Finland). Acta Bot. Fennica 114. 150 p.

[55] Hautala A, Rautiainen L (1998) Oulanka – Paanajärvi. Photos: Hautala H, Rautiainen, L, texts: Systra Y, Viramo, J. Kuusamo, Articmedia. 144 p.

[56] Halonen P (1993) The lichen flora of the Paanajärvi National Park. Oulanka Reports 12, 45-54.

[57] Borodulina GS, Systra YJ (2001) Springs along the nature path in Paanajärvi National Park. Oulanka Reports 25, 5-8.

[58] Ponge J-F (2003) Humus forms in terrestrial ecosystems: a framework to biodiversity. Soil Biol.& Biochemistry 35, 935-945.

[59] Lukina NV, Poljanskaja LM & Orlova MA (2008) Nourishing regime of soils of the northern taiga forests. Moscow, Nauka. 342p. (in Russian).

[60] Kravchenko AV, Kuznetsov OL (2003) The role of protected areas in Karelia's border zone in the conservation of floristic biodiversity. In: Gromtsev, A.N., Kitaev, S.P., Krutov, V.I. et al. (eds.). 2003. Biotic diversity of Karelia: conditions of formation, communities and species. pp.69-76. Russian Academy of Sciences, Karelian Research Centre, Petrozavodsk.

[61] Ladoga Lake 2002) Atlas. Sankt-Peterburg. 129 p. (in Russian).

[62] Lehtovaara JJ (1994a) Geological map of Finland. Sheet 1823, Kilpisjärvi., Pre-Quaternary rocks, Geol.Survey of Finland.

[63] Lehtovaara JJ (1994b) Geological map of Finland, Sheet 1842 Halti, Pre-Quaternary rocks.Geolo. Survey of Finland.

[64] Lehtovaara JJ (1995) Kilpisjärven ja Haltin kartta-alueiden kallioperä. Summary: Pre-Quaternary rocks of the Kilpisjärvi and Halti map-sheet areas. Geol. Survey of Finland, Espoo 1995. 64 p.

[65] Pirrus E (2009) Big stones of Estonia. Story of large erratic boulders. Tallinn, Teaduste Akadeemia kirjastus (in Estonian, summary in English).

[66] Golovin A, Krinochkin L (2004) Geochemical specialization of bedrock and soils as indicator of regional geochemical endemicity. Geologija 48, 22-28, Vilnius.

[67] Gilder SSB (1964) The London Letter. Canad.Med.Ass.J. v. 91, 1081-1082.

[68] Thornton I, Farago ME, Thims CR et al (2008) Urban geochemistry: research strategies to assist risk assessment and remediation of brownfield sites in urban areas. Environ.Geochem. Health 30, 565-576.

Chapter 10

Image Processing for Pollen Classification

Marcos del Pozo-Baños, Jaime R. Ticay-Rivas, Jousé Cabrera-Falcón,
Jorge Arroyo, Carlos M. Travieso-González, Luis Sánchez-Chavez,
Santiago T. Pérez, Jesús B. Alonso and Melvín Ramírez-Bogantes

Additional information is available at the end of the chapter

1. Introduction

Palynology - *"The study of pollen grains and other spores, especially as found in archaeological or geological deposits. Pollen extracted from such deposits may be used for radiocarbon dating and for studying past climates and environments by identifying plants then growing."* [1]

Over 20% of all the world's plants are already at the edge of becoming extinct [2]. Saving earth's biodiversity for future generations is an important global task [3] and as many methods as available must be combined to achieve this goal. This involves mapping plants distribution by collecting pollen and identifying them in a laboratory environment.

Pollen grain classification has been an expensive qualitative process, involving observation and discrimination of features by a highly qualified palynologist. It is still the most accurate and effective method. But it certainly limits research progress, taking considerable amounts of time and resources [4].

Automatic recognition of pollen grains can overcome these problems, producing purely objective results faster. Such a tool would provide invaluable in the studies of flora. This advantages were obvious for Flenley [5] [6], who proposed the implementation of an automatic pollen grain classification system in 1968. However, the idea was intractable at that time. Mainly, because of technology restrictions. Nowadays, technology is not a barrier any more, and the discussed system is a reality thanks to computer vision.

This chapter presents the latest results obtained by the authors in the field of automatic pollen grain classification. This will be done by introducing a developed system, paying special attention to the phases of *preprocessing* (section 3.1) and *feature extraction* (section 4). Results for a 17 pollen species database obtained with the commented system will also be shown (section 6).

2. Related work

The begins of automatic pollen identification were based on scanning electron microscope (SEM) images. Langford applied statistical classifiers on texture parameters on 1988, reporting a 94.30% of accuracy on a six pollen class database [7]. Later, artificial neural networks (ANN) were used on the classification task, achieving a success rate of 100% with 3 classes [8].

However, SEM images are expensive and difficult to produce and the use of light microscope (LM) images were explored in 1998 [9]. Again, first attempts were not fruitful due to the low quality images provided by the technology of the time. But recent works has demonstrated that the use of LMs images is, in fact, possible.

For example, [10] reported a 100% of success with a small database containing 4 classes. Moreover, it was one of the first works using artificial neural networks for the classification phase, along with texture parameters. Again, [11] used artificial neural networks for classification. This time, brightness and shape descriptors were extracted as pollen features. A 90% of accuracy with a 3 class database was reported.

[12] and [13] presented a more complex work, combining shape and ornamentation of the grains; using simple geometric measures, and concurrence matrices applied for the measurement of texture. Again, artificial neural networks were used for classification. These works reported a 87.7% recognition rate for a 5 classes database and a 97.7% for a three class database respectively.

[14] describes an automatic optical recognition and classification of pollen grains system. This is able to locate pollen grains on slides, focus and photograph them before identify the species applying a trained neural network. The system achieved a 90% of recognition rate with a 3 class database.

Other works use more sophisticate capture methods, achieving 3 dimensional representations of the pollen grains. [15] presented a combination of statistical reasoning, feature learning and expertise knowledge. A feature extraction algorithm was applied alternating 2D and 3D representations. Iterative refinement of hypotheses was used during the classification process. This work reported a 77% of accurate rate in a database with 30 classes and 97% when only 4 classes were used. An other example, [16], which used a confocal laser microscope to create the 3D models, achieved a 90% recognition rate with 3 classes database.

3. Pollen extraction

At the actual development stage of the system, the detection of pollen grain is highly but not fully automatic. This should not be of any surprise, as the task of pollen location inside sample images is itself a different problem, which is as much complicated as the problem studied in this chapter.

Thus, users should first select and area with a pollen grain inside. Preferably, an area, as small as possible, where an isolated pollen grain is located. This user selected region of interest (ROI) is then automatically preprocessed to detect the contour of the grain (see figure 1).

Figure 1. An example of a pollen grain manually selected by the user.

3.1. Preprocessing

This section introduces the automatic preprocessing algorithm used for pollen extraction and preparation. It is important to remind that this process is applied to the image area manually selected by the user, like that showed on figure 1. The preprocessing steps are (see figure 2):

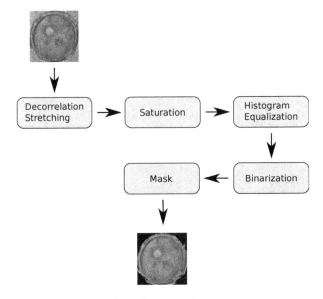

Figure 2. Automatic preprocessing steps for pollen extraction.

1. *Decorrelation stretching*: This process aims to reduce the autocorrelation of the information contained in the image [17]. This is done as a three steps process:

    (a) The original bands are transformed to their principal components.

    (b) The principal components are then stretched separately.

    (c) The resulting data is transformed back to the original space applying the inverse of the principal component transformation.

    The results is a linear transformation of the spectral bands, resulting in uncorrelated variables with unit variance, and enhancing displays. The result can be seen in figure 3.

2. *Saturation*:The saturation channel of the image represents the amount of colour used at each pixel, i.e. the lower the saturation is the greyer the pixel is. This channel is actually extracted from the HSV image representation [18].

In particular, the saturation channel is computed as:

$$S = \begin{cases} 0, & \text{if } MAX = 0 \\ 1 - \frac{MIN}{MAX} & \text{otherwise} \end{cases} \tag{1}$$

The result of computing the saturation channel of the docorrelation stretched image is shown on figure 3. The simplification of the task of differentiating pollen and background is obvious.

3. *Histogram equalization*: Equalizing the histogram of an image aims to obtain a uniform distribution of the pixel values. This maximizes the contrast without loosing structural information, i.e., conserving the entropy [19].

4. *Binarization*: The binarization of an image consist of transform each pixel's value to '0' or '1' depending on whether it has a value lower or higher (respectively) than a set threshold. This results on a simple image containing pure geometric information.

5. *Mask*: Finally, in a bid to obtain a clear mask of the pollen grain, several image processing functions are applied such as "imfill" and "bwareaopen" provided by the Image Processing Toolbox of Matlab [20].

The resulting mask can be either used for feature extraction or to remove the background of the pollen grain image. The result of applying each preprocessing step can be seen on figure 3

4. Feature extraction

Pollen images by their own does not prove to be a high quality information for the task of automatic pollen grain classification. Although they contain the necessary information, this information is hidden and diffused around the image and behind other unimportant data. In order to extract the relevant information from raw samples, they need to be further processed by the *feature extractor*.

A total of 50 features are extracted from the pollen images. I.e. the output of the *feature extraction* block is a vector with length 50. These 50 features corresponds to 24 geometric parameters carrying information regarding size and basic shape, and 26 texture parameters with information about how pixel intensities are distributed on the image. A detailed view of each of these features will be given here.

Certainly, colour may be an attractive source of information. However, since the preparation of pollen grain samples imply the use of a stain, it is not recommended to use it. Moreover, the stain effects is not constant along time and the colour of the same sample may change.

4.1. Geometric parameters

Geometric parameters contain information about the size and the basic shape of the pollen grains. The 24 geometric parameters extracted in the systems presented in this chapter are:

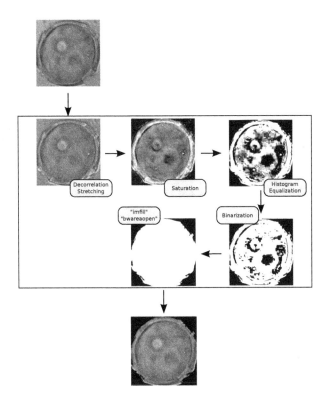

Figure 3. The result of applying each preprocessing step to a pollen grain image. Note that the sequence followed is the same as in figure 2.

- *Area*: Refers to the amount of pixels with level '1' in the pollen mask.

- *BoundingBox*: Smallest rectangle enclosing the pollen. In particular, parameters width and hight are used as:

$$BoundingBox(1) = \text{width}$$
$$BoundingBox(2) = \text{hight} \tag{2}$$

- *Centroide*: Refers to the mass centre of the pollen grain. Coordinates (x, y).

- *MajorAxisLength*: Length of the major axis of the ellipse with the same second order normalized central moment of the object.

- *MinorAxisLength*: Length of the minor axis of the ellipse with the same second order normalized central moment of the object.

- *ConvexArea*: Area of the smallest convex shape enclosing the object.

- *EquivDiameter*: Diameter of the circle with the same area as the object.

$$EquivDiameter = \sqrt{\frac{4 \times Area}{\pi}} \qquad (3)$$

- *Solidity*: Portion of the area of the convex region contained in the pollen.

$$Solidity = \frac{Area}{ConvexArea} \qquad (4)$$

- *Perimeter*: Length of the perimeter of the mask image.
- *Extent*: Portion of the area of the bounding box contained in the pollen.

$$Extent = \frac{Area}{Area_{BoundingBox}} \qquad (5)$$

- *Eccentricity*: Relation between the distance of the focus of the ellipse and the length of the principal axis.
- *WeightedCentroid*: This is a centroid computing weighted by the pixel values of the grey-scale image.
- *Shape*: Measures how circular is the pollen. Its values are in the range [0,1], where 1 corresponds to a perfect circle.

$$Shape = \frac{4 \times \pi \times Area}{Perimeter^2} \qquad (6)$$

- *Thickness*: This is the number of times that the mask has to be eroded with a 3x3 square filter, until it disappears, e.i. the image gets black.
- *Box*: These are the coordinates of an inner rectangle area computed from the *BoundingBox* parameters as:

$$Box(1) = \frac{BoundingBox(1)}{4}$$
$$Box(2) = \frac{BoundingBox(2)}{4}$$
$$Box(3) = \frac{BoundingBox(1)}{2}$$
$$Box(4) = \frac{BoundingBox(2)}{2} \qquad (7)$$

- *Hight*: Length of the largest line enclosed in the pollen.
- *Width*: Length of the largest line enclosed in the pollen and perpendicular to *Hight*.

4.2. Texture parameters

Texture parameters provide information regarding how pixels are distributed on the image, such as contour changes or objects inside the pollen grain.

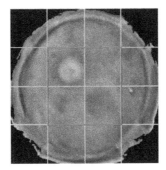

Figure 4. Example of the inner rectangle area computing from the *BoudingBox*.

The first 4 of the 26 texture parameters introduced in this section are computed using the grey level co-occurrence matrix (GLCM). This matrix gives information about the frequency of pixel value pairs combinations. In particular, the value of GLCM(i,j) is the number of times that a pixel with value 'j' sits next and at the left of a pixel with value 'i'. Figure 5 shows and example of this.

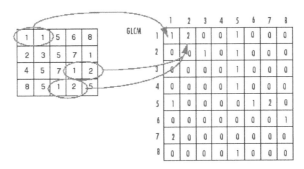

Figure 5. Example of a grey level co-occurrence matrix.

- *Contrast*: Mean intensity difference between a pixel and its neighbours. This value is computed as:

$$Contrast = \sum_{i,j} |i - j|^2 p(i, j) \tag{8}$$

- *Correlation*: Measures how must correlated it a pixel with respect to its neighbours. This value is computed as:

$$Correlation = \sum_{i,j} \frac{(i - \mu_i)(j - \mu_j)p(i, j)}{\sigma_i \sigma_j} \tag{9}$$

- *Energy*: Sum of the squared elements of the GLCM. This is:

$$Energy = \sum_{i,j} p(i, j) \tag{10}$$

- *Homogeneity*: Measures how close the distribution of objects of the GLCM are to the diagonal of the GLCM. This is:

$$Homogeneity = \sum_{i,j} \frac{p(i,j)}{1 + |i - j|} \tag{11}$$

- *Entropy*: This measure is applied to six different images derived from the original pollen grain image. These images are the the the outer and inner bounding box (*BoundingBox* and *Box*) of the blue channel of the RGB representation, the saturation and the value channels of the HSV representation. A representation can be seen in figure 6.

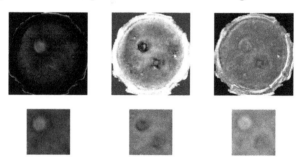

Figure 6. Images used to compute the *entropy* measures. They correspond to channels blue, saturation and value (left-right) and outer and inner bounding box (up-down).

Entropies are are scalar values representing a statistical measure of the randomness of the pixel values. Each value is computed as:

$$Entropy = \sum p \log_2 p, \tag{12}$$

where p is the histogram count of the corresponding image.

- *Fourier Descriptors*: These measures are based on the analysis of the pollen contour points, and it provides information about the pollen shape. It is worth it to mention that a major property of the *fourier descriptors* is its invariance to geometric transformations, such as rotation, scale and sift.

To compute these parameters, the complex representation of the contour $z_i = x_i + jy_i$ is used, where $i = 0, 1, 2..., N_c - 1$ with N_c the number of points of the contour. Moreover, the contour is sampled every 2 degrees. Now, the discrete Fourier transform (DFT) of z is:

$$a(u) = \frac{1}{N_c} \sum_{i=0}^{N_c-1} z_i e^{-j2\pi u/N_b} \quad u = 0, 1, 2, ..., N_b - 1 \tag{13}$$

The resultant complex coefficients $a(u)$ are transformer in a power spectrum $|a(u)|^2$. Finally, the discrete cosine transform (DCT) is applied to reduce the dimensionality of the vector, ending up with a vector of length 5.

- *Relative areas*: This is a 5 elements vector which values correspond to the number of active pixels (pixels with value '1') after binarizing the pollen image with different thresholds. In particular, the thresholds used are 0.3, 0.4, 0.5, 0.6 and 0.7. Figure 7 shows an example of this.

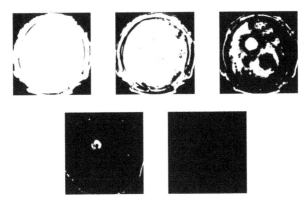

Figure 7. Results of applying thresholds 0.3, 0.4, 0.5, 0.6 and 0.7 respectively to a pollen image.

- *Relative objects*: In this case the number of objects (group of connected pixels with value '1' and surrounded of pixels with value '0') contained inside the pollen grain are counted, using an inverted and masked version of the binarized images computed the *relative areas*. See figure 8 for an example.

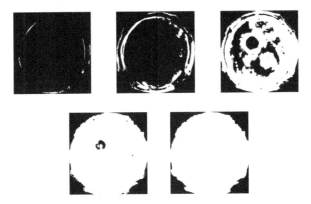

Figure 8. Images used to compute the *relative objects*.

5. Classification

Several works such as [10], [22], [11] and [13] used artificial neural networks (ANNs) as classifiers. These algorithms works as follow:

- Parameters are computed from a set of training samples.

- The computed parameters are passed to the ANN so that it gets trained. This means that the ANN automatically adjusts its parameters to solve the problem of classify the parameters in different classes.
- After the training process, a new testing parameters vector can be passed to the ANN and it will produce an output regarding the sample class.

An ANN is a mathematical model inspired in the structure and functional aspects of the biological neural networks. It could be defined as a set of simple computational elements massively interconnected following a hierarchical organization [21].

In this case, a multilayer perceptron architecture trained by a back propagation algorithm (MLP-BP) is proposed. The principal characteristic of this algorithm is its ability to solve non-lineal problems. Its architecture is composed of several layers. Each layer corresponds to a set of neurons receiving data from the previous layer and transmit data to the next layer. This layer can be divided in "input layer", "hidden layer" and "output layer" as shown in figure 9. In this case, the number of hidden layers is set to one.

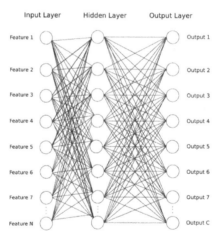

Figure 9. Architecture of the multilayer perceptron.

It is important note that the training process of the ANN contain an aleatory factor which determines the solution found. In other words, the training process does not avoid local minimums. To overcome this limitation, the proposed classifier implements 11 individual ANNs and sum their resulting scores to obtain a final response. The idea behind this fusion is that the set of computed solutions complement each other, i.e. some solutions correct the errors produced by others.

6. Experimentation methodology, results and discussion

A system were implemented in order to test the quality of the proposed approach. This system uses all the techniques introduced in previous sections (preprocessing, feature extraction and classification). This section gives the details about the database used and the experimental procedure, along with a detailed explanation of the obtained results.

6.1. Database

The database used for the experimentation contains 345 images of 17 different pollen grain classes. Images has been captured with a 2 mega-pixels digital camera connected to a microscope set to apply a 40 times zoom.

More precisely, these images correspond to 17 sub-genders and species of 11 different families of tropical honey plants situated in Costa Rica (Central America). Table 1 shows the exact information about family, gender and specie.

Class	Family	Gender	Specie	Samples
1	Asteraceae	Baltimora	Recta	24
2	Asteraceae	Tridats	Procumbels	47
3	Asteraceae	Critonia	Morifolia	21
4	Asteraceae	Elephentopus	Mollis	17
5	Bombacaceae	Bombacptis	Quinata	18
6	Caesalpinaceae	Cassea	Gradis	35
7	Combretaceae	Combretum	Fructicosum	25
8	Comvulvulaceae	Ipomea	Batatas	15
9	Fabaceae	Aeschynomene	Sensitiva	24
10	Fabaceae	Cassia	Fistula	36
11	Fabaceae	Miroespermyn	Frutesens	18
12	Fabaceae	Enterolobium	Cyclocarpun	18
13	Myrsinaceae	Ardisia	Revoluta	18
14	Malpighiaceae	Bunchosin	Cornifolia	36
15	Saphindaceae	Cardioesperman	Grandiflorus	20
16	Saphindaceae	Melicocca	Bijuga	26
17	Verbenaceae	Lantana	Camara	25

Table 1. The exact information about family, gender and specie of the 17 classes included in the DDBB used. The last column expresses the number of samples of pollen grains extracted from the database.

Applying the pollen grain extraction algorithm introduced in section 3, a total of 423 pollen images distributed on all species were obtained. The number of samples extracted for each sample was greater than one. This was possible thanks to images such as that shown in figure 10 where more than one pollen grain could be extracted. Figure 11 shows a sample of each pollen specie included in the DDBB.

6.2. Experiments

First, remember from section 5 that the design of the classifier include 30 ANNs fused at the score level. Thus, the number of hidden units on the ANNs had to be specified. To do so, a set of experiments with different configurations were executed to find the optimal value. To obtain a valid measure of the performance of the system, 30 iterations of a hold 50% out cross-validation procedure was executed. Results will be shown and discussed in sections 6.3

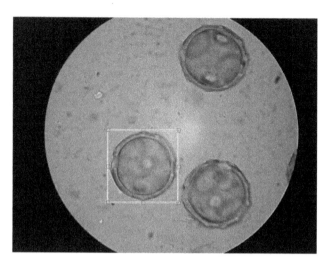

Figure 10. Database image sample. Note that more than one pollen grain can be extracted from this image.

Class	Sample	Nº	Class	Sample	Nº	Class	Sample	Nº
1	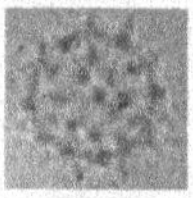	24	7		25	13	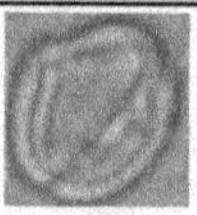	18
2	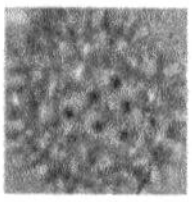	47	8		15	14	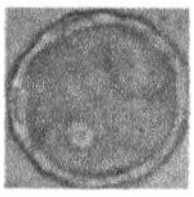	36
3		21	9		24	15		20
4		17	10		36	16		26
5		18	11	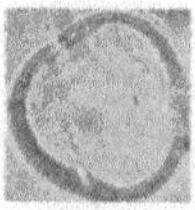	18	17	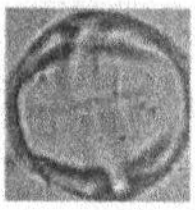	25
6	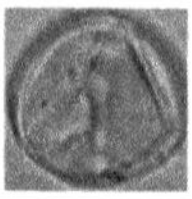	35	12		18			

Figure 11. Samples of the 17 different pollen grain species.

and 6.4 respectively. For now, it is enough to note that the optimal value were found with a 30 neurons hidden layer.

Thus, using this optimal configuration of the ANNs, further experiments were executed to evaluate the performance of the designed system. In this case, 30 iterations of a K-folds cross-validation procedure were applied with values of 'K' equal 3, 5, 7 and 10.

Note that the set of all experimental procedures (hold-50%-out and 3, 5, 7 and 10 folds) are based on divisions of the database in disjoint training and test sets. Moreover, this experiments can be seen as using different proportions of the database of training, i.e. using a different number of samples for training. In particular, the proportions of samples used for training are 1/2, 2/3, 4/5, 6/7 and 9/10 respectively.

6.3. Results

It is important to note that every experiment was repeated 30 times in order to obtain a valid measure of the system's performance. Therefore, results are given in terms of mean percentage and standard deviation (mean % and std).

The first experiment tested different configurations of the ANNs. Figure 12 shows the progress of the success rate when the number of units in the hidden layer increased from 10 to 150. A highest rate of 90.54% of success rate were obtained with 30 units (see table 2).

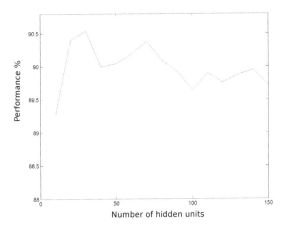

Figure 12. Performance progress for different number of units in the hidden layer of the ANN.

A second group of experiments aimed to measure the system's performance with different number of samples for training. Table 3 shows the results obtained for 3, 5, 7 and 10 folds. Note that the success rate increased with the number of training samples (from 90.54% to 92.81%), while the std decreased (from 1.29 to 0.74).

6.4. Discussion

It can be argued that the number of hidden units of the ANNs could be further optimized executing a finer search around the point found. However, based on the similar accuracy measures obtained between 10 and 80 units and stds higher than the range of accuracy

Neurons in the hidden unit	Mean % ± std
10	89.29% ± 2.11
20	90.40% ± 1.69
30	**90.54% ± 1.29**
40	90.00% ± 1.66
50	90.05% ± 1.78
60	90.19% ± 1.33
70	90.38% ± 1.52
80	90.09% ± 1.37
90	89.92% ± 1.42
100	89.64% ± 1.49
110	89.90% ± 1.34
120	89.76% ± 1.55
130	89.87% ± 1.37
140	89.95% ± 1.42
150	89.72% ± 1.64

Table 2. Performance progress for different number of units in the hidden layer of the ANN.

Experiment	Mean % ± std
Hold-50%-Out	90.54% ± 1.29
3 k-folds	91.40% ± 1.05
5 k-folds	92.38% ± 0.75
7 k-folds	92.43% ± 0.82
10 k-folds	92.81% ± 0.74

Table 3. Results for 30 iterations of different experiments.

percentages, paying the cost of running a finer search for a minimal increment of performance was not worth it. Therefore, 30 units were chosen as the optimal point.

On the other hand, the results obtained for the second round of experiments show an increasing in both system's performance and stability. This seems to indicate that the performance of the system may increase with a bigger training database.

7. Conclusions

This chapter has introduced the problem of automatic pollen grain classification, which is vital for biologists and flora researches among others. As pointed out in section 3, the task of automatically detecting the pollen grains from samples is a complex problem itself and fall beyond the scope of this chapter. Thus, a semi-automatic algorithm for pollen extraction was explored instead,

The chapter mainly focused its attention in giving a fair amount of both geometric and texture parameters. Moreover, the extraction of this parameters relied on the good work performed by the preprocessing block during the pollen's perimeter definition.

Finally, these parameters were tested implementing a completed system. In particular, the system used the semi-automatic pollen detection and preprocessing algorithms introduced, along with the mentioned feature extraction techniques and a classifier based on the fusion of

11 ANNs at the score level. The system was tested executing a number of experiments using different hold-out and k-folds cross-validation procedures. The results showed success rates between 90.54% and 92,81%, pointing out the quality of the presented parameters for pollen grain classification. Moreover, these results improve those achieved by other authors such as [10], [22], [11] and [13], even though the number of classified species was significantly larger.

Acknowledgements

This work has been supported by Spanish Government, in particular by "Agencia Española de Cooperación Internacional para el Desarrollo" under funds from D/027406/09 for 2010, D/033858/10 for 2011, and A1/039531/11 for 2012. To M.Sc. Luis A. Sánchez Chaves, Tropical Bee Research Center (CINAT) at Universidad Nacional de Costa Rica, for provide the database images and palynology knowledge.

Author details

del Pozo-Baños Marcos, Ticay-Rivas Jaime R., Cabrera-Falcón Jousé, Travieso Carlos M., Pérez Santiago T. and Alonso Jesús B.
Signal and Communication Department Institute for Technological Development and Innovation in Communications University of Las Palmas de Gran Canaria IdeTIC-ULPGC. Las Palmas de Gran Canaria, Spain

Arroyo Jorge, Sánchez-Chavez Luis and Ramírez-Bogantes Melvin
Escuela de Matemáticas, Universidad Nacional, Costa Rica

8. References

[1] Definition of "palynology" by the Oxford Dictionaies site (http://oxford dictionaries.com/definition/palynology). Visited last time on April 2011.

[2] Plants under pressure: a global assessment. The first report of the IUCN Sampled Red List Index for Plants. Royal Botanic Gardens, Kew, UK (2010)

[3] Sytnik KM (2010) Preservation of biological diversity: Top-priority tasks of society and state. Ukrainian Journal of Physical Optics 11(suppl. 1), S2-S10

[4] Stillman EC, Flenley JR (1996) The Needs and Prospects for Automation in Palynology. Quaternary Science Reviews 15, 1-5.

[5] Flenley JR (1968) The problem of pollen recognition. In: Problems in Picture Interpretation, (ed. M. B. Clowes and J. P. Penny), pp. 141-145. CSIRO, Canberra.

[6] Flenley JR (1990) Some prospects for palynology in the South- West Pacific Region. Massey University Faculty of Social Sciences Occasional Papers, No. 1.

[7] Langford M, Taylor GE, Flenley JR (1990) Computerized identification of pollen grains by texture analysis. Review of Palaeobotany and Palynology, Volume 64, Issues 1-4, 23 October 1990, pp.197-203.

[8] Treloar WJ (1992) Digital image processing techniques and their application to the automation of palynology. Ph.D. Thes., University of Hull, Hull UK.

[9] Treloar WJ and Flenley JR (1996) An investigation into the potential of light microscopy for the automatic identification of pollen grains by the analysis of their surface texture. Poster Presentation, 9th International Palynological Congress, Houston TX.

[10] Li P and Flenley J (199) Pollen texture identification using neural networks. Grana 38(1), pp.59-64.

[11] Rodriguez-Damian M, Cernadas E, Formella A, Sa-Otero R (2004) Pollen classification using brightness-based and shape-based descriptors. Proceedings of the 17th International Conference on Pattern Recognition, ICPR 2004, August 23-26, vol.2, pp.23-26.

[12] Zhang Y, Fountain DW, Hodgson RM, Flenley JR and Gunetileke S (2004) Towards automation of palynology 3: pollen pattern recognition using Gabor transforms and digital moments. Journals of quaternary science, vol.19, 2004, pp.763-768.

[13] Rodriguez-Damian M, Cernadas E, Formella A, Fernandez-Delgado M and De Sa-Otero P (2005) Automatic detection and classification of grains of pollen based on shape and texture. IEEE Trans. on Systems, Man, and Cybernetics, Part C: Applications and Reviews 36(4), 531-542.

[14] Allen GP, Hodgson RM, Marsland SR and Flenley JR (2008) Machine vision for automated optical recognition and classification of pollen grains or other singulated microscopic objects. Mechatronics and Machine Vision in Practice, 15th International Conference on , 2008, pp.221-226.

[15] Boucher A and Thonnat M (2002) Object recognition from 3D blurred images, 16th International Conference on Pattern Recognition, vol.1, pp.800-803.

[16] Ronneberger, Burkhardt O, Schultz H, General E (2002) Purpose object recognition in 3D volume data sets using gray-scale invariants-classification of airborne pollen-grains recorded with a confocal laser scanning microscope. Proceedings on Pattern Recognition, 2002. 16th International Conference, vol.2, pp.290- 295.

[17] Campbell NA (1996) The decorrelation stretch transformation. INT. J. Remote Sensing, vol.17, no.10, pp.1939-1949.

[18] Agoston MK (2005) Computer graphics and geometric modeling: implementation and algorithms. Springer Verlag.

[19] Baxes GA (1994) Digital Image Processing: Principles and Applications. John Wiley & Sons.

[20] Matlab documentation at Mathworks. http://www.mathworks.com. Visited last time on April 2011.

[21] Hopfield JJ (1982) Neural networks and physical systems with emergent collective computational abilities. Proc. NatL Acad. Sci. USA Biophysics, vol.79, pp.2554-2558.

[22] France I, Duller A, Duller G and Lamb H (2000) A new approach to automated pollen analysis. Quaternary Science Reviews 18, pp.536-537.

Permissions

The contributors of this book come from diverse backgrounds, making this book a truly international effort. This book will bring forth new frontiers with its revolutionizing research information and detailed analysis of the nascent developments around the world.

We would like to thank Gbolagade Akeem Lameed, for lending his expertise to make the book truly unique. He has played a crucial role in the development of this book. Without his invaluable contribution this book wouldn't have been possible. He has made vital efforts to compile up to date information on the varied aspects of this subject to make this book a valuable addition to the collection of many professionals and students.

This book was conceptualized with the vision of imparting up-to-date information and advanced data in this field. To ensure the same, a matchless editorial board was set up. Every individual on the board went through rigorous rounds of assessment to prove their worth. After which they invested a large part of their time researching and compiling the most relevant data for our readers. Conferences and sessions were held from time to time between the editorial board and the contributing authors to present the data in the most comprehensible form. The editorial team has worked tirelessly to provide valuable and valid information to help people across the globe.

Every chapter published in this book has been scrutinized by our experts. Their significance has been extensively debated. The topics covered herein carry significant findings which will fuel the growth of the discipline. They may even be implemented as practical applications or may be referred to as a beginning point for another development. Chapters in this book were first published by InTech; hereby published with permission under the Creative Commons Attribution License or equivalent.

The editorial board has been involved in producing this book since its inception. They have spent rigorous hours researching and exploring the diverse topics which have resulted in the successful publishing of this book. They have passed on their knowledge of decades through this book. To expedite this challenging task, the publisher supported the team at every step. A small team of assistant editors was also appointed to further simplify the editing procedure and attain best results for the readers.

Our editorial team has been hand-picked from every corner of the world. Their multi-ethnicity adds dynamic inputs to the discussions which result in innovative outcomes. These outcomes are then further discussed with the researchers and contributors who give their valuable feedback and opinion regarding the same. The feedback is then collaborated with the researches and they are edited in a comprehensive manner to aid the understanding of the subject.

Apart from the editorial board, the designing team has also invested a significant amount of their time in understanding the subject and creating the most relevant covers. They scrutinized every image to scout for the most suitable representation of the subject and create an appropriate cover for the book.

The publishing team has been involved in this book since its early stages. They were actively engaged in every process, be it collecting the data, connecting with the contributors or procuring relevant information. The team has been an ardent support to the editorial, designing and production team. Their endless efforts to recruit the best for this project, has resulted in the accomplishment of this book. They are a veteran in the field of academics and their pool of knowledge is as vast as their experience in printing. Their expertise and guidance has proved useful at every step. Their uncompromising quality standards have made this book an exceptional effort. Their encouragement from time to time has been an inspiration for everyone.

The publisher and the editorial board hope that this book will prove to be a valuable piece of knowledge for researchers, students, practitioners and scholars across the globe.

List of Contributors

Suzane M. Fank-de-Carvalho
National Council for Scientific and Technological Development - CNPq, Brasília, Distrito Federal, Brazil
Universidade de Brasília – UnB, Biological Institute, Electron Microscopy Laboratory, Brasília, Distrito Federal, Brazil

Sônia N. Báo
Universidade de Brasília – UnB, Biological Institute, Electron Microscopy Laboratory, Brasília, Distrito Federal, Brazil

Maria Salete Marchioretto
Instituto Anchietano de Pesquisas/UNISINOS, PACA Herbarium, São Leopoldo, Rio Grande do Sul, Brazil

Andrzej Kędziora, Krzysztof Kujawa, Hanna Gołdyn, Jerzy Karg, Zdzisław Bernacki, Anna Kujawa, Stanisław Bałazy, Maria Oleszczuk, Mariusz Rybacki, Ewa Arczyńska-Chudy, Rafał Łęcki, Maria Szyszkiewicz-Golis, Piotr Pińskwar, Dariusz Sobczyk and Joanna Andrusiak
Institute for Agricultural and Forest Environment, Polish Academy of Sciences, Poznań, Poland

Cezary Tkaczuk
Siedlce University of Natural Sciences and Humanities, Siedlce, Poland

Stefanos Kalogirou
Hellenic Centre for Marine Research, Hydrobiological Station of Rhodes, Greece

Ernesto Azzurro
ISPRA, National Institute for Environmental Protection and Research, Livorno, Italy

Michel Bariche
Department of Biology, American University of Beirut, Beirut, Lebanon

Tan Jing, Tian Yan, Wang Shang-Wu
Sichuan Agriculture University, Sichuan, China

Feng Jie
Forestry Bureau Li County, Sichuan, China

Gabriella Buffa, Edy Fantinato and Leonardo Pizzo
DAIS - Dept. of Environmental Sciences, Informatics and Statistics, Ca' Foscari University, Italy

Heimo Mikkola
University of Eastern Finland, Kuopio Campus, Kuopio, Finland

Guilherme Fernando Gomes Destro, Tatiana Lucena Pimentel, Raquel Monti Sabaini, Roberto Cabral Borges and Raquel Barreto
Coordination of Enforcement Operations, Brazilian Institute of Environment and Renewable Natural Resources – IBAMA, SCEN, Trecho II, Ed. Sede, Brasília/DF, Brazil

Mohd Nazip Suratman
Faculty of Applied Sciences, University of Technology MARA, Shah Alam, Malaysia

Ylo Joann Systra
Department of Mining, Chair of Applied Geology, Tallinn University of Technology, Tallinn, Estonia

del Pozo-Baños Marcos, Ticay-Rivas Jaime R., Cabrera-Falcón Jousé, Travieso Carlos M., Pérez Santiago T. and Alonso Jesús B.
Signal and Communication Department Institute for Technological Development and Innovation in Communications University of Las Palmas de Gran Canaria IdeTIC-ULPGC, Las Palmas de Gran Canaria, Spain

Arroyo Jorge, Sánchez-Chavez Luis and Ramírez-Bogantes Melvin
Escuela de Matemáticas, Universidad Nacional, Costa Rica

Printed in the USA
CPSIA information can be obtained
at www.ICGtesting.com
JSHW011454221024
72173JS00005B/1075